Springer Series in Reliability Engineering

For further volumes:
http://www.springer.com/series/6917

George Anders · Alfredo Vaccaro
Editors

Innovations in Power Systems Reliability

Springer

Editors
Prof. George Anders
Department of Microelectronics and Computer Science
Technical University of Lodz
Lodz
Poland
e-mail: george.anders@p.lodz.pl

Prof. Alfredo Vaccaro
Department of Engineering
Università degli studi del Sannio
Piazza Roma 21
82100 Benevento
Italy
e-mail: vaccaro@unisannio.it

ISSN 1614-7839

ISBN 978-0-85729-087-8 e-ISBN 978-0-85729-088-5

DOI 10.1007/978-0-85729-088-5

Springer London Dordrecht Heidelberg New York

British Library Cataloguing in Publication Data
A catalogue record for this book is available from the British Library

Cover design: eStudio Calamar, Berlin/Figueres

Printed on acid-free paper

Springer is part of Springer Science+Business Media (www.springer.com)

Preface

Electric power networks are, in general, among the most complex yet most reliable systems engineered by man. However, these large interconnected systems often operate under high stresses because of the increasing demand for electric energy and the difficulty of improving the infrastructure due to economic and environmental constraints. The major challenges facing the electric power industry today include the need for balancing resource adequacy, safety, network reliability, stability, economics, environmental and other public objectives to optimize resources while satisfying the growing demand. This optimization must be performed with due consideration of such constraints as having to meet reliability criteria and stability margins.

Armed with such a vision, this book covers a wide spectrum of issues ranging from methods for balancing the resources to various reliability and security aspects of the electrical grid. These topics are presented by a number of prominent researchers, scientists and practitioners from many countries. While the book focuses on the technological breakthroughs and roadmaps for implementing new technologies, it also presents a much needed forum for sharing of the best practices.

Chapter 1 describes some strategies for meeting challenges to the grid's ability to provide reliable power delivery. Solutions are offered through applications of modern technologies, such as advanced feedback control schemes using wide area measurements, wide-area visualization techniques, and intelligent operational tools applying Standard IEC-61850 and information semantics. The goal is to provide a vision for a comprehensive and systematic approach to meet the criteria for grid safety and reliability through new information services. Some of the concepts suggested in this chapter, such as advanced information services combined with the new computational paradigms for maintenance and on-line power equipment diagnostic, and integrating data from a myriad of sensors, will be instrumental in achieving the reliability, efficiency, and financial soundness of future power grids.

In the present business environment of competition and re-regulation, the determination of asset values and methods of reaching the best investment decisions are of increasing interest. Traditional approaches to establishing maintenance and replacement expenditures can no longer satisfy regulators or bottom-line-driven decision-makers. Chapter 2 addresses the subject of optimal maintenance policies for power equipment. In the emerging operating environment of deregulation and market-based competition, every management decision involves a certain amount of risk. These risks need to be evaluated and courses of actions selected so that the risks are minimized. Quantitative methods are needed that combine technical aspects with financial and business risk factors. For quantitative risk evaluations, analytical tools are necessary.

For the maintenance (or asset sustainment) function at an electric utility, the following question is of particular interest: Faced with multiple options for re-investment in equipment maintenance, what is the best course of action to take in order to maximize reliability at minimum cost? The decision-maker can use several criteria for selecting the best reinvestment policy. In the past, engineers operating an electric power system were mainly concerned about equipment reliability, with the financial aspect playing a secondary role. However, in the new economic environment, the reliability and financial aspects of system operation will be equally important. Hence, both reliability and cost should be considered in the selection of maintenance alternatives. With this in mind, a substantial effort has been made to develop suitable mathematical models and decision support tools to address the issue of maintenance/refurbishment/investment option selection.

Mathematical models can be deterministic or probabilistic. Because maintenance models are used for predicting the effects of maintenance in the future, probabilistic methods are more appropriate than deterministic ones, even if the price for their use is increased complexity and a consequent loss in transparency. For these reasons, the use of such methods is spreading only slowly. Examples of simpler mathematical models, still based on fixed maintenance intervals (scheduled maintenance) are compared with more complex ones that incorporate the idea of condition monitoring, where decisions about the timing and amount of maintenance are dependent on the actual condition of the device (predictive maintenance). Such policies can be optimized with respect to any of the model parameters, such as the frequency of inspections or the life-cycle costs. This chapter also discusses model optimization, both through a sensitivity analysis and through a mathematical formalism.

The development of theoretical and practical aids for selecting reliability models for power system equipment represents a very active area of research. This topic is addressed in Chap. 3. The main purpose of this chapter is to present an up-to-date review of the basic theoretical and practical aspects of the major reliability models. The presentation also includes a review of some models that are rarely discussed in the literature, but that, in the authors' opinion, can be very useful. Some new models or new ways of justifying the usefulness of the older models are also presented. These aspects are illustrated with practical examples that show how to perform rational model selection. The authors stress that the analyst should spend a sufficient amount of time in performing the preliminary analysis to fully understand the consequences of adopting—sometimes without sufficient information—a given model.

The second purpose of this chapter, closely related to the first, is to highlight the rationale behind a selection of the models that are based on the phenomenological and physical characteristics of the aging of the power system equipment. These models consider the probabilistic laws governing the stresses acting on the devices and the degradation (deterioration) processes that the equipment or its components are often subjected to. The authors argue that this "technological" approach, which is also referred to in the recent literature as an "indirect reliability assessment", might be in practice the only feasible tactic available to the researcher in the presence of a limited amount of data, as is typically the case in the field of modern power systems. The chapter also addresses the relationship between purely mathematical models and those exploiting the physical characteristics of the devices.

Equipment reliability models also constitute a basic building block in the analysis of the entire electric power network. As a matter of fact, the development of mathematical models representing the reliability characteristics of the electrical devices went hand-in-hand with the development of power system reliability evaluation techniques. New technological developments in power system generation and transmission warrant a new look at the well-established reliability assessment methods. In particular, renewable energy resources are receiving considerable attention in the continued deployment, growth and development of bulk electric power systems. At the present time, the most promising new source of electrical energy is wind power and the governments around the world are making commitments to add considerable amount of this new generation resource to the existing power grids. This is discussed in Chap. 4. The increasing use of wind power clearly indicates the importance of reassessing the traditional models for the reliability evaluation of the composite power systems containing significant amounts of wind energy.

To address this issue, Chap. 4 proposes advanced methodologies for adequacy assessment of integrated composite generation and transmission systems containing wind generators. Adequacy evaluation of composite generation and transmission systems is a complex task that includes detailed modelling of the generation and transmission facilities. The emergence of wind generation as an important electrical energy source creates some challenging complications in evaluating the adequacy of composite systems, because wind power behaves quite differently from the conventional electric power generating facilities. This chapter discusses the general area of composite system adequacy evaluation and some of the new techniques that can be utilized to incorporate wind power in a system assessment. Wind power modelling in both generation and composite system adequacy evaluation is discussed and illustrated by application of the proposed method to the two well-known reliability test systems. The studies presented utilize sequential and non-sequential Monte Carlo simulation and illustrate the effects of addition of independent and correlated wind generation to the two test systems.

One of the goals of a bulk electric system reliability evaluation is the determination of a need for new investments in the grid infrastructure. This is discussed in Chap. 5. Utilities may be asked by regulators (such as the North American Electric Reliability Corporation [NERC] in the United States) to identify their transmission line and substation facilities that are critical for operation of the power network. Generators, transmission lines, and power transformers represent the major portion of bulk transmission systems, and these assets should perform well in order to achieve a high level of reliability for the bulk power system. Such high performance can be achieved if there are no restrictions on budgets. In many cases, budget constraints have been imposed, and the owner of the system has to set priorities with regard to the work that needs to be done on those transmission components. Such work may involve large or small projects, depending on the objective and the importance of each project. Depending on the network configuration, and on how loads and generators are connected to the network, the consequences of generation and transmission component outages could be more or less significant. One of the goals of a bulk electric system reliability evaluation is the determination of a need for new investments in the grid infrastructure. This is discussed in Chap. 5. Utilities may be asked by regulators (such as the North American Electric Reliability Corporation [NERC] in the United States) to identify their transmission line and substation facilities that are critical for operation of the power network. Generators, transmission lines, and power transformers represent the major portion of bulk transmission systems, and these assets should perform well in order to achieve a high level of reliability for the bulk power system. Such high performance can be achieved if there are no restrictions on budgets. In many cases, budget constraints have been imposed, and the owner of the system has to set priorities with regard to the work that needs to be done on those transmission components. Such work may involve large or small tasks depending on the objective and the importance of each project. Depending on the network configuration, and on how loads and generators are connected to the network, the consequences of generation and transmission component outages could be more or less significant.

A facility-ranking procedure would help utilities perform critical facility assessment. A documented assessment procedure for identifying and ranking facilities will help utilities justify assessment results. Additionally, scarcity of resources, both financial and human, requires that the available funds are directed to the places that would benefit the network the most. For all these reasons, robust and accurate ranking procedures could be very helpful. This topic is addressed in Chap. 5. The proposed approach utilizes some concepts of the spectral graph theory to rank electric power substations in a high-voltage network. This approach is only one of the possible approaches to finding the most critical facilities in the network. Other approaches could include additional information related to the reliability of the system components. This extension of the ideas presented in Chap. 5 will undoubtedly be the subject of new research in the near future.

Continuing with the theme of identifying the critical infrastructure in an electric power grid, a logical extension of this topic is the question how to reinforce the grid so that the networks are adequately fulfilling their design tasks. This question is addressed by transmission system planners. The main objective of multistage transmission expansion planning (TEP) is to define where, when, and what reinforcements should be placed in the network to ensure an adequate quality level of energy supply to customers. In a competitive energy market, TEP is a complex optimization task to ensure that the power system will meet the predicted demand and the security criteria, along the planning horizon, while minimizing investment, operational, and interruption costs. This practice is the only rational response to conflicting customer and regulatory demands.

Several approaches for solving the TEP challenges can be found in the literature, however, only a few have considered the multi-stage nature of the TEP problem. The multi-stage nature of the TEP problem requires consideration of multiple time periods, determining possible sequences of transmission reinforcements. To deal with the multi-stage nature of the problem, simplified studies (also known as static analyses) can determine, for just one stage, where new transmission facilities should be installed. Different from most approaches to static planning, the ideas presented in this chapter solve the TEP problem considering the chronology of reinforcements. The goal of the suggested approach is not only to define what reinforcements should be placed in the electrical network and their corresponding locations, but also when they should be added within the planning horizon to ensure an adequate level of energy supply to the customers. In the end, the best expansion plans must be selected in order to minimize the present value costs defined in the objective function.

Whereas the transmission expansion planning is a purely system planning problem, modeling of the day-to-day operation of the power system is of interest also to researchers involved with issues of system operation. In the restructured environment, the improvement of the economic efficiency of the electricity markets has been the focus of several studies. Central to these efforts is a better understanding of the nature of the tight coupling between market and system operations. An important aspect of this coupling is the dependence of market outcomes on the way the system is operated. A key driver in system operations is the security criterion, with which compliance must be ensured. Chap. 7 focuses on the dependence of market performance on system security. In this chapter, the authors propose an approach to quantify market performance as a function of a specified security criterion for both single- and multi-settlement environments.

The chapter investigates the interactions between the system security criterion and the associated economics in terms of the marginal costing—used to determine the security prices—and the evaluation of expected system security costs. The problem is analyzed in various contexts by both empirical and analytical means. The empirical studies investigate the adverse impacts of market participants' behaviours on the performance of electricity markets. The analytical studies, on the other hand, focus on the impacts of constrained system operations on markets to determine the unavoidable losses in the economic efficiency of electricity markets.

The expected system security costs are evaluated, taking explicitly into account the random nature of the outages and the costs of the required security control actions to deal with them. The authors argue that there is a clear need, in the restructured environment, to quantify market performance as a function of system security in a way that appropriately reflects the regional transmission operations. This quantification further requires the consideration of different market and system conditions that may exist within a period in order to capture the range of impacts under such conditions. One approach to address this issue is a cost/benefit analysis, taking into account the expected costs of operating the system and the expected outage costs. Such an approach may be viewed as the application of the notion of "value of reliability", which obviously is a topic of great interest to the market participants.

It is a well-known fact that the increase in energy demand and the advent of the deregulated market mean that both the static and the dynamic system limits must be considered in a modern power systems reliability analysis. Chapter 8 discusses a general analytical method for the probabilistic evaluation of power system transient stability. The chapter also reviews some of the basic contributions available in the relevant literature and previous results obtained by the authors. The first part of the chapter introduces the basic concepts required for the calculation of the probability of system stability. The chapter characterizes the random variables that enter this analysis (e.g. system load, fault clearing time, and critical clearing time) and discusses the methods of analytical or numerical calculations. The values of these parameters are uncertain, and the discussion in this chapter shows that ignoring the uncertainty may lead to a serious underestimation of the probability of system instability.

A Bayesian statistical inference approach is then proposed for the probabilistic transient stability assessment; in particular, the chapter discusses both point and interval estimation of the transient instability probability of a given system. A Bayesian approach is particularly useful in this context, because the parameters affecting transient stability probability (e.g., mean values and variances of the above random variables) are not generally known and have to be estimated. The authors propose application of the well-established system models for the description of the load evolution in time.

The second part of the chapter investigates a new aspect of the on-line statistical estimation of the transient instability probability. The goal is to predict whether the system will become unstable with the help of advanced modelling tools, including a new Bayesian approach utilizing the Dynamic Linear Model for the stochastic evolution of the system load.

A major hurdle in the widespread application of the probabilistic transient stability analysis was the low computational efficiency of the classical models. In the numerical studies discussed in this chapter, the authors show that computations involving "tracking" of the transient stability versus time can be performed very fast.

The reported results could be very important in a modern, liberalized market in which fast and large variations of load and generation are expected to have a significant effect on the transient stability probability. To conclude the chapter, some results on the robustness of the estimation procedure are also briefly discussed. The discussion demonstrates that the assumptions regarding the system parameter distributions do not affect the efficiency of the proposed approach.

To perform efficient and accurate probabilistic analysis of the system transient stability, network parameters must be well defined. This conclusion became even more obvious after a number of reports analyzing the 2003 blackouts events pointed out that several national and transnational grids had been managed without sufficient real-time data, particularly in the presence of a large number of new uncertainties. Reliable real-time data, oriented to the monitoring of system dynamics, were not available, and the operators did not have enough time to take decisive and appropriate remedial actions. What has become clear is that, after the blackouts of 2003, despite the revolution driven by electric industry restructuring and energy market re-regulation, the general approach to power system security has not changed. A key feature is still the lack of dynamic data concerning key system parameters.

Chapter 9 discusses estimation of dynamic system parameters. In this chapter, the authors propose an optimization method, utilizing a nonlinear programming algorithm, to obtain such estimates. In the proposed procedure, time domain simulation trajectories are compared with on-line measurements in order to update or estimate dynamic parameters. The main advantage of this method is its flexibility because it can be adopted for estimating parameters such as synchronous machine constants, external network equivalents or the constants for frequency or voltage dependent loads. The authors show that the methodology can be applied during the on-line power system operation and provides a more reliable database for real-time dynamic security and control. In fact, a frequent update of the power system dynamic model can guarantee more reliable simulations and, consequently, more effective control.

Finally, Chap. 10 addresses advanced methodologies for reliable power flow analysis in the presence of data uncertainties. Power flow analysis is used to determine the steady state of the power system for a specified set of load and generation values. It is one of the most extensively used tools in various power engineering applications, including network optimization, voltage control, state estimation, and market studies.

The most common formulation of the power flow problem—the deterministic power flow—has all input data specified from the snapshot corresponding to a selected point in time. Alternatively, the analyst can construct data to reflect the required assumption about the expected generation/load profiles for a certain peak demand condition. The solution for the study is deemed representative for a limited set of system conditions. However, when the input conditions are uncertain, there is a need to analyze numerous scenarios to cover the range of uncertainty. Under such conditions, reliable solution algorithms, incorporating the effect of data uncertainty into the power flow analysis, are, therefore, required. Reliable power flow solution algorithms allow the analyst to estimate both the uncertainty in the input data and in the solution tolerance. In this way, the uncertainty propagation effect is explicitly represented, and the level of confidence of power flow studies can be assessed.

Acknowledgment We would like to thank Dr. Giosuè di Franco for his valuable support in editing the book.

George Anders
Alfredo Vaccaro

Contents

List of Contributors

George Anders, Department of Electrical and Computer Engineering, Technical University of Lodz, Lodz, Poland, Kinectrics Inc., 800 Kipling Avenue, Toronto, ON, M8Z 6C4, Canada

Roy Billinton, Power System Research Group, University of Saskatchewan, 57 Campus Dr, Saskatoon, SK, S7N 5A9, Canada

Sergio Bruno, Dipartimento di Elettrotecnica ed Elettronica (DEE), Politecnico di Bari, via Re David 200, 70125, Bari, Italy

Elio Chiodo, Department of Electrical Engineering, University of Naples, "Federico II" Via Claudio 21, 80125, Naples, Italy

Ron Coutu, ISO New England Inc., 1 Sullivan Road, Holyoke, MA, 01040, USA

Aleksandar D. Dimitrovski, Oak Ridge National Laboratory, Power & Energy Systems Group, ETSD, One Bethel Valley Road, P.O. Box 2008, Oak Ridge, TN, 37831-6070, USA

Yi Gao, Power System Research Group, University of Saskatchewan, 57 Campus Dr, Saskatoon, SK, S7N 5A9, Canada

George Gross, University of Illinois at Urbana Champaign, 1406 W. Green, Urbana, IL, 61801, USA

Teoman Güler, University of Illinois at Urbana Champaign, 1406 W. Green, Urbana, IL, 61801, USA

Dange Huang, Power System Research Group, University of Saskatchewan, 57 Campus Dr, Saskatoon, SK, S7N 5A9, Canada

Rajesh Karki, Power System Research Group, University of Saskatchewan, 57 Campus Dr, Saskatoon, SK, S7N 5A9, Canada

Roger L. King, Mississippi State University, Box 5405, Mississippi, Mississippi State, MS, USA

Massimo LaScala, Dipartimento di Elettrotecnica ed Elettronica (DEE), Politecnico di Bari, via Re David 200, 70125, Bari, Italy

Davide Lauria, Electrical Engineering Department, University of Naples Federico II, Via Claudio 21, I-80125, Naples, Italy

Armando M. Leite da Silva, Institute of Electric Systems and Energy, Federal University of Itajubá—UNIFEI, Itajubá, MG, Brazil

Eugene Litvinov, ISO New England Inc., 1 Sullivan Road, Holyoke, MA, 01040, USA

Vahid Madani, Pacific Gas and Electric, 1919 Webster Street, Oakland, CA, 94612, USA

Luiz Antônio F. Manso, Department of Electrical Engineering, Federal University of São João Del Rei—UFSJ, São João Del Rei, MG, Brazil

Giovanni Mazzanti, Department of Electrical Engineering, University of Bologna, Viale Risorgimento 2, 40136, Bologna, Italy

Leandro S. Rezende, Institute of Electric Systems and Energy, Federal University of Itajubá—UNIFEI, Itajubá, MG, Brazil

Kevin Tomsovic, Department of Electrical Engineering and Computer Science, University of Tennessee, 414 Ferris Hall, Knoxville, TN, 37996, USA

Alvaro Torres, Dept. of Electrical and Electronics Engineering, University of Los Andes, Bogotá, Colombia, SNC Lavalin, Cra. 10 97A-13, Of. 207, Bogotá, Colombia

Alfredo Vaccaro, Department of Engineering, University of Sannio, Piazza Roma 21, 82100, Benevento, Italy

Strategies and Roadmaps to Meet Grid Challenges for Safety and Reliability

Vahid Madani and Roger L. King

1 Introduction

When reliable systems or operations are envisioned, there is an expectation that the entity being considered will produce consistent results, preferably meeting or exceeding some standard of operation, within some uncertainty. Often, reliability also has a temporal component associated with it.

The reliability of providing consistent service to end users is becoming a challenge as the demand and end users' expectations for energy increase. At times, this has resulted in power grids approaching their limits, and in the severest cases blackouts have occurred in parts of the grid. Figure 1 shows the frequency of transmission outages based on data from the NERC Disturbance Analysis Working Group (DAWG). The figure shows approximately 24 outages per year in the US with curtailments in the 100–1,000 MW range, about 5 outages in the 1,000–10,000 MW range, and one outage every 4 years at 10,000+ MW [1]. The large-scale outages are not unique to one country or a specific region or part of the world [2], and could be triggered by mechanical failures in the power grid networks or by external forces such as natural calamities (earthquakes, hurricanes, etc.) and more recently the threat from human-induced damages (e.g., cyber attacks).

The electric grid is designed and operated to withstand any single (and often double) contingency by its protection and controls system. Therefore, a common phrase used during a postmortem event analysis is "Relays cannot start a disturbance". However, this is not exactly telling the whole story. There can be a

V. Madani (✉)
Pacific Gas and Electric, 1919 Webster Street, Oakland, CA 94612, USA
e-mail: VxM6@PGE.COM

R. L. King
Mississippi State University, Box 5405, Mississippi State, MS, USA
e-mail: rking@cavs.msstate.edu

G. Anders and A. Vaccaro (eds.), *Innovations in Power Systems Reliability*,
Springer Series in Reliability Engineering, DOI: 10.1007/978-0-85729-088-5_1,

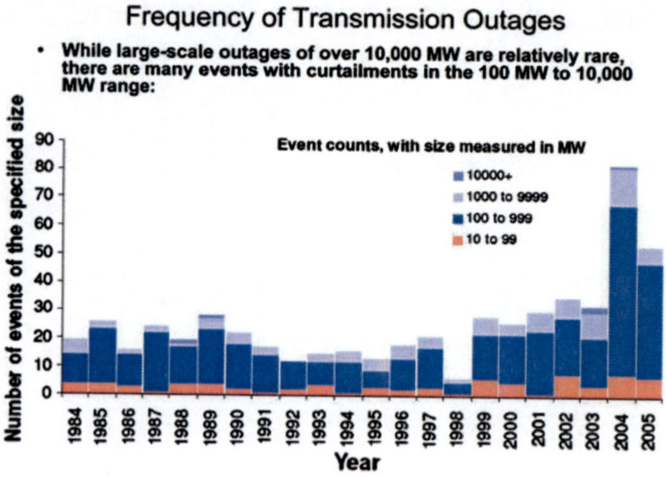

Fig. 1 Blackout frequencies, 1984–2005

significant impact to the grid's operation and reliability if a relay misoperation occurs during a contingency (i.e., a fault). Incorrect operation is both failure to detect or unplanned action. In most widespread disturbances, there is usually a misoperation that aggravates the events.

In the last few years there have seen several events in North America, and abroad, that might have been contained if it were not for unanticipated and unexpected protection system performance. Often, the test of properly functioning relaying system is during contingencies or faults. Protection systems (protective relays and associated relay systems) are expected to perform reliably during a grid disturbance. This expectation places a priority in maintaining the highest degree of reliability in the protection systems for ensured safe and reliable operation of the grid.

2 The Reliable Grid Challenge

Understanding the complexities of the interconnected power grid, need for proper planning, good maintenance, and sound operating practices are key components of an effective strategy in grid reliability. Formulation of reliable power system strategies begins with accurate modeling and system analysis of strengths, weaknesses, limitations, expectations, and the interactions. Today's grid requires a multi-scaled system approach to define technology-based solutions for reliable operation. These scales include:

- *The big picture* Interconnected grid system with defined boundaries, and forward looking solutions for the overall grid and telecommunication infrastructures.
- *The providers of power* Individual power producers and power companies and associated support over life cycle.

- *The component level* Individual elements affecting the system such as generation siting, substation capacity, or the end user.
- Move toward standardization and use of open solutions to support harmonization, transparency and interoperability.

The result of such an approach is the ability to provide explicit, normative explanations, and working definitions for common information being used amongst different groups in operation, planning, and protection.

Evolutions in protection and restoration principles for the smart grid are being made possible by wide-area measurement systems (e.g., PMUs). Real-time adjustment of the protection system's security-dependability is within reach given the advancements in technology and investments in communication system infrastructures. Further improvements with standardization, processing relay settings, event recordings, and distributed data sources can also be achieved through ontology or knowledge-based semantics.

Resource and transmission adequacy are necessary components of a reliable and economic supply. Though the reliability and market economics are driven by different policies and incentives, they cannot be separated when the objective is reliability and availability. Today, grid planning faces an extremely difficult task given the challenge to achieve resource adequacy in today's restructured industry, where market economics and local concerns often drive many of the decisions.

It is important to take a fresh and balanced approach to viewing the system as a whole by implementing various planning, operations, maintenance, and regulatory measures and weighing the costs, performance impact and risks associated with each measure.

This chapter provides a vision and the roadmap for creating actionable intelligence for reliable and real-time grid operation by capturing the knowledge and experience of the power system operation and control personnel and merging this knowledge with real-time data from disparate sources. Elements of this approach include formulation of a multi-disciplinary team, knowledge discovery methods that encompass electrical and computer engineering and industrial and systems expertise through industry collaborators. Many of the building blocks of this approach have already been established and results have been demonstrated in literatures. See [3] and other references in support of the roadmaps and building blocks. It is anticipated that this new perspective will yield results to improve system reliability and overall system performance.

3 A Technology-Based Solution

Early energy management systems, such as Supervisory Control and Data Acquisition (SCADA), were specially developed electronic devices running specialized operating systems providing only fundamental functions like real-time data processing and collecting, but nearly no real-world applications. Also, proprietary

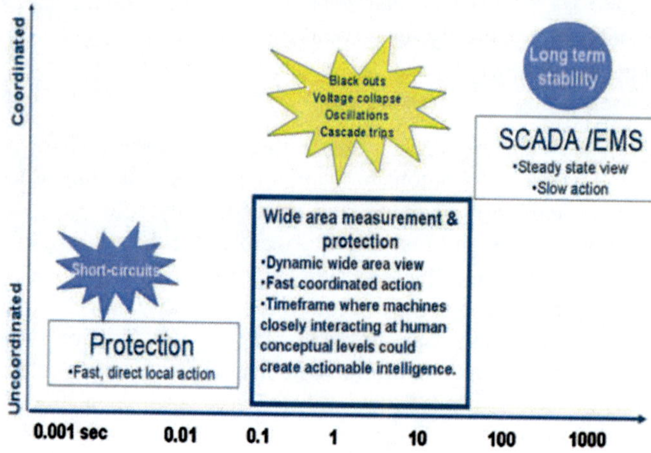

Fig. 2 Critical timing and reaction times for power grid operations

databases/interfaces restricted access by third parties to design innovative Information Technology (IT) solutions that overcome data exchange problems. Several applications in the areas of Customer Relationship Management (CRM), Geographic Information Systems (GIS), SCADA/ERM (Extended Runtime Modules), Distribution Management Systems (DMS), Outage Management Systems (OMS), Customer Information Systems (CIS), need to integrate and exchange data and information seamlessly (i.e., interoperability).

Figure 2 [4] shows the timing of events in the electric power grid and the reaction times commonly available for either an automated or operator intervention. At one end of the spectrum are the actions taken by the power system protection equipment to take an automated response based on measured quantities (e.g., frequency, current). At the other end of the spectrum, the operator controls the system in a steady-state mode using data acquired from a host of sensors via a SCADA system. Actions may be automated or are more often made based upon an operator's visual interpretation of the data presented through a variety of meters and display devices.

In steady-state operations, an operator normally has adequate time to consider the data, consult text-based help guides, or seek another operator's opinion before having to make a decision. Between these two ends of the spectrum is a time in which operators may have to make decisions based simply on heuristics or past experiences. Obviously, these actions may not result in the best outcome for reliability. This is the critical time period in which immediate actions must be made by system operators to prevent wide area collapses of the grid.

To ensure the secure and stable operation of the power system across the temporal spectrum, it is required to develop and apply new decision support tools that provide actionable intelligence in the required timeframe. In the aspects of secure and stable control, we need to think of an *automatic pilot* power system concept, representing a trend to improve Energy Management Systems (EMS).

To reach reliable real-time operation in auto-pilot, we need many tools and services that should be configured to work cohesively, including Operator Training System (OTS), Dynamic Security Analysis (DSA), Optimal Power flow (OPF), short circuit, etc. These tools need to operate harmoniously and independently, effectively organize the analysis results and share information amongst the various layers of the information service.

In other aspects, such as emergency control, restoration control and, etc., multiple services are required to be harmoniously interconnected into a multi-agent system to perform calculations, analyses, and be able to create actionable intelligence to support auto-pilot operation.

3.1 Novel Applications and Analysis

Adaptive protection and controls have been studied and advanced applications have been implemented in many modern day devices [5]. Advanced concepts by adjusting the need for more dependability or security as applicable, through the application of fuzzy logic and based on a set of real-time factors, such as system state index, nodal price, and equipment outage impact index have also been explored [6] and today's technologies such as real-time simulation tools support the concepts to go beyond theoretical and even demonstration stages to "proof of concepts".

Wide-area monitoring, protection, and control (WAMPAC) systems are emerging as a cost-effective solution to improve system planning, operation and maintenance. WAMPAC systems can take advantage of the latest advances in sensing, communication, computing, visualization, and algorithmic techniques and technologies.

Synchronized phasor measurement (PMU) technology and applications are an important element and enabler of WAMPAC, which has been receiving considerable attentions from the power industry. With its precise time synchronization, the measurement from the different locations in the system can be collected and compared in real time. It is hence ideal for monitoring and controlling the dynamic performance of a power system [7].

Implementation of this functionality gained significant attention after the 14 August 2003, Northeast Blackout event. One of the major challenges faced during the investigation was the rather limited availability of time-synchronized event recordings in the affected regions. As a result, one of the North American Electric Reliability Corporation (NERC) recommendations states: "Facilities owners shall, in accordance with regional criteria, upgrade existing dynamic recorders to include GPS time synchronization and, as necessary, install additional dynamic recorders" [8].

The applications area of phasor measurement units (PMU) can vary from visualization, postmortem analysis, state estimation improvements, congestion management, controlled system islanding, angular and voltage stability alarming and automated control, adaptive protection and relay settings, intelligent load-shedding, system restoration, etc. Application of PMUs in association with novel applications and analysis methodologies could help improve system reliability by mitigating undesirable responses of devices due to hidden failures, by monitoring system changes affecting settings and thus provide early warnings [9, 10]. Additional system developments to improve reliability include:

Novel algorithms for disturbance monitoring devices in order to calculate the proximity to voltage instability. These smart devices are installed at designated locations such as major load centers. They use local measurements to estimate the voltage stability margin and send alarm signals to a control center when it detects a local weak condition.

Installation of PMUs over the critical transmission corridors to monitor reactive power transfer to the load center.

Transmitting computation results from different geographical locations via a communication network to a control center. This will enable on-line wide-area voltage stability monitoring and control.

While microprocessor relays have gained full industry acceptance, there is a large number of legacy non-intelligent relays still in operation resulting in decreased reliability due to increased maintenance and failure costs or concerns that those costs will increase. The number of electromechanical and solid-state protection devices is estimated to about 70% in the US. It is not trivial to determine end of life and probability of failure for electromechanical relays. In addition, reduction in the skill set and knowledge familiar with troubleshooting, testing, and repairs of the old technology is diminishing across the power industry spectrum (see Sect. 3.4).

There are methods to assess probability of failure and replacement needs using criteria such as: age, maintenance practices and records, industry experience with certain relay types, and criticality of failure to name a few.

However, it is not easy to justify relay upgrades entirely based on age. Reluctance to upgrade to microprocessor relays is further emphasized by the complexity associated with increased functionality (e.g. settings), need for firmware upgrades, short life span of computer technology, and overall need to change the protection system philosophy and design. In addition, while digital relays provide a wealth of data, users may be faced with data overload. It is often the case that even data already available are not used or even collected (see Sect. 3.3).

New generations of microprocessor protective relays and substation/distribution automation systems are offering lower installed cost, integrated flow of rich information for operations and management, and improved performance of system protection and security. Those benefits cannot be achieved by one-for-one replacement of old devices with new ones.

A successful strategy should focus on how to successfully integrate devices that once operated in isolation, and how to use the new functional characteristics of the latest product generations to meet new operating challenges while lowering costs and improving operations. New designs must reflect innovative ways of combining the proven functions and elements, building on what has been demonstrated to achieve enhanced operating and cost benefits. The successful strategy, leading to the most cost-effective results, needs to identify order and speed in which relays need to be upgraded [11, 12].

To protect investments for future use, it is necessary to evaluate new technology in the time frame of the upgrades. Cost savings are also achieved through technology management and standardization. For example, introducing IEC 61850 as quickly as is practical can result in future-proof solutions with additional benefits. Use of IEC 61850 could help replace control wiring, simplify integration and data flow, allow for easier engineering and design changes, and reduction in installation and O&M costs [12].

3.2 Standardization

Standardization is the key to meeting every aspect of today's reliability and power delivery on the smart grid. Over decades of incremental upgrades, stranded asset uncertainties, social and environmental policy and regulations, and an absence of authority to enforce regulatory measures, the grid infrastructure expansions were kept to a minimum. Now, standards should be applied in all aspects of the infrastructure; from the substation design to bus configuration to control building equipment.

There is a direct relationship between grid reliability and protection and control justifying investments in standardized infrastructure system upgrades. Long-term vitality and viability is an important strategic requirement of standardization. Some of the elements of efficient and effective systematic upgrades to meet customer demand include:

- *Regulatory compliances* Considerations of reliability and potential impact to bulk interconnected power system. Regional Reliability Council discussions and the resulting directions and guidelines.
- Requirements of high-level internal strategic directions to eliminate discrete components such as control switches, interposing auxiliary devices, and metering instrumentation.
- Flexibility, and adaptability to moving technology.
- Familiarity and maintaining core competency skills, cohesive resource training, and resource management.
- Benchmarking, trend settings.
- Implementation of new technologies that bring about processes and other efficiencies, and decreases backlogged maintenance work.

Combining the notion of wide-area monitoring with standardization gives rise to the emerging technology area of sensor webs. Sensor web enablement (SWE) technology is a service oriented open standard developed by Open Geospatial Consortium (OGC) for discovery and acquisition of sensor data. SWE can integrate sensor data irrespective of physical/logical characteristics of the sensors, providing a platform for interoperability, essential in achieving seamless inter-utility communication [13].

Proper monitoring and critical information exchange in real time is a key for reliable operation in the grid. The disparity in protocols used in the power industry and lack of infrastructure of information exchange are hindrances to achieving reliability. Therefore, using the standards-based sensor web technology is one approach available for achieving interoperability in the power systems. Sensor web enablement (SWE) and common information model (CIM) provide a solution to heterogeneity of data and lack of central repository of the sensor data for proper action, in case of a contingency. The sensor data from utilities, published in CIM format, can be exposed via a sensor observation service (SOS). This provides a standard method for discovering and accessing sensor data between utilities, which facilitates the rapid response to handle contingences. In addition, the application of SWE in power industry pushes power industry one step closer towards auto-pilot operation.

3.3 Information Service

Since it is not possible to completely prevent blackouts, then effective and fast power system restoration is necessary to minimize the impact of major disturbances. This requires rapid decisions in a data-rich, but information-limited environment. The streams of data from a variety of sensors do not provide system operators with the necessary information to act on in the timeframes necessary to minimize the impact of a disturbance. Even if there are fast models that can convert the data into information, the system operator must deal with the challenge of not having a full understanding of the context of the information and, therefore, the information content cannot be used with any high degree of confidence. Some of the key elements for response in smart grids are:

- Well-defined procedures that require overall coordination within the affected area, as well as with the neighboring grids.
- Reliable and efficient software tools to aid operators and area coordinators in executing dynamic control procedures and in making the right decisions.
- Control solutions reducing the overload and instability risks during recovery.

Today's technology allows improved processes and smart systems to aid in decision-making to minimize impacts of outages (spatially and temporally). Standard operating procedures, based on pre-defined system conditions and operating parameters, can be provided via a set of power system information services. For example, rapid restoration or minimizations of outages by selected islanding are options for consideration in minimizing the consequences of an outage to a user.

Information services are focused on providing the right information at the right moment to the right decision maker. High-level operational information services (i.e., actionable intelligence) are often needed along with supportive sensor data or trends to provide context. The information services required by grid operators could vary from scenario development to estimates of socio-economic impacts of failures to quantitative statistics, trends and forecasts. These services also must be available in a geospatial context and at various temporal scales to support the needs of system operators, planners, and regulatory agencies. Information services must be characterized by a strong integration of grid data with ancillary data and information, and this will require a knowledge-based approach for capturing the best practices of utilities and regulators. The complexity of these information services will require a network of partners who will contribute to the production of the services. To facilitate these services it will be incumbent upon the power research community to develop tools to facilitate operational data acquisition and handling in interoperable formats and to create information products through a coordinated process chain. The successful conversion of power sensor data into actionable intelligence will require the integration of power system expertise in modeling, data management and service delivery to describe the state of the grid and to predict responses to actual and potential change.

3.4 Education

The continuation of the technology explosion of the second half of the twentieth century requires the availability of a diverse and highly capable technical workforce. Unfortunately, the education of engineers has not kept pace with the global demand. As a result there is a tremendous shortage of technical personnel all around the world. In the context of globalization this is a complex challenge and the cooperative efforts among stakeholders are required [14].

The US Department of Energy (DOE) and the North American Electric Reliability Corp. (NERC) identified the aging workforce as a critical challenge facing the electric power industry and the educational system that supports it. If not managed properly, the loss of experience and expertise will affect reliability, safety, productivity, innovation, and the capability to solve pressing issues, such as grid modernization and climate change.

> The aging of the American workforce has emerged as a critical issue facing American productivity in the 21st century. As the so-called "Baby Boomer Generation" reaches retirement eligibility, the impact will be felt across both the public and private sectors. These 78 million individuals born between 1946 and 1964 have accumulated a wealth of experience and knowledge, and represent 44% of America's workforce. For electric utilities, whose service quality and reliability depends on maintaining an adequate, knowledgeable workforce, managing the upcoming retirement transition is a particular challenge [15].
>
> The reliability of the North American electric utility grid is dependent on the accumulated experience and technical expertise of those who design and operate the system.

As the rapidly aging workforce leaves the industry over the next five to ten years, the challenge to the electric utility industry will be to fill this void... [16].

The education of engineers has not kept pace with the technological developments. The universities cover very few classes in power systems in undergraduate programs and practical experience in signal processing and advance feedback control systems are needed to bring the practical knowledge to the universities. Though it is a science that can be covered through sound basic principles, its actual implementation permits alternatives. The alternative that is selected depends upon the power engineer's experience and the traditions of the electric utility company. Indeed, the entire power engineering education curriculum is at a crossroads and needs complete rejuvenation. Experience to date has shown that students can be attracted to and retained in power programs if they are exposed early to the joys of creation through design, discovery through research and invention through hands-on experimentation [17].

The paper by [14] gives several examples of how universities are working with industry and government to develop novel approaches in fostering power engineering education which is a lynchpin in the grid reliability quest.

4 Next Steps

This chapter presents a vision and transformation blueprint for meeting the protection and control needs of the twenty-first century to generate and deliver reliable power in the smart grid. The roadmap includes new concepts in use of modern tools and techniques as well as hardware and applications. Protection and control markers such as resource and asset management, process for harmonization of different plans and disciplines to a united vision, and justification strategies and benefits of investments are highlighted. Use of modern technology ad methods of testing and detecting equipment or design failures are highlighted.

Some of the concepts suggested in this paper about utilizing information services and integrating data from a myriad of sensors will be required to maintain social and environmental obligations for the electric utility industry. The protection and control will be instrumental in achieving reliability, efficiency, and financial aspects of the twenty-first century grid.

Exchange of information stemming from the worldwide experiences and the innovations in technology shed new lights on the current conditions, procedures, regulations and design of power systems of the future. Examination of the root causes for blackouts, for example, the resulting effects on neighboring systems, and implementation of proven solutions to help prevent propagation of such large-scale events should help design reliable power delivery infrastructures for today and in the future.

References

1. Hines P, Apt J, Liao H, Talukdar S (2006) The frequency of large blackouts in the United States electrical transmission system: an empirical study. Carnegie Mellon
2. Madani V, Novosel D (2005) Getting a grip on the grid. IEEE Spectr Mag 42(12):42–47 (December)
3. Madani V, Higinbotham W (2007) Advantages in Modern Communications as applied to Remedial Action Schemes and Control Centers, International Institute for Research and Education in Power Systems, August
4. Novosel D (2005) IEE international conference on energy trading and risk management, UK, November
5. Madani V, Novosel D, Apostolov A, Corsi S (2004) Innovative solutions for preventing wide area disturbance propagation. In: International institute for research and education in power systems (IREP) symposium proceedings, August
6. Khorashadi-Zadeh H, Zuyi L, Madani V Artificial intelligence and fuzzy logic concepts for adaptive dependable and secure protection systems (for electric power system)
7. Horowitz SH, Phadke AG (2003) Boosting immunity to blackouts. Power and Energy Magazine 1(5):47–53 (September/October)
8. NERC Recommendations to August 14, 2003 Blackout—prevent and mitigate the impacts of future cascading blackouts. http://www.NERC.com
9. Phadke A (2009) Wide area measurements for improved protection of power systems. In: Proceedings of the innovations in protection and control for greater reliability infrastructure development (i-PCGRID-2009), March, San Francisco
10. Thorp J, Bernabeu E (2009) Adaptive security and dependability with PMUs. In: Proceedings of the innovations in protection and control for Greater Reliability Infrastructure Development (i-PCGRID-2009), March, San Francisco
11. Madani V, Novosel D, Zhang P, Meliopoulos A, King R (2006) Vision in protection and control area. In: Meeting the challenges of 21st century, IEEE PSCE
12. Myrda P, Tates D, Udren E, and Novosel D (2006) Optimal strategies for system-wide protection and control replacement programs. CIGRE, Paris, August
13. Dahal N, Mohan V, Durbha S, Srivastava A, King R, Younan N, and Schulz N (2009) Wide area monitoring using common information model and sensor web. In: Power systems conference and exposition, March
14. Reder W, Madani V, King R, Venkata M (2007) The challenges and opportunities to meet the workforce demand in the electric power and energy profession. In: IEEE VDE conference on meeting the growing demand for engineers and their educators 2010–2020, November
15. Department of Energy, Workforce Trends in the Electric Utility Industry: A Report to the United States Congress Pursuant to Section 1101 of the Energy Policy Act of 2005, August 2006, http://www.oe.energy.gov/DocumentsandMedia/Workforce_Trends_Report_090706_FINAL.pdf
16. North American Electric Reliability Corp (2006) Long-term reliability assessment, October 2006. http://www.nerc.com/files/LTRA2006.pdf
17. Athreya D (2007) Shortages of qualified faculty. In: First Indo-US Collaboration for engineering education, Mysore, India, 5 June

Optimal Maintenance Policies for Power Equipment

George J. Anders

1 Introduction

Maintenance is defined in an IEEE/PES Task Force report [1] as an activity "wherein an unfailed device has, from time to time, its deterioration arrested, reduced or eliminated." It is an important part of asset management. As deterioration increases, the asset value (condition) of a device is reducing; the connection between asset value, time, maintenance and reliability is shown in Fig. 1. The curves in the figure are called *life curves*. Since they are derived from probabilistic information, the times shown represent means.

The maintenance policy is aimed at achieving failure-free operation of the system and prolonging the remaining life of equipment. The remaining lifetime of a device depends to a large extent on two factors—frequency of making inspections (technical surveys) and the quality of repairs (for given part of a device either the most crucial and necessary repairs can be made or a complete overhaul can be provided). Defining both, times when the inspections should be performed and which components should be repaired, are difficult tasks. Usually when an inspection takes place, the equipment is temporarily unavailable (that results in additional costs). As a result, utilization costs can be overestimated due to the fact that inspections are made too frequently.

To address this problem, we will start this chapter with a discussion how the life curves could be used to find an optimal maintenance policy. The presentation will show that once the life curves are generated for various maintenance policies, a

G. Anders (✉)
Department of Electrical and Computer Engineering, Technical University of Lodz, Lodz, Poland
e-mail: george.anders@kinectrics.com, george.anders@attglobal.net

G. Anders
Kinectrics Inc., 800 Kipling Avenue, Toronto, ON, M8Z 6C4, Canada

G. Anders and A. Vaccaro (eds.), *Innovations in Power Systems Reliability*,
Springer Series in Reliability Engineering, DOI: 10.1007/978-0-85729-088-5_2,

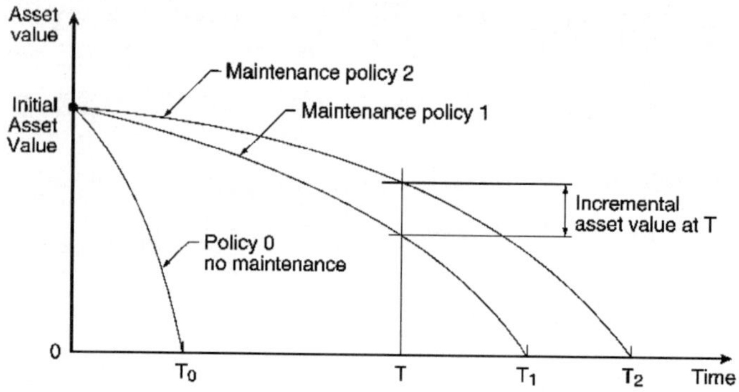

Fig. 1 Life curves

most advantageous one from the life extension point of view can be selected performing a sensitivity analysis. However, for finding a true optimized policy, we need a mathematical model. We will discuss briefly deterministic and probabilistic modeling of maintenance activities and we will focus on a Markov model that could be optimized with respect not only to the remaining life of the equipment but also to its availability and the maintenance costs.

Maintenance activities on power system equipment are not taken in isolation from the system performance requirements. For the Maintenance or Asset Sustainment function at an electric utility, the following aspect of the decision-making process is of a particular interest:

Faced with multiple options for re-investment on a particular set of equipment like breakers, disconnects, transformers, etc. (e.g., do nothing, continue with current maintenance practice, refurbish, replace, monitor, and so on), what is the best course of action to maximize reliability at minimum cost?

The effects of changes in maintenance policy are difficult to foresee since there is usually no historical data reflecting the performance of the component subject to the revised policy. Here, mathematical models offer an invaluable help and, as mentioned above, one such model utilizing Markov chain will be investigated in this chapter.

Changes in a component maintenance policy are usually undertaken in order to improve reliability of supply to the customers most affected by the performance of this component and, occasionally, to generate savings in the system operation and maintenance for the utility. This aspect of the problem is seldom modeled mathematically, mostly because of a lack of easy-to-use and reliable tools modeling complex operation of a substation or a small area with several substations. This problem will be addressed in more detail in this chapter as well.

Occasionally, changes in a component maintenance policy may have a profound effect on the reliability of the larger area or even on the entire system. To measure this effect, one needs to model reliability of the entire power system under consideration.

Recognizing the interdependence of a component maintenance policy with the area and system reliability, a new paradigm in reliability analysis by combining the notion of component, small area and system reliability concepts into a single application was proposed by [2] and will be summarized here. This is a conceptual leap in the traditional thinking where each aspect of system operation is analyzed separately. The concepts reflecting this new way of thinking were implemented in a computer platform that allows an analysis of a component maintenance policy in the context of a customer, area and system needs. An example of a study with this platform will be discussed. A thought of linking component maintenance with a small area distribution reliability analysis has been explored in [3, 4]; however, there are no analytical tools involved in this analysis.

Looking at a component maintenance policy from a larger perspective brings one additional important aspect into play. Namely, with a limited maintenance budget a question arises which component in the system should we maintain first? Traditionally, maintenance policies followed a time-based pattern suggested by the equipment manufacturers. Recently, the Reliability Centered Maintenance (RCM) has been applied in many electric utilities.

The cornerstone of the RCM methodology is a classification of component importance in the system operation. This aspect is also implemented in the approach described in this chapter. A numerical example illustrating these concepts on the 24-bus IEEE reliability test system is presented in the final part of the chapter.

2 Selecting the Best Maintenance Alternative

From a system point of view, the governing thought is that starting with a prescribed budget we have to identify the assets that should be put on the priority list of the components whose maintenance policy will affect the key performance indicators the most. The initial step in the analysis is, therefore, bulk electric system reliability study. The study involves analysis of the effects of failures of major components such as lines, transformers and generators. The methods for bulk power system reliability modeling are well established and since the literature is vast, the reader is referred to one of the many papers listed in [5–9]. Since all commercial programs to assess the BES reliability are geared to analysis of very large systems, the buses have zero failure rates at this stage. As a result of this step, the most vulnerable load buses (delivery points) are identified. From this analysis, a single bus or a set of neighboring buses is selected for further studies.

The selected buses form a small area that is now analyzed in more detail. A network diagram of the selected area shows all the important components that affect customer reliability in this region. Such an area can have several hundred components that are maintained and are subject to failure. A reliability analysis of the reduced system is now performed taking into account constituent components' reliability characteristics. This includes modeling of protection system operation, common mode outages, maintenance-dependent outages and others. The component reliability indices may be either based on the utility's historical experience or can be taken from the available external databases. As a result of this analysis, the reliability indices computed at this stage are assigned to the bus(es) representing the station(s) in the BES study or to the components connected to the buses of interest. These indices are now entered in the bulk electric system reliability program and the first step is repeated. There are several possible ways the station indices could be transferred to the BES reliability evaluation program. The new reliability indices reflect the performance of individual components forming the substations in the selected area. This will be our base case scenario.

It is important to mention that this part of the proposed approach is philosophically different from the methods published in the literature dealing with the evaluation of the BES reliability taking into account the station-originated outages [5, 10–13]. Paper [10] proposed some models to take station-originated failures into account in the bulk electric system reliability evaluation. Also, some computational techniques have been proposed in the remaining references listed above that evaluate station-related failures. However, these papers have concentrated on the concepts and effects of station-originated outages and not on methods of identifying them. Paper [14] simulated various failure modes of station components and computed the reliability indices of connected lines and generators. The approach described in this chapter is a variation of this concept with a more comprehensive modeling of a substation operation.

Even though we have "homed" on the area of interest, there are still too many components for which the maintenance policies should be analyzed. The next step is then a prioritization of all components in the selected region. There are two types of prioritization lists, one ranking the components on the basis of their structural importance, the other on the basis of the reliability importance. These two lists may result in a quite different ranking of components as discussed in the numerical example presented here. Components at the top of either list are selected for further studies.

A model for the component deterioration process taking into account the presently applied maintenance policy is now built. One of the outcomes of this process is the evaluation of the component failure rate. If the computed failure rate is much different than the one used in the base case study, the above two steps could be repeated with the new failure rate of the component and a new base case scenario established.

We are now ready to contemplate changes in the maintenance policy of the selected component(s). The analysis results in new failure rate for this component. The area and system studies are now repeated with the new information and the effect of the new maintenance policy analyzed. The diagram in Fig. 2 summarizes the procedure described above. In this figure, the programs used by the author are named, namely REAL for the BES reliability evaluation, WinAREP for small area reliability analysis, Asset Management Planner (AMP) and RiBAM for component maintenance investigations. The programs are described in more detail later in the chapter.

The proposed approach required a construction of a computer platform that allows seamless transfer of data and results between various computer programs forming the constituent parts of the platform. The computational engines

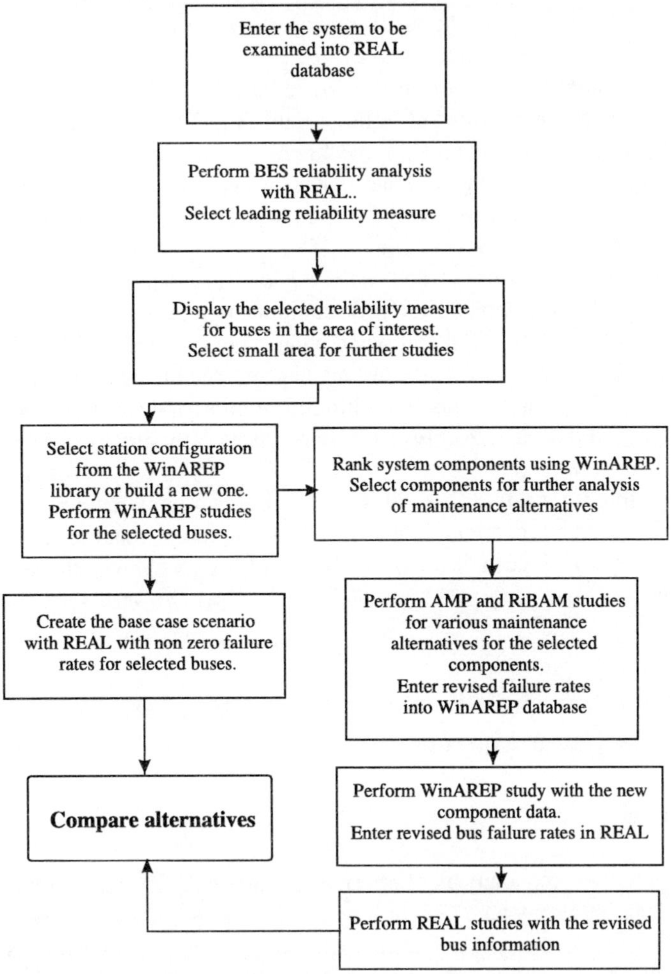

Fig. 2 Flowchart for maintenance strategy optimization

employed in the platform are only briefly described here. References to published articles describing their features are included.

The presentation in this chapter will start with the analysis of a component maintenance policies and their optimization and will conclude with a review of the effect of the component policy on the system reliability.

3 Maintenance Optimization with Life Curves [2]

Conditions for three maintenance policies are illustrated in Fig. 1, including Policy 0 where no maintenance is performed at all, and Policies 1 and 2 where maintenance is performed according to different rules.

Let failure be defined as the asset condition where asset value becomes zero, and lifetime, as the mean time it takes to reach this condition; furthermore, let reliability be linked with the mean time to failure. Now, life extensions T_0 to T_1 when Policy 1 is applied instead of Policy 0, and T_1 to T_2 when Policy 1 is replaced by Policy 2 can be clearly seen in the figure. So are the changes in the asset condition (value) at any time T. Note that in a given study failure, lifetime and reliability can be defined differently; e.g., failure could be tied to any asset condition which is deemed unacceptable.

As far as reliability is concerned, Policy 2 is clearly superior to Policy 1. It is also obvious that maintenance affects component and system reliability. But maintenance has its own costs, and when comparing policies, this has to be taken into account. The increasing costs of carrying out maintenance more frequently must be balanced against the gains resulting from improved reliability. When costs are also considered, Policy 2 in Fig. 1 may be very costly and, therefore, may not be superior to Policy 1.

It is possible to study life curves for finding optimal maintenance policy. However, since no mathematical model exists to represent relationships shown in Fig. 1, the way to proceed is to do a case analysis as shown in the following example.

3.1 Example for High-Voltage Air-Blast Breakers Using the Life Curve Concept

3.1.1 General

This study involves the analysis of several breakers with a total operating history of about 100 breaker-years. According to the current policy, three types of maintenance are routinely performed on each breaker. About every 8 months, minor maintenance (timing adjustments, lubrication) is carried out at a cost of about $700. Its average duration is 0.25 day. Approximately every 10 years, medium maintenance is performed involving replacement of some parts, taking on the average 2 days, at a cost of about $6,000. Major maintenance involving breaker overhaul takes place every 15 years, with an average duration of 22 days and a cost of about $75,000.

In the study, four alternative maintenance policies are compared. The first option is to continue with the present maintenance policy. The second is to do nothing, i.e., to run the equipment in the future without any maintenance. The third is to perform major overhaul, followed by a slightly modified version of the original policy. The last option is to replace the equipment with a new one ($90,000) and continue with the modified maintenance policy. The modified policy differs from the original one in that the minor maintenance after overhaul or replacement is scheduled every 15 months instead of every 8 months, a reduced maintenance policy.

3.1.2 Life Curves

Figure 3 shows life curves for the three "basic" policies. Considering the four options above, the life curves for two of them, the one involving replacement and the other to stop maintenance, are shown in Fig. 3. These curves were derived with the AMP and RiBAM programs, described later, using the following assumptions: (a) the "present" moment when the choice is made among the options is 20 years into the life of the breakers, (b) if the option chosen requires action (replacement, overhaul), there is a delay of 3 years before the action is implemented, and (c) upon failure, repair is performed which brings the device to an assumed 90% of its original condition.

3.1.3 Cost Studies

In a financial evaluation, a *time horizon* must be selected which usually starts at the present, when the study is made, and includes a predetermined number of years for which the costs of the various operating and maintenance options are calculated and compared. For the present study, a time horizon of 10 years was selected. This is also shown in Fig. 3 by a horizontal line between the 20th and 30th year marks.

Cost computations involve the calculation of the expected number of failures, and of the various types of maintenance activities, during the specified time horizon. The cost of each maintenance activity is expressed by its present value. The costs are then expressed as functions of the delay.

Figure 4 illustrates the present costs for all options with a 3-year delay for each. The diagram shows that in the given case the best option (of those considered) is to continue with the original maintenance policy. The expected cost of this is $100,000 for the 10-year time horizon. The costs are highest for the "Stop All Maintenance option" because the probability of failure is much higher than for the

Fig. 3 Life curves with **a** replacement, **b** no maintenance, after a 3-year delay

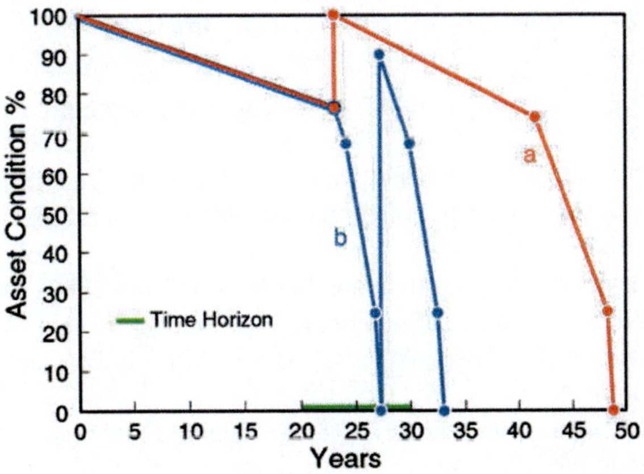

Fig. 4 Cost diagram for the various options

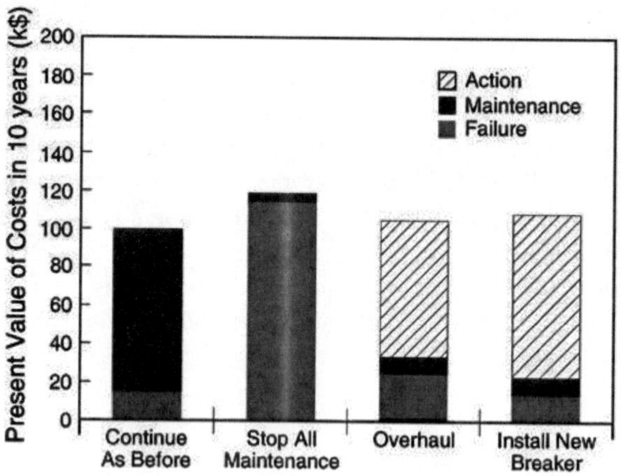

other options. The maintenance cost is high for the "Continue as Before" policy because minor maintenance is performed quite often.

To summarize, while for highest reliability the "Install New Breaker" policy is the choice, for lowest costs the "Continue as Before" option must be selected.

3.1.4 Sensitivity Studies

It is quite possible to perform optimization of each option with regard to, say, maintenance frequency. Such studies were not carried out. Instead, sensitivity studies were performed to find out how "robust" the findings are if some of the input values are subject to uncertainty. While details are not discussed here, the results show that the cost of the option "Continue as Before" appears to be little affected if several of the input data are varied around their assumed values.

3.2 Review of Maintenance Approaches

3.2.1 Regular Versus "As Needed" Maintenance

Maintenance has been performed for a long time on a great variety of devices, machines and structures. Traditionally, maintenance policies have been chosen either on the basis of long-time experience or by following the recommendations of manuals issued by manufacturers. In both cases, maintenance has been carried out at regular, fixed intervals. This practice is also called *scheduled* maintenance and, to this day, this is the maintenance policy most frequently used by electric utilities.

It was found, however, that scheduled maintenance may be quite costly in the long run, and may not extend component lifetime sufficiently. For the last 15 years or so, variations of a new approach have been tried and implemented by many industrial undertakings and several electric power utilities. The essence of this approach is that maintenance should be undertaken not regularly but only when needed. Such an approach is called *predictive* maintenance. To find out when maintenance is needed, condition monitoring—periodic or continuous—and appropriate criteria for triggering action are required.

3.2.2 Improvement Versus Replacement

Maintenance activities may result in the restoration of a device to conditions better than those it was found in, or in its replacement with a new one. However, for a long time, it had been assumed that even restoration would result in "as new" conditions, which clearly is not what happens in practice. Most often, only limited improvement would take place; however, this is very difficult to take into account.

A large number of replacement policies are described in the literature; in fact, most of the literature concerns itself with replacement only, neglecting the possibility that maintenance may result in smaller improvements at smaller costs. Maintenance policies involving limited condition improvement are mostly based on experience, and such empirical approaches cannot predict and compare changes in reliability as a result of applying various maintenance policies.

3.2.3 Empirical Approaches Versus Mathematical Models

Empirical approaches are based on experience and manufacturers' recommendations. This does not mean that they are necessarily very simple. The method called Reliability Centered Maintenance (RCM), introduced about 18 years ago [15], is empirical, yet quite sophisticated. It is based on condition monitoring (and, therefore, may not follow rigid maintenance schedules), failure cause analysis and an investigation of operating needs and priorities. From this information, it selects the critical components in a system (those which are dominant contributors to system failure or to the resulting financial loss) and initiates more stringent maintenance programs for these components. It assists in deciding where the next dollar budgeted for maintenance should go.

An important advantage of the RCM approach is that it also considers external, non-deterioration-originated failures (e.g., those caused by weather, animals and humans). A good example is the case of overhead lines in distribution systems. According to fault and interruption statistics in the UK, the percentages of failure causes of such lines are the following (since only the dominant failure causes are shown, the percentages are rounded and do not add up to 100):

- Weather 55%;
- Damage from animals 5%;
- Human damage 3%;

- Trees 11%;
- Aging 14%.

The conclusion appears to be that the maintenance budget for overhead lines should be divided almost equally between internal and external programs. The external budget would be spent mostly on tree trimming and some design changes, such as the erection of barriers and fences.

Maintenance policies based on *mathematical models* are much more flexible than heuristic policies. Mathematical models can incorporate a wide variety of assumptions and constraints, but in the process they can become quite complex. A great advantage of the mathematical approach is that the outcomes can be optimized. Optimization with regard to changes in some basic model parameter can be carried out for maximal reliability or minimal costs.

Mathematical models can be *deterministic* or *probabilistic*. Since maintenance models are used for predicting the effects of maintenance in the future, probabilistic methods are more appropriate than deterministic ones, even if the price for their use is increased complexity and a consequent loss in transparency. For these reasons, the use of such methods is spreading only slowly.

The simpler mathematical models are still based on fixed maintenance intervals (scheduled maintenance), and optimization will be carried out, in most cases, through sensitivity analysis, by varying, say, the frequency of maintenance. More complex models incorporate the idea of condition monitoring where decisions about the timing and amount of maintenance are dependent on the actual condition of the device (predictive maintenance). Such policies can be optimized with respect to any of the model parameters, such as the frequency of inspections.

3.3 Linking Component Reliability and Maintenance: A Probabilistic Approach

3.3.1 Basic Models

A simple failure-repair process for a deteriorating device is shown in Fig. 5: The various states in the diagram are explained. The deterioration process is represented by a sequence of stages of increasing wear, finally leading to equipment failure. Deterioration is, of course, a continuous process in time, and only for easier modeling it is considered in discrete steps.

The number of deterioration stages may vary, and so do their definitions. In most applications, the stages are defined through physical signs such as markers on

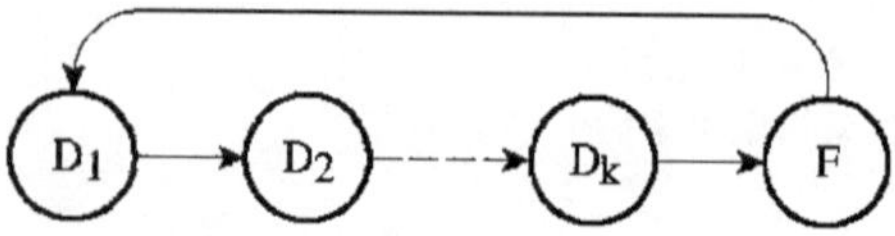

Fig. 5 State diagram including (D1, D2, ...); *F* failure state

wear or corrosion. This, of course, makes periodic inspections necessary to determine the stage of deterioration the device has reached. The mean times of the stages are usually uneven, and are selected from performance data or by judgment based on experience.

The process in Fig. 5 can be readily represented by a probabilistic mathematical model. If the rates of transitions shown between the states can be assumed time-independent, the mathematical models describing such processes are known as Markov models. Well-known techniques exist for the solution of these models [16]. It can be proven that in a Markov model the times of transitions between states are exponentially distributed. This property and the constant-rate property follow from each other.

3.3.2 The Effect of Maintenance

One way of incorporating maintenance into the model in Fig. 5 is shown in Fig. 6. It is immediately clear that in this arrangement there is no assumption made that maintenance would produce "new" conditions; in fact, the effect of maintenance can now be limited: it is assumed that it will improve the device's condition to that which existed in the previous stage of deterioration. This contrasts with many strategies described in the literature where maintenance is considered equivalent to replacement.

If a failure has external causes (e.g., inclement weather), there is a single step from the working to the failed state. Now, the constant failure-rate assumption leads to the result that maintenance cannot produce any improvement because the chances of failure in any future time interval are the same with or without maintenance (a property of the exponential distribution). That maintenance will not do any good in such cases agrees with experience as expressed by the oft-quoted piece of wisdom: "If it ain't broke, don't fix it!" The situation is quite different for deterioration processes where the times from new conditions to failure are not exponentially distributed even if the times between subsequent stages of deterioration are (this can be rigorously proven). In such a process, maintenance *will* bring about improvement, and one can conclude that if failures are the consequence of aging, maintenance has an important role to play.

In Fig. 6, the dotted-line transitions to and from state M1 indicate that maintenance while in state D1 should really not be performed because it would lead

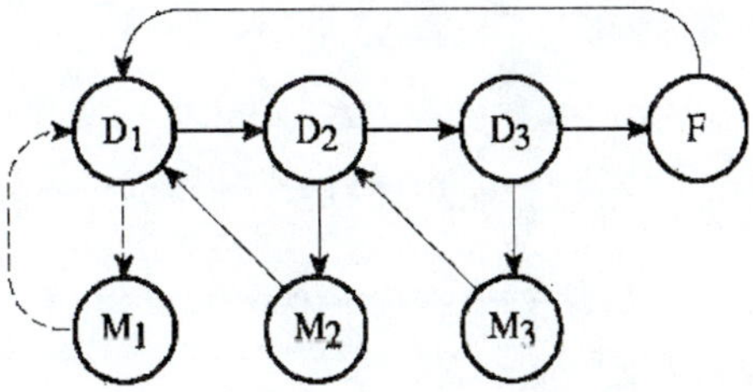

Fig. 6 State diagram including three stages of deterioration stages and maintenance (*F* failure state)

back to state D1 and, therefore, it would be meaningless. State M1 could be omitted if the maintainer knew that the deterioration process was still in its first stage and, therefore, no maintenance was necessary. Otherwise, maintenance must be carried out regularly from the beginning, and state M1 must be part of the diagram.

It should be observed that this and similar models solve the problem of linking maintenance and reliability. Upon changing any of the maintenance parameters, the effect on reliability (say, the mean time to failure) can be readily computed.

3.3.3 A Practical Model

A more sophisticated model [17] based on the scheme in Fig. 6 and tested in practical applications is shown in Fig. 7. A program, called Asset Management Planner (AMP), using this model, was developed by Kinectrics Inc. in Toronto, Canada. It computes the probabilities, frequencies and mean durations of the states of a component exposed to deterioration but undergoing regular inspections and receiving preventive maintenance.

Without maintenance, the path from the onset (entering D1) would run through the stages of deterioration to the failure state F. With maintenance, this straight path to failure is regularly deflected by inspection and maintenance.

According to the diagram, in all stages of deterioration, regular inspections take place (I1, I2, I3), possibly several times, and at the end of each inspection a decision is made to continue with minor (M) or major (MM) maintenance, or forgo maintenance and return the device to the state of deterioration it was in before the inspection. Another point of decision is after minor maintenance when, if the results are considered unsatisfactory, major maintenance can be initiated.

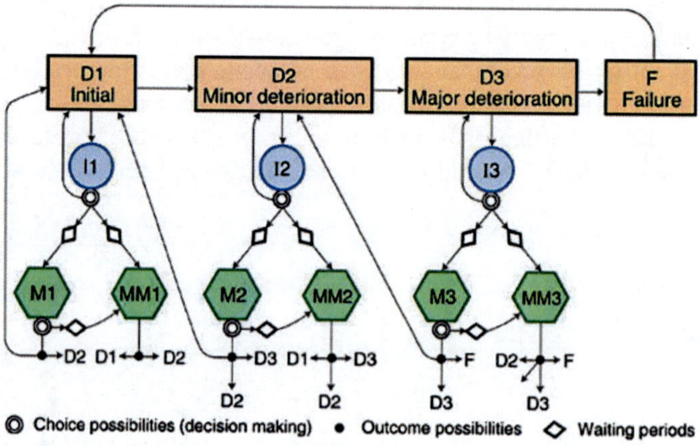

Fig. 7 The AMP model

The result of all maintenance activities is expected to be a single-step improvement in the deterioration chain, following the principle shown in Fig. 7. However, allowances are made for instances when no improvement is achieved or even when some damage is done during maintenance, the latter resulting in the next stage of deterioration.

The choice probabilities (at the points of decision-making) and the probabilities associated with the various possible outcomes are based on user input and are estimated from historical records.

Another technique, developed for computing the so-called first passage times (FPTs) between states, will provide the average times of first reaching any state from any other state. Although not shown, the technique is implemented in the AMP model. If the end state is F, the FPTs are the mean remaining lifetimes from any of the initiating states.

This information is necessary for constructing life curves. It can be observed that the AMP model can handle both scheduled (regular) and predictive (as needed) maintenance policies.

Figure 6 shows an arrangement for scheduled maintenance: the rate of starting maintenances is always the same (this rate is the reciprocal of the *mean* time to maintenance; the actual times constitute a random variable).

The scheme in Fig. 7 incorporates an arrangement for predictive maintenance. Condition monitoring is done through regular inspections, and if it is found that no maintenance is needed the device is returned to the "main line" without undergoing maintenance.

Mathematical Description of the Model in Fig. 7

Transition Rates. Assuming that transition rates between states are known (computed from historical data), transition rate matrix **Q** can be built with components λ_{ij} denoting a transition rate from state i to j and:

$$\lambda_{ii} = -\sum_{j,j\neq i} \lambda_{ij}. \tag{1}$$

The transition rates from states Dx to Ix are computed as the reciprocal of the time to inspections, while the transition rates from states Dx to Dy are reciprocals of the times when the device reaches another stage of deterioration without any maintenance.

The repair states are characterized by two parameters: duration and the probabilities of departure to other states. Duration of a state can be determined from historical records for both the Ix and Mx states. In the first case, it is the average duration of inspections, in the second case, the time of performing the repairs. The departure rate from state i to j is then defined as

$$\lambda_{ij} - \frac{p_{ij}}{d_i} \tag{2}$$

where p_{ij} is the probability of transition from state i to j.

We also have

$$\sum_j p_{ij} = 1 \quad \text{for } i = 1, 2, \ldots, n \tag{3}$$

where n is the number of the repair and inspection states.

This definition of the transition rate matrix describes a semi-Markov process [18].

Cost of a State. In addition, for every state, one can define a cost of residing in this state. It is especially important for the Mx states because it symbolizes the costs of repairs.

Both values (costs and duration) can be written as two vectors: cost **C** and duration **D**.

4 Optimal Maintenance Policies for Power Equipment

There are many maintenance optimization models utilizing simplified deterministic mathematics [19]. One such model is presented as an illustration in the next section. A more sophisticated probabilistic optimization model is discussed afterwards.

4.1 A Simple Deterministic Model [19]

Consider a device that breaks down from time to time. To reduce the number of breakdowns, inspections are made n times a year when minor modifications may be carried out. The optimal number of inspections is to be determined which minimizes the total yearly outage time, consisting of the repair times after failures and the inspection durations.

Let the failure rate be $\lambda(n)$ occurrences per year, where λ is independent of time but is a function of the inspection frequency. Therefore, the total downtime $T(n)$ is also a function of n. Further, let it be assumed that

$$\lambda(n) = \frac{n}{k+1} \tag{4}$$

where the numerical value of k indicates the failure frequency when no inspections are made. If t_r is the average duration of one repair and t_i the average duration of one inspection, then

$$T(n) = \lambda(n)t_r + nt_i. \tag{5}$$

Substituting (4), taking the derivative of $T(n)$ with respect to n, and equating it with zero,

$$\frac{dT(n)}{dn} = \frac{-kt_r}{(n+1)^2} + t_i = 0. \tag{6}$$

From the second statement, the optimal value of n becomes

$$n_{opt} = \left(\frac{kt_r}{t_i}\right)^{0.5} - 1 \tag{7}$$

with $k = 5$ per year, $t_r = 6$ h and $t_i = 0.6$ h, one obtains that $n_{opt} = 6.07$ per year, or the optimal inspection frequency is about one in every 2 months. The total outage time is $T(6) = 7.9$ h/year, whereas without inspections it would be $T(0) = 30$ h/year.

As can be seen, optimization is easily included in mathematical models. On the other hand, modeling the relation between maintenance (inspection) and reliability (failure rate) is still a problem. In the example above, this relation is given by (4). It should be observed that this relation is *assumed*, and not a result of calculations. What is missing is a mathematical model where this relation is part of the model itself, and the effect of maintenance on reliability is part of the solution.

4.2 Maintenance Optimization with a Probabilistic Model

The following section describes a mathematical model for the selection of an optimal maintenance policy [20]. The original model described above presented a method of calculation of the remaining life of equipment without suggestions on how the maintenance policy modeled could be optimized. Here, we will define several possible optimization procedures to find out the best maintenance policy. The optimization process will be illustrated with an optimization algorithm for Markov models utilizing a simulating annealing approach in a practical numerical example involving high-voltage circuit breakers.

4.2.1 The Objective Function

In the optimization procedure discussed here, the quantities of interest are: (1) the *Remaining Life of Equipment* represented in the model as the FPT from the current deterioration state to the failure state [21], (2) *the Life Cycle Costs* represented as the cost of maintenance and failure, and (3) equipment *Unavailability*. Our goal is thus to define an optimization model that would minimize a function of these three parameters, i.e.:

$$F(r) = \min f(\text{total\_cost}, -\text{FPT}, \text{unavailability}). \tag{8}$$

Vector **r** symbolizes parameters of the model that can be varied and is described later in this chapter. To transform the multi-objective optimization problem described by (8) into a more practical single optimization formulation, f is defined as a special function that transforms three parameters to be expressed in the same units of measurement and is described below.

The nature of the problem that we are aiming to solve leads us to the decision that one could use an algorithm based on simulated annealing [20] to find an optimal solution.

A brief review of the way that the three input parameters are evaluated is given as follows.

FPT

In Markov theory, the FPT, T_{ij} represents the time when the model (starting from state i) will reach state j for the first time. In the case of the considered model, the most interesting is the time when the device will reach state F. T_{ij} will be equivalent to the remaining lifetime of the equipment. FPT will be measured in years.

Unavailability

During both inspection and repair, the device is temporarily out of service. The proposed model enables computation of the equipment unavailability. This value is usually measured in days per year.

In the model shown in Fig. 7, several values can be treated as parameters that can be modified in order to find the optimal solution. These parameters are:

- frequency of making inspections (time to inspections). These parameters correspond to transitions from states Dx to Ix;
- funds spent on maintenance (cost of states Mx);
- durations of the repair states.

The last two items define the depth and the speed of repairs. Each of these quantities can be varied independently or simultaneously. The possible optimization scenarios will be described next.

4.2.2 Parameters of the Maintenance Optimization Problem

The term "optimal maintenance policy" implies a selection of maintenance parameters for which the function in (8) will reach its minimum. The parameters that can be varied are described as follows.

Time to Inspection Optimization (TTI)

Knowing that during inspections a device is temporarily unavailable and being aware of the fact that every inspection is connected with additional costs, the main aim of the TTI optimization is to find the best points in time to perform the inspections.

Parameters that can be optimized are transition rates between deterioration states (**Dx**) and inspection states (**Ix**). This type of optimization changes the values of the elements of the transition rate matrix **Q**.

Cost Optimization

The second group of parameters that can be optimized are the costs of the states that represent the repairs (cost of the **Mx** states). After making a decision to spend additional funds on repair, one can expect the following effects:

- time spent in the repair stage will be shorter;
- better (deeper) repair—the equipment will be reconditioned with more care and it will end up in a "better" deterioration state. It does not necessarily mean though, that the repair time will be shorter.

Maintenance Time Optimization

In this optimization, we assume that the time of repair (elements of the duration vector **D**) is a function of the funds spent on repair and that the probability of transition from state **Mx** to **Dx** is constant (the probability matrix **P** does not change). Parameters that are optimized are elements of the cost vector **C** representing repair states. On the basis of this cost, duration of every state is computed (elements of the duration vector **D**) and then, the elements of the transition rate matrix **Q** are recalculated using (2).

Maintenance Depth Optimization

This type of optimization assumes that the probability of transition from a repair state to any other state is a function of the funds spent on repair, i.e., $p_{i,j} = f(c_i)$. The rationale for this thinking is that the more funds are spent on maintenance, the more likely it is that the equipment will end up in a higher (better) deterioration state than before the repairs. Parameters that are optimized are the elements of the cost vector **C**, but in this case, duration of the repair is constant (duration vector **D** does not change). After modification of the elements of the probability matrix **P**, the elements of the probability matrix **P** and the elements of the transition rate matrix **Q** are recalculated using (2).

4.2.3 Constraints

The constraints in this problem relate to the permissible changes in the components of the cost, probability and duration vectors. Thus, there are lower and upper limits on the amount of money available for maintenance and minimum and maximum times between inspections. Section 4.3 presents boundary conditions used in the numerical example. The optimization problems defined above will be solved using a simulated annealing algorithm [20].

4.2.4 Definition of the Optimization Function

Since the quantities to be optimized are expressed in different units and are of a different order of magnitude, it would be very difficult to formulate the objective function that would be just an algebraic sum of these variables. To address this problem, [20] proposed to use a notion of *utility* from a multi-attribute utility theory (MAUT) [22].

MAUT is one of the methods that form a multi-criteria decision analysis (MCDA) for quantifying the value of something (e.g., a project) based on its characteristics, impacts, and other relevant "attributes". It is useful for project prioritization because it provides a relatively simple and defensible way to capture all sources of project value, including non-financial (or "intangible") components of value.

More precisely, MAUT is an approach for deriving a "*utility function*" that, according to decision theory, quantifies a decision-maker's preferences over the available alternatives to a decision. The utility function, u, is such that the best alternative is the one that optimizes u.

In order to evaluate optimal parameters of the maintenance policy, there is a need to compare three values expressed in different measures: FTP—expressed in years, cost—expressed in thousands of dollars per year and unavailability—expressed in days per year, respectively. This is achieved by introducing a suitable utility function, which is described next.

Utility Functions

Calculation of utility requires a definition of a utility function. The form of this function determines the ability of taking a higher risk of failure in order to find a better solution. A utility function can be constructed to reflect the risk preference of the analyst: From this viewpoint, the analyst can be classified as:

- risk-seeker,
- risk-averse,
- risk-neutral.

Fig. 8 Characteristics of different utility functions

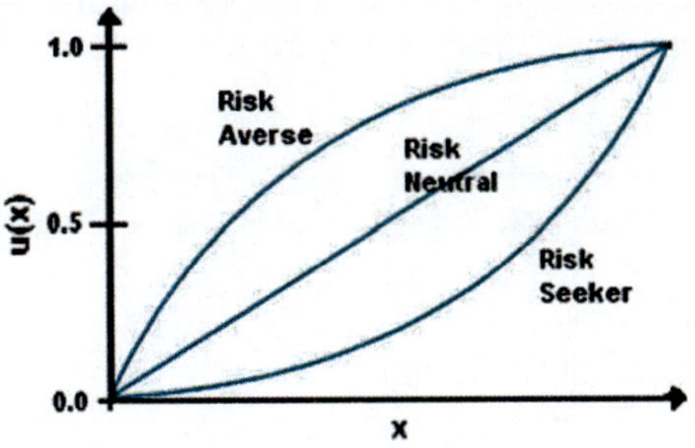

Their characteristics are presented in Fig. 8.

One of the commonly used utility functions is a power expression shown as follows:

$$u(x) = \frac{(x-a)^R}{(b-a)^R}. \tag{9}$$

Parameter R is responsible for the definition of a risk-acceptance attitude with a risk-seeker characterized by the value greater than one and a risk-averse person with the parameter smaller than one. A risk-neutral analyst will be assigned the value of $R = 1$. The constants a and b in (9) represent the minimum and the maximum value of the variable x, respectively.

A very useful characteristic of the utility function in (9) is the fact that calculated utility values are between 0 and 1. In our optimization problem, each of the three optimized parameters is represented by (9) with the same parameter R.

4.3 Numerical Example

The ideas described above will be demonstrated with a model of maintenance policy for high-voltage, air-blast circuit breakers with real historical values of model parameters.

4.3.1 Model Parameters

Figure 9 shows the model with the transition rates and probability of transitions between the states indicated on the arrows. These parameters are the same as used in the numerical example discussed in [17].

After applying (1)–(3), the transition rate matrix **Q** is shown in Table 1.

The input values of the duration and costs of the states, obtained from historical data supplied by a large utility in Canada, are summarized in Table 2.

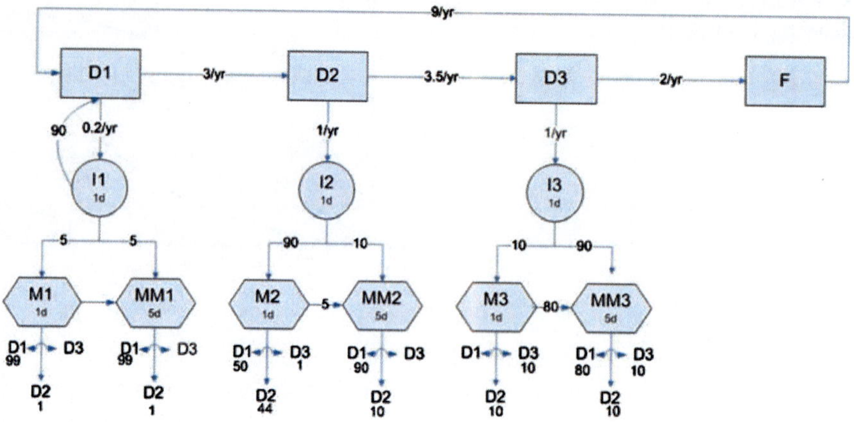

Fig. 9 Model of maintenance policy with transition rates between the states (*d* days, *y* years) and probability of transitions between them

4.3.2 Base Case Results

The steady-state probabilities and other model parameters computed from the standard Markov equations and transition rate matrix are shown in Tables 3 and 4.

In reality, the states with the probability equal to zero in Table 3 have non-zero values of this parameter but the values are smaller than 0.0001 and not shown here.

4.3.3 Model Optimization

The assumption made in the numerical analysis presented here states that an increase in the amount of money spent on maintenance can either result in shorter durations of the repairs or a greater depth of the repairs leading to the higher probability of the equipment landing in a better state. Therefore, it is assumed that the durations of the repairs and the probability of a transition from a repair state to another state are functions of the cost of the state, i.e.,

$$(d_i, p_{i,j}) = g(c_i) \tag{10}$$

Function g has a form $g(x) = \alpha x^d$, $g(x) = \beta - \gamma \ln(x)$ for the duration and probability variables, respectively, with the values of α, β, γ and d different for each value of i and j. Our goal is to find such values of the model parameters for which (8) is minimized. We will assume that the analyst is risk-averse and assign the value of $R = 0.2$.

The objective function f in (8) is an algebraic sum of utility functions (9) for the variables FPT, unavailability and total_cost and is given by

Table 1 Transition rate matrix **Q** for the model in Fig. 9

	D1	D2	D3	F	I1	I2	I3	M1	MM1	M2	MM2	M3	MM3
D1	−0.01	0	0	0	0.01	0	0	0	0	0	0	0	0
D2	0	0	0	0	0	0	0	0	0	0	0	0	0
D3	0	0	0	0	0	0	0	0	0	0	0	0	0
F	3.044E−04	0	0	−3.044E−04	0	0	0	0	0	0	0	0	0
I1	90.0	0	0	0	−100.0	0	0	5.0	5.0	0	0	0	0
I2	0	0	0	0	0	−100.0	0	0	0	90.0	10.0	0	0
I3	0	0	0	0	0	0	−100.0	0	0	0	0	10.0	90.0
M1	99.0	1.0	0	0	0	0	0	−100.0	0	0	0	0	0
MM1	19.8	0.2	0	0	0	0	0	0	−20.0	0	0	0	0
M2	50.0	44.0	1.0	0	0	0	0	0	0	−95.0	0	0	0
MM2	18.0	2.0	0	0	0	0	0	0	0	0	−20.0	0	0
M3	0	10.0	10.0	0	0	0	0	0	0	0	0	−100.0	80.0
MM3	18.0	2.0	0	0	0	0	0	0	0	0	0	0	−20.0

Table 2 Duration and cost of each state

	Cost ($)	Duration (days)
D1	0	
D2	0	
D3	0	
F	144,000	
I1	200	1
I2	200	1
I3	200	1
M1	1,200	1
MM1	14,400	5
M2	1,200	1
MM2	14,400	5
M3	1,200	1
MM3	14,400	5

Table 3 Steady-state probability of each state

D1	D2	D3	F	I1	I2	I3	M1	MM1	M2	MM2	M3	MM3
0.542	0.219	0.043	0.195	0	0	0	0	0	0	0	0	0

Table 4 Solution of the base case maintenance policy model

Unavailability	0.23 day/year
Total cost	$33,380
FPT (years)	26.87

$$f(c) = w_1 u(\text{total\_cost}) + w_2 u(\text{unav}) - w_3 u(\text{FPT}) \tag{11}$$

with weights $w_1 = w_3 = 1$ and $w_2 = 0.5$ assigned arbitrarily by the author.

4.3.4 Constraints

The following limiting values were adopted for the model parameters. The lower bound of the costs is that of the present value used by the utility and the upper is the cost of the next level of repair:

$$1/\lambda_{i,j} \in \langle 1\text{d}, 365\text{d} \rangle \tag{12}$$

$$\begin{aligned} &c_{\text{M1}}, c_{\text{M2}}, c_{\text{M3}} \in \langle \$100, \$10,000 \rangle \\ &c_{\text{MM1}}, c_{\text{MM2}}, c_{\text{MM3}} \in \langle \$10,000, \$100,000 \rangle \end{aligned} \tag{13}$$

This defines completely the optimization problem. The results are discussed later.

Table 5 Optimal cost of each state

	Cost ($)	Duration (days)
D1	0	
D2	0	
D3	0	
F	144,000	
I1	200	1
I2	200	1
I3	200	1
M1	3,497	0.20
MM1	32,556	3.86
M2	4,861	0.12
MM2	59,750	3.18
M3	4,732	0.13
MM3	69,306	3.04

Table 6 Parameters of maintenance policy model

Unavailability	0.16 day/year
Total cost	$23,700
FPT (years)	42.73

4.3.5 Simulation Results

After a series of simulations, the SA algorithm gives the following results: with the optimized cost of breaker states shown in Table 5 and other computed parameters shown in Table 6.

All three parameters were improved by applying the optimal maintenance policy. The unavailability and the total utilization cost of the breaker were reduced by about 30% each and the expected remaining life was increased by about 60%. This is the effect of increasing the expenditures on all maintenance activities. At the same time, the probability of moving from the maintenance state to a higher state has increased compared to the base case, hence the overall improvement in all components of the objective function. For example, the probability of moving from states MM2 and MM3 to D1 increased by about 10% accompanied by a substantial increase in the transition probability from the minor maintenance states M2 and M3 to states D1 and D2, respectively.

Finally, a sensitivity study was performed to determine the effect of the value of the parameter R. Figure 10 presents the dependence of the objective function $-u(x)$ at the optimum value of the vector $x = x^*$ on R. The values of the independent variable have been normalized as follows:

$$x' = \frac{x}{x_{\max} - x_{\min}}. \tag{14}$$

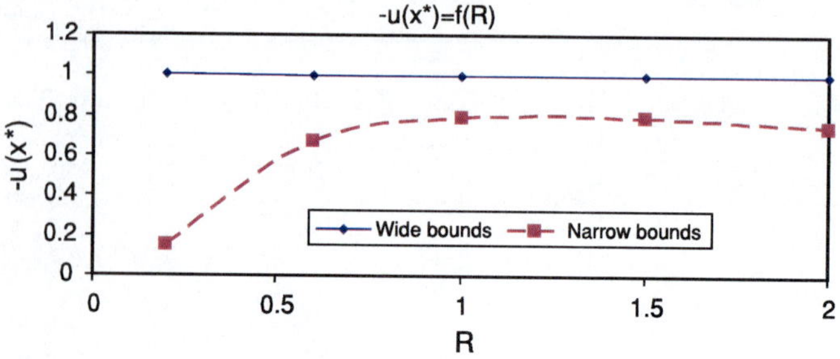

Fig. 10 Results of the sensitivity analysis

Two cost ranges were considered. In addition to the one given by (13), a reduced range was defined as follows: for minor maintenance between $100 and $5,000 and for major maintenance between $30,000 and $70,000.

The first observation is that the parameter R plays a role only for a narrow interval case. In this case, the best values of the objective function are obtained for the risk-averse decision-maker, with virtually no distinction between the risk-neutral and risk taking persons.

5 System Effect of a Component Maintenance

5.1 Bulk Power System Reliability Evaluation

The first step of the analysis performed by the computer platform is the evaluation of the reliability of a bulk electric system. A general approach adopted in many computer programs for this type of analysis is presented in Fig. 11. In this figure, a particular implementation in the software called REAL [23, 24] is shown.

A brief characteristic of the blocks in Fig. 11 is given as follows.

- Sequential or pseudo-chronological Monte Carlo simulation is used to select system states [25].
- DC power flow model is used to analyze the system states.
- Linear programming (LP) is used to solve, by redispatching and load shedding, system problems (i.e., overloads).
- Failure/repair rates are considered for both generation and transmission equipment.

Since reliability computations often involve an analysis of large systems, two measures are introduced to deal with the efficiency of the computations. First, we employ a pseudo-chronological Monte Carlo simulation during the reliability evaluation process [25]. Second measure involves a division of the entire network into three parts [23].

The first part of the network (i.e., Equipment Outage Area) involves a full representation of random behavior of transmission and generation elements. The

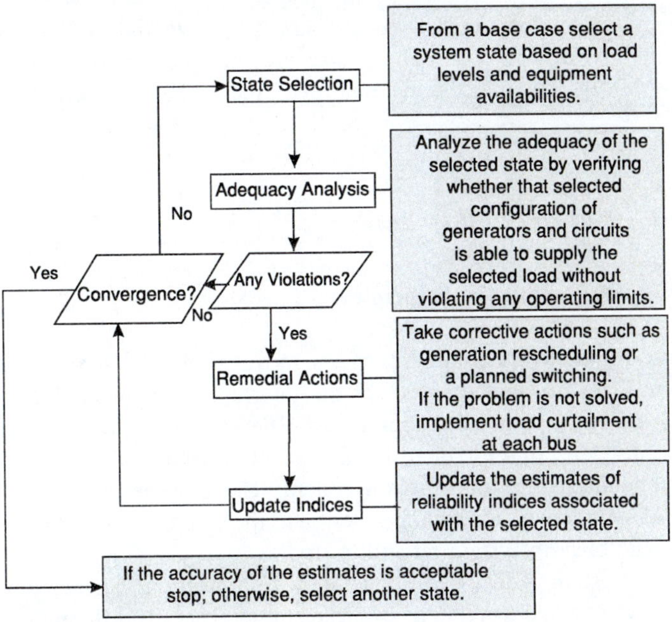

Fig. 11 Simplified flowchart of the REAL model

Fig. 12 Network representation for outage scheduling

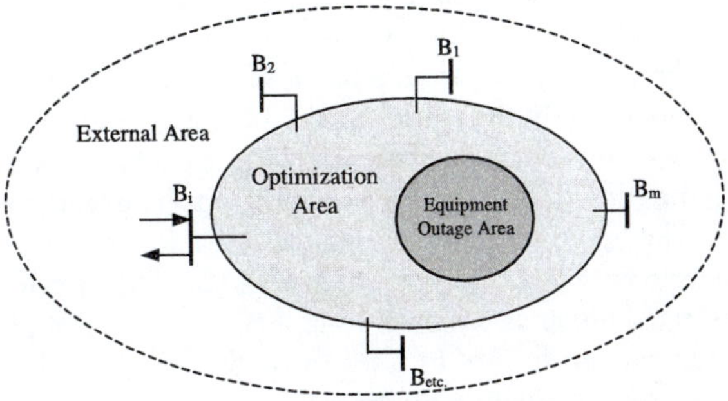

second, larger, network (i.e., Optimization Area) involves representation of all its elements for load flow and remedial action analysis. The elements in the second network that do not belong to the first network are not allowed to fail, but generators may be redispatched and load can be cut, if necessary. Finally, the third network (i.e., External Area) includes both previous networks and equivalent representation of the remaining components of the original load flow file. The idea of performing outage simulation on a part of the entire network was first introduced in [26]. The concept introduced there would be equivalent to using only the outage and optimization areas in our approach. The representation of the networks is illustrated schematically in Fig. 12.

The output of this part of the analysis is a set of standard reliability indices including the loss of load costs. The indices are computed for the system and for each bus. From the bus indices, the area for further analysis is selected.

5.2 Small Area Reliability Study

5.2.1 Reliability Evaluation Principles

Area supply reliability is commonly measured in terms of BES delivery point interruptions. The key indices used are the interruption frequency, duration and probability. Delivery point interruptions in a BES system can occur as a result of several reasons. Most of these interruptions are attributable to facilities' outages or security problems in the transmission system. The proposed approach uses the continuity of supply only as a failure criterion as the equipment overload issues are tackled in bulk electric system reliability studies.

The power system is designed to operate with protection schemes to minimize the effects of the component outage events resulting from the different phenomena described below. Often these effects are localized to an operating area such that widespread outages in the power system will not occur. Therefore, the reliability indices of a delivery point can be studied by modeling component outages within an area containing the delivery point and a few buses away.

A delivery point interruption, which is referred to as a system failure in this part of the chapter, is seldom a result of a single outage event. Overlapping of outage events is likely the cause. Since the outage events are contained to the initiating faulted components and are not wide spreading, good results can still be achieved by limiting the study of overlapping outage events to three or fewer components.

The approach is based on the Area Reliability Evaluation Program (AREP) developed by Hydro One Networks [27, 28] to calculate reliability indices of a selected group of customers supplied by a series of power sources. The method of minimal cuts is used to assess a continuity of supply from sources to sinks and to evaluate the reliability indices.

5.2.2 System Modeling and Component Data

The various phenomena modeled in the area reliability evaluation include:

- independent outages caused by faults,
- independent outages caused by false trips,
- common mode outages caused by faults,
- common mode outages caused by false trips,
- breaker failure (active and passive),
- protection-dependent failure,

- maintenance-dependent outages,
- repair-dependent outages,
- maintenance events,
- normally open breakers,
- operation of various protection zones.

Calculation of the frequency and duration of various outages involving the above phenomena are discussed in [27]. The effect of adverse weather is also included.

5.2.3 Classification of Outages

A delivery point is assumed to be interrupted if and only if all the electrical paths between the delivery point and all source points are interrupted. Interruptions are grouped into four types of system failures with different interruption durations as follows:

- *Permanent*: if the interrupted delivery point(s) can only be restored by repairing the corresponding component(s) on permanent outage.
- *Switching*: if the interrupted delivery point(s) can only be restored by isolating the corresponding component(s) on permanent outage.
- *Temporary*: if the interrupted delivery point(s) can only be restored by restoring the outage component(s) to service via manual reclosing disconnects and circuit breakers.
- *Transient*: if the interrupted delivery point(s) can only be restored by auto reclosure.

5.2.4 Calculation of Reliability Indices

As mentioned above, the reliability indices for small area studies are calculated using a minimal cut set approach. This method, although it is an approximation, yields very accurate results and is much more practical with larger systems than the Markov process. The frequency and duration equations can be found in [16, 29, 30].

5.2.5 Area Network Representation

In order to perform area supply reliability analysis, the pattern of the power flow from the sources to the delivery points has to be established. Therefore, a direction of power flow has to be assigned to every connection specified in the network. The connection elements used to connect two adjacent components can be either one or bi-directional.

System components are normally protected by the nearest breakers. The protection zone for each component is established using this rule. These are defined as standard protection zones. As an alternative, a nonstandard protection zone for component(s) or breaker(s) can be established to override the standard protection zone defined above.

The approach assumes that all components in the electrical system being analyzed are self-switched; i.e., each component can be isolated from the electrical system without also isolating of another component. Nevertheless, the one may specify a component to be switched out with other components, i.e., a nonstandard switching zone.

In many substations, certain breakers can be normally open. When a permanent system failure is identified by the software, attempts will be made to restore the system by closing the normally open breakers one by one in a specified order. If none of the closings are successful in restoring the system, then the classification of the failure remains as permanent. If, however, any one closing restores the system, then the failure is classified as switching, and frequency and duration calculations are performed accordingly.

A standard failure criterion would state that the supply of M out of N delivery points must be interrupted in order to have a system failure.

5.2.6 Study of Independent and Dependent Outages

Independent and dependent component outages are studied in the area reliability evaluation using failure modes and effect analysis. Although the general approach used is the same, different techniques are used to perform analysis for these two types of outages.

Independent Study

The minimal cut technique is used to simulate independent outage events and determine if they result in delivery point interruptions. The basic idea of this technique is that once an independent outage event or overlapping of two independent outage events causes an interruption, it will not be combined further with other outage events in the failure modes and effect analysis.

Dependent Study

While studying dependent outages, the three-component rule (maximum three components can be out of service at the same time) as well as transitional analysis are used to perform failure effect analysis. Unlike independent outages, the minimal cut set approach is not used to perform failure effect analysis when studying dependent outages. This is due to the fact that when dealing with dependent outages, system states that result in failures do not have to be minimal cut states. For this reason, transitional analysis is used to carry out failure effect analysis using a state transition diagram [27].

5.2.7 Ranking of Components

A crucial functionality introduced in ASSP relates to the selection of those components for the in-depth maintenance policy analysis that are the most important from the system operation point of view. A component's contribution to the system failure is termed its *importance.* It is a function of failure characteristics and system structure. An importance analysis is akin to a sensitivity analysis and is thus useful for system design, operation, and optimization. For example, we can estimate possible variations in system failure probability caused by uncertainties in component reliability parameters. Inspection, maintenance, and failure detection can be carried out in their order of importance for components, and systems can be upgraded by improving components with relatively large importance.

We will consider two ranking methods of system components. Both are based on the calculation of the derivative of the system failure probability with respect to the component probability of failure. This derivative, which will be used as one of the importance indicators, gives a measure of the sensitivity of the system failure probability with respect to the given component reliability. A given component can be important because this derivative takes a high value. This normally is the case when the system fails when this component fails, or this component appears in one or more minimal cuts that have a small number of other components.

The partial derivative considered here can have a high value even when the component has very small probability of failure. In order to take into account a contribution of the component failure probability to the system probability of failure, we will introduce the second measure called criticality importance.

The basic mathematical principles used in the development of the ranking tables are given in [16]. The implementation in the computer platform is an extension of this approach and is briefly discussed below.

Structural Importance

This is the simplest of the importance criteria and is merely the partial derivative (the classical sensitivity) of the probability of system failure p_{F} with respect to a component failure probability p_j. Thus, for the jth component, we have

$$\mathrm{IST}_j = \frac{\partial p_{\mathrm{F}}}{\partial p_j}. \tag{15}$$

Since the probability of system failure is a linear function of the component failure probability, the expression for the system probability of failure can be written as

$$p_{\mathrm{F}} = p_j K_j - (1 - p_j) L_j + H_j. \tag{16}$$

Constant K_j is equal to the sum of all components in the expression for the probability of failure that contain factor p_j with p_j excluded, L_j contains the terms with factor $q_j = 1 - p_j$, with q_j excluded and H_j contains the terms that have neither p_j nor q_j.

From (15), we have

$$\mathrm{IST}_j = K_j - L_j. \tag{17}$$

Thus, the structural importance can be easily evaluated if a mathematical expression for system failure can be written in form (16). Since, in a usual area reliability problem, there can be hundreds of minimal cuts, construction a symbolic expression for system failure probability can be a formidable task. As part of this development, a very efficient algorithm to accomplish this task has been programmed in the computer platform described here.

The structural importance of components can be used to evaluate the effect of an improvement in component reliability on the delivery point(s) reliability, as follows: By the chain rule of differentiation, we have

$$\frac{\partial p_{\mathrm{F}}}{\partial t} = \sum_{j=1}^{m} (K_j - L_j) \frac{\mathrm{d}p_j}{\mathrm{d}t} = \sum_{j=1}^{m} \mathrm{IST}_j \frac{\mathrm{d}p_j}{\mathrm{d}t} \tag{18}$$

where t is a common parameter—say, the time elapsed since the system development began. Thus, the rate at which system failure probability decreases is a weighted combination of the rates at which component probabilities of failure decrease, where the weights are the structural importance numbers.

From (18), we may also obtain

$$\Delta p_{\mathrm{F}} = \sum_{j=1}^{m} (K_j - L_j) \Delta p_j = \sum_{j=1}^{m} \mathrm{IST}_j \Delta p_j \tag{19}$$

where Δp_{F} is the perturbation in system failure probability corresponding to perturbations Δp_j in component failure probabilities.

Criticality Importance

The criticality importance considers the fact that it is more difficult to improve the more reliable components than to improve the less reliable ones. Dividing both sides of (19) by p_{F}, we obtain

$$\frac{\Delta p_{\mathrm{F}}}{p_{\mathrm{F}}} = \sum_{j=1}^{m} \frac{p_j}{p_{\mathrm{F}}} \mathrm{IST}_j \frac{\Delta p_j}{p_j}. \tag{20}$$

The criticality importance of the *j*th component is defined as

$$\mathrm{ICR}_j = \frac{p_j \mathrm{IST}_j}{p_{\mathrm{F}}}. \tag{21}$$

Thus, the criticality importance is a fractional sensitivity.

5.3 Analysis of Component Maintenance Policy

Having selected the components for further analysis, the analysis of the effect of changes of the component maintenance policy on its remaining life and failure rate are called from the computer platform. The most important features of the proposed approach are described below.

5.3.1 Calculation of the Remaining Life of the Equipment

The remaining life of the equipment is computed using the Markov model as shown in Fig. 8. The calculations use the notion of the FPT discussed later.

FPT

Let T_{ij} be the FPT from state i to state j in a finite-state, continuous-time Markov chain with continuous parameter (CTMC) $\{Z(t),\ t \geq 0\}$ with state space $\Omega = \{1, 2, \ldots, n\}$. The continuous parameter is often time. The transition rate matrix is defined as $\mathbf{A} = [\lambda_{ij}]$, where λ_{ij} $(i \neq j)$ represents the transition rate from state i to state j and the diagonal elements $\lambda_{ii} = -\sum_{j \neq i} \lambda_{ij}$. We let $\eta = \max|\lambda_{ij}|$. Let C represent the set of absorbing states[1] and B $(=\Omega - C)$ the set of the transient states in the CTMC. From the matrix $\mathbf{A}$, a new matrix $\mathbf{A}_{\mathrm{B}}$ of size $|B| \times |B|$, where $|B|$ is the cardinality of the set B, can be constructed by restricting $\mathbf{A}$ to only the states in B.

Since $Z(t)$ is distributionwise equivalent to a Poisson process, we have [18]:

$$T_{ij} = \sum_{k=0}^{N_{ij}} V_k(t) \quad \text{where } V_k(t) = \eta\, \mathrm{e}^{-\eta t} \quad \text{for all } k. \tag{22}$$

[1] A state is considered an absorbing state if there are no outgoing transitions from that state. Therefore, for an absorbing state i, $\lambda_{ij} = 0$ for all j.

Thus, $E(T_{ij}) = E(N_{ij})\frac{1}{\eta}$, where N_{ij} is the FPT of a discrete parameter Markov chain. But (see page 167 of [16]):

$$E(N_{ij}) = \sum_{k=1}^{n} n_{ik} = \left(I - \frac{A_B}{\eta} - I\right)^{-1} = -\eta \sum_{k=1}^{n} A_B^{-1}(i,k) \tag{23}$$

where only the transient states are in $\mathbf{A_B}$. Hence,

$$E(T_{ij}) = \sum_{k=1}^{n} -A_B^{-1}(i,k). \tag{24}$$

Those FPTs are used to generate the life curve of the equipment.

Life Curves of the Equipment

The concepts of a life curve and discounted costs are useful to show the effect of equipment aging with time and were discussed at the beginning of this chapter. Figure 1 shows an example of two life curves for the same type of equipment under two different maintenance strategies. If a "better" maintenance strategy is selected, or if the operating conditions are more favorable, the equipment will last longer and at a particular point in time, its condition (or asset value) will be higher.

The generation of a life curve requires several steps. They are described in the following.

This happens in several steps, as explained below with the help of Fig. 13:

- First, the borderlines between the deterioration stages D1, D2 and D3, expressed in terms of percentages of equipment condition, are marked on the vertical axis and entered into the program.

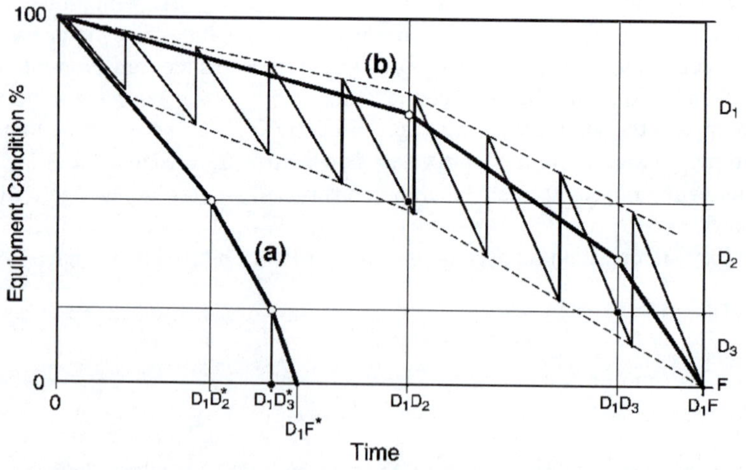

Fig. 13 Development of life curves **a** without maintenance, **b** with maintenance

- Next, AMP/FPT calculations are carried out by the program, to determine the FPTs between states D1 and D2, D1 and D3, and D1 and F. These are entered on the time axis of Fig. 13. Using the AMP model, the effects of maintenance are already incorporated.
- If there was no maintenance, the FPTs D1D2*, D1D3* and D1F* would be obtained and the corresponding life curve would run as shown. With maintenance, the life curve is no longer a smooth line but a rugged one indicating the deterioration between maintenances and the improvements caused by them. A crude realization of the process is shown in Fig. 13. It is a deterministic approximation that does not consider all possibilities inherent in the AMP model; nevertheless, it helps to visualize how an equivalent smooth life curve is constructed.
- The equivalent smooth life curve is drawn by observing the following simple rules. At time 0 it must be at 100%, at D1F it must be 0. At the remaining two ordinates, by arbitrary decision, it should be near the lower quarter of the respective domains. (In Fig. 13, the midpoints are used, an earlier convention.)

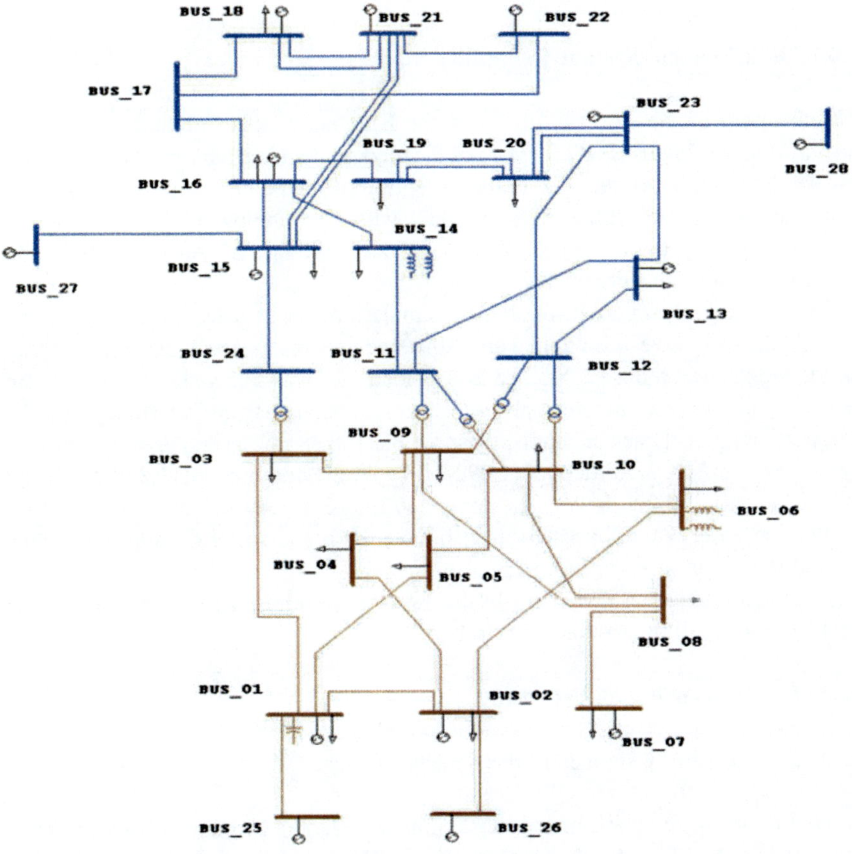

Fig. 14 The diagram of the 24-bus RTS system

5.4 Numerical Example

To illustrate how the complete study is performed, let us assume that a maintenance budget for the high-voltage breakers in the system under consideration is specified and our task is to examine several asset sustainment options for these pieces of equipment. For illustration purposes, we have selected the IEEE Modified Reliability Test System (MRTS). MRTS is a modification of the IEEE RTS [31], with the objective of stressing the transmission network. Bearing in mind this objective, the original generating capacities and peak loads are multiplied by two. The system has 24 buses, 38 circuits and 14 plants (32 generating units). The total installed capacity is 6,810 MW, with a peak load of 5,650 MW. Even though the computer platform can handle very large power systems, as described in [23], a relatively small system was selected for illustrative purposes because (1) it is familiar to many power system engineers, and (2) it allows a better understanding of the procedures adopted in the platform.

The following sections describe the procedure indicated in Fig. 2.

5.4.1 Bulk Electric System Reliability Study

The first step in the study is to set up the BES network information. Figure 14 displays the 24-bus IEEE RTS [31]. All the electrical and reliability parameters are as specified in this reference with the exception of load and generation quantities, which are doubled from the original values with a chronological load model with the load curve represented by 8,760 hourly values. Outages of generators, lines and transformers are considered.

In this study, a pseudo-chronological simulation was selected with 50,000 samples. A flat load curve was applied and, since the network is small, the entire system was selected as an outage area. The loss of load cost was selected as the governing reliability index. The unit interruption cost curves are taken from Ontario Hydro [32]. The participation of each consumer class per bus is the same as that used in [25]. The system total is: 19.2% residential (1,092 MW), 24.2% commercial (1,379 MW) and 56.6% industrial (3,229 MW). The computed LOLC values are shown in Fig. 15.

We can observe that buses 14 and 16 have the highest LOLC values. The area around these buses is shown in Fig. 16.

The three stations represented by buses 14, 15 and 16 will be represented in detail in the small area reliability study.

5.4.2 Small Area Reliability Study

Assigning Reliability Indices to the Selected Buses

A diagram showing details of the substations represented by buses 14, 15 and 16 is given in Fig. 17. The bus and breaker failure rates and repair times are not a part of the RTS database. The adopted parameters are summarized in Table 7.

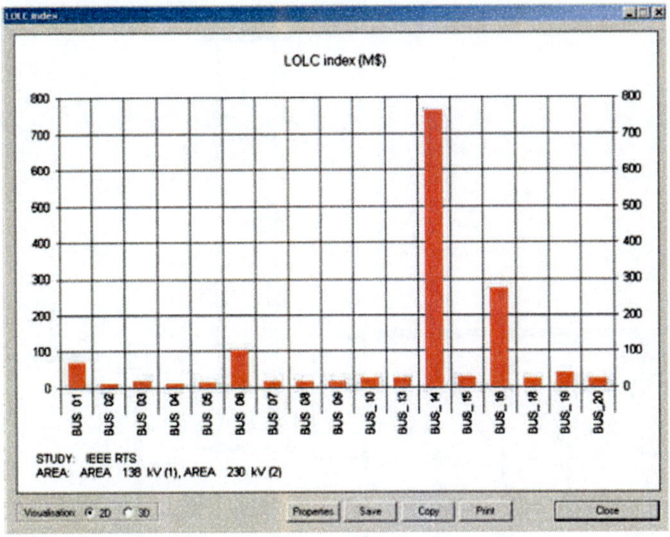

Fig. 15 LOLC values for selected buses

Fig. 16 Area selected for detailed study

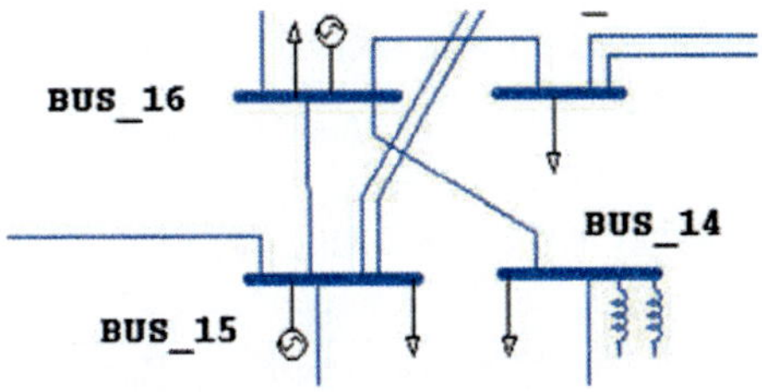

The duration of a permanent outage of a breaker is set at 6 weeks. In addition to a fault, a failure to open or close is also modeled for breakers and the probability of a stuck breaker is equal to 0.006. Failure rate for transient outages is assumed to be equal to 0. The duration of the temporary outages is assumed to be equal 30 min for all components. Each of the three stations has one node denoted as a sink to represent the load connected at the station. Stations representing buses 15 and 16 in Fig. 17 have 2 (nodes 14 and 22) and 3 (nodes 27, 47 and 48) source nodes, respectively, representing possible power inflows to the station either from the generator or the external lines connected to this station but not represented in Fig. 16. In this example, each of the substations in Fig. 17 is analyzed individually. When the substation representing bus 16 is analyzed in this way, it has 4 sources (one additional source represents a possible inflow from bus 15) and 2 sinks (one its own load and the other representing the supply to bus 14).

Table 8 summarizes the reliability indices for the station with the 6-diameter arrangement in Fig. 9 representing bus 15 in Fig. 16.

The analysis is now repeated for the remaining two stations and the results from the last row in tables similar to Table 2 are used now in the BES reliability study.

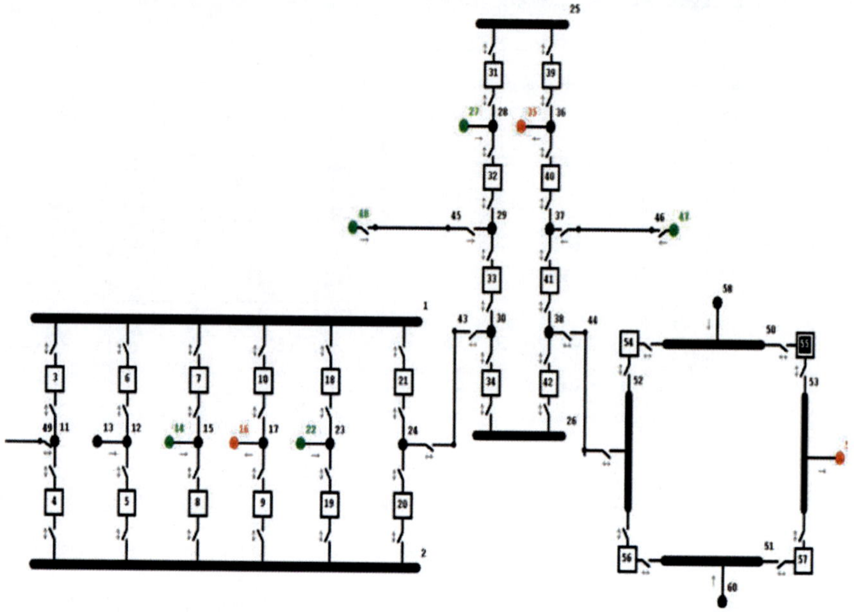

Fig. 17 Three stations representing buses 14, 15 and 16 in Fig. 16. The relative location is the same as in Fig. 14

Table 7 Component reliability parameters

Element	Permanent outages		Temporary outages	
	Failure rate (1/year)	Duration (h)	Failure rate (1/year)	Duration (h)
Breaker	0.025	1008	0.01	0.5
Bus	0.006	12.3	0.009	0.4
Line	As per IEEE	As per IEEE	0.5	0.4

Table 8 Reliability indices for the station representing Bus_15

Failure type	Number of failures	Freq. (1/year)	Duration (h)	Prob. (h/year)	MTTF (1/year)
Permanent	12	0.0042	21.12	0.0891	236.97
Temporary	56	0.0122	0.196	0.0024	81.94
Switching	58	0.0504	0.501	0.0253	19.82
Total	128	0.0668	1.746	0.1168	14.96

A new base case is now established. The results are similar to the ones shown in Fig. 15. Figure 18 shows the differences in the LOLC between the base case and the case where all buses are 100% reliable. Only the buses with load in this system are shown.

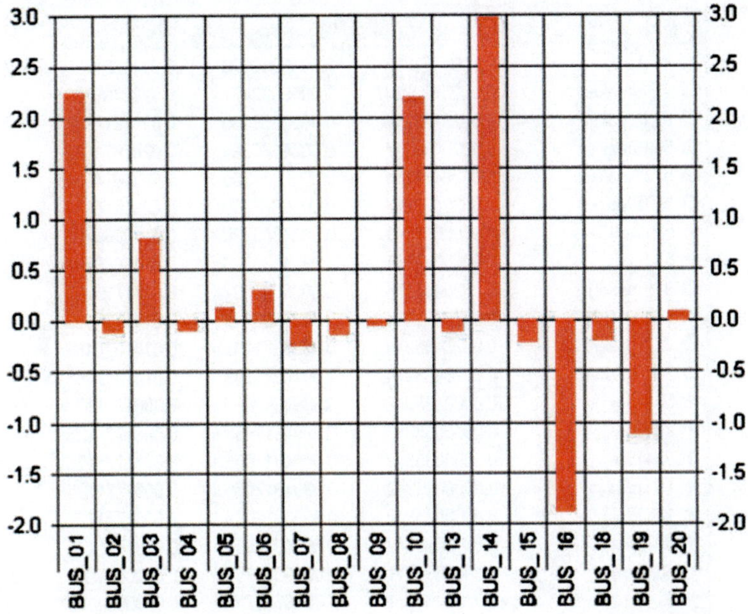

Fig. 18 The difference in the LOLC between the base case and the studies in Fig. 15

We can observe that buses 1, 10 and 14 show a substantial increase in the LOLC values and bus 16 shows a comparative decrease in the customer interruption cost.

The next step in the analysis is the creation of the component ranking tables.

Criticality Ranking of Components

There are 24 high-voltage breakers in the 3 stations shown in Fig. 17. A visual inspection of the station configurations immediately points out that the breakers in the ring bus are very important for the station with this configuration. The criticality of the other breakers is not so apparent. The results of the criticality analysis are summarized in Fig. 19.

The elements in this table are ordered according to their criticality importance. Several interesting observations can be drawn analyzing this table.

The two methods rank the system components quite differently. For example, all the breakers in the ring bus substation and three other breakers are ranked in the list of the first 10 most critical components.

However, from a structural point of view, only buses (and line 44) enter this list.

Out of 24 breakers in this system only 11 are important either from a structural or criticality point of view. This is because, in this analysis, only single and

Component	Probability of failure	Structural import...	Criticality import...
44 (Line)	0.0004771689	0.9999052970	0.8343851850
41 (Breaker)	0.0028433214	0.0056872471	0.0282789700
54 (Breaker)	0.0028433214	0.0056676380	0.0281814667
56 (Breaker)	0.0028433214	0.0056676380	0.0281814667
57 (Breaker)	0.0028433214	0.0056676380	0.0281814667
55 (Breaker)	0.0028433214	0.0056676380	0.0281814667
52 (Bus)	0.0000084041	0.9994365726	0.0146886703
53 (Bus)	0.0000084041	0.9994365726	0.0146886703
9 (Breaker)	0.0028433214	0.0028500702	0.0141715399
10 (Breaker)	0.0028433214	0.0028500702	0.0141715399
42 (Breaker)	0.0028433214	0.0028371199	0.0141071468
34 (Breaker)	0.0028433214	0.0028371199	0.0141071468
59 (JBus)	0.0000035342	0.9994317055	0.0061771125
38 (JBus)	0.0000035342	0.9994317055	0.0061771125
36 (JBus)	0.0000035342	0.9994317055	0.0061771125
16 (JBus)	0.0000035342	0.9994317055	0.0061771125
17 (JBus)	0.0000035342	0.9994317055	0.0061771125
35 (JBus)	0.0000035342	0.9994317055	0.0061771125
50 (Bus)	0.0000084041	0.0056515705	0.0000830609
51 (Bus)	0.0000084041	0.0056515705	0.0000830609
37 (JBus)	0.0000035342	0.0113260006	0.0000700018
2 (Bus)	0.0000084041	0.0028419904	0.0000417686
1 (Bus)	0.0000084041	0.0028419904	0.0000417686
26 (Bus)	0.0000084041	0.0028290769	0.0000415788
30 (JBus)	0.0000035342	0.0028290631	0.0000174854
31 (Breaker)	0.0028433214	0.0000035022	0.0000174139
39 (Breaker)	0.0028433214	0.0000035022	0.0000174139
25 (Bus)	0.0000084041	0.0000034922	0.0000000513
28 (JBus)	0.0000035342	0.0000034922	0.0000000216

Fig. 19 Criticality ranking of system components

two-element minimal cut sets were considered and the remaining breakers did not appear in any of these cuts.

Out of all the breakers in this system, breaker 41 is the most important from the criticality point of view. Both methods would rank the breakers in the same way.

Since a group of breakers at the ring bus (bus 14 in Fig. 14) is at the top of the list, the maintenance activities at this station are reviewed in more detail in the next section.

5.4.3 Analysis of the Breaker Maintenance Policy

We will analyze the performance of the high-voltage breakers. We will start by reviewing the present maintenance policy.

Present Maintenance Policy

The present maintenance policy is described earlier in the chapter in the numerical example discussing the application of the life curves. We will recall that, reflecting a general utility practice, we assumed that three types of maintenance are routinely performed on each breaker.

About every 8 months to a year, a minor maintenance is performed involving timing adjustments and lubrication at a cost of about $700. Medium maintenance involving replacement of some parts is performed approximately every 10 years and costs about $6,000. A major maintenance involving breaker overhaul takes place every 15 years and costs about $75,000.

In the breaker example, three types of maintenance were modeled. The equipment could be in one of four possible states: "as new—D1", "slightly deteriorated—D2", in "major deterioration state—D3" or "failure—F". The output of this analysis yields information shown in Fig. 20.

As the result of the study, we obtain the expected time to failure from various deterioration states (these times range from 40.5 to 27.1 years from "as new" to "badly deteriorated" breaker, respectively) and the percentage of the lifetime that the breaker is expected to be in each deterioration state.

Alternative Maintenance Policy

The next step in the analysis is the selection of a maintenance procedure by either staying with the present policy (as described above) or performing major refurbishment or replacing the breaker with a new, possibly of a different type. These studies will be performed by analyzing the life curves of the equipment and the associated costs. A description of the computer program used to perform this study can be found in [33]. The time horizon for the calculation of the discounted costs is set at 10 years with the inflation and discount rates of 3 and 5%, respectively. The system and penalty costs associated with equipment failure are equal to $10,000 each in this example.

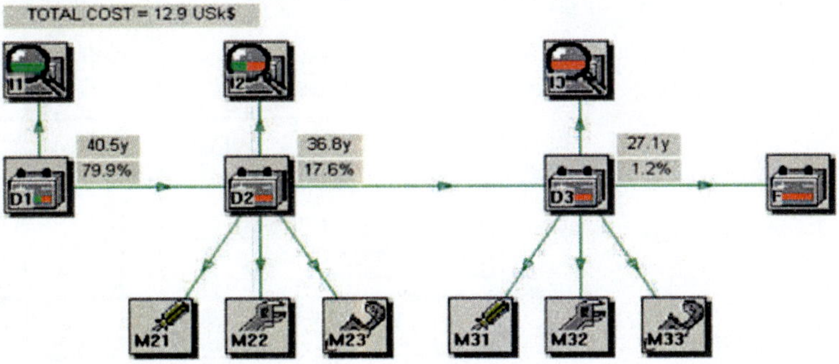

Fig. 20 Reliability indices for the air-blast breakers

In order to calculate the effect of the revised maintenance policy, we need to specify the present asset condition (assumed at 80% in this example) or asset value. This information determines where the equipment is located on the life curve.

In order to analyze maintenance alternatives, possible actions need to be defined. We will consider three possible maintenance actions for the breakers in the station represented by bus no. 14 with the ring construction. In addition to continuing present maintenance policy, we will consider a major refurbishment of the air-blast breakers at a cost of $75k per breaker or a replacement with a newer design (e.g., SF6 construction). The cost of a new breaker is set at $150k. The refurbished breakers will have the same maintenance policy as the current system whereas the new design will have the minor maintenance performed once a year rather than every eight months and medium repairs every 5 years. With the financial and engineering data specified, the calculations of reliability and cost information can now proceed.

5.4.4 Life Curves

Figure 21 shows the life curves of the selected breaker under various maintenance policies considered in this example. The two new maintenance alternatives will be introduced after a 3-year delay. In the replacement action, the equipment is assumed to return to as new condition.

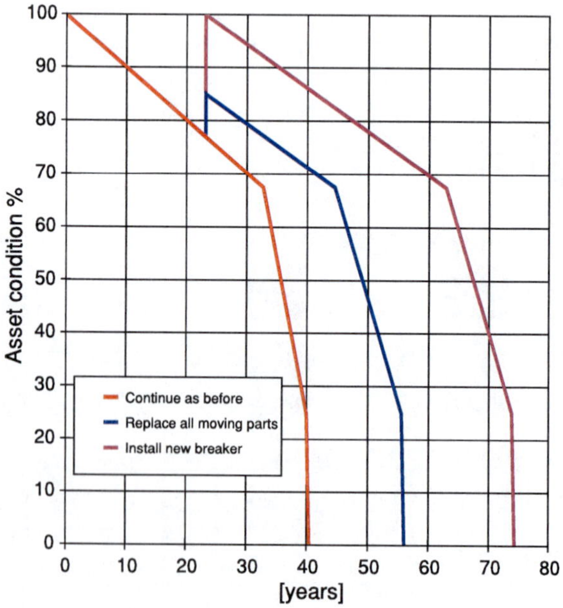

Fig. 21 Life curves for various maintenance policies

5.4.5 Cost Curves

Cost computations involve calculation of the number of different types of repairs during the specified time horizon. The number of repairs in the period before and after the action is taken is computed separately. The cost of each repair is then expressed by its present value. Failure and associated costs are computed as expected values. The probabilities of failure before and after the action are computed by the program.

The cost curves are presented as functions of the delay. Since each action can have different delay, the delay time can be specified either in years or as a percentage of the specified delay time. The last option allows display of the curves in one screen without the necessity of providing a separate delay scale for each action.

Figure 22 shows the cost diagram for all actions with a 3-year delay for each (100% of assumed value).

If a 3-year delay was contemplated in the application of any action, the best policy would be to either stay with the present policy or to perform a major refurbishment with the resulting cost of about $95,000. The major portion of this cost is due to the refurbishment action itself. We can observe that for the 10-year time horizon, the expected cost of failure (computed during the analysis but not shown here) is fairly small in all the cases since the probability of failure is small for these actions. The maintenance cost is high for the present maintenance policy because minor repairs are performed quite often and, during the 10-year horizon, one medium and one major repair will be performed.

5.4.6 Comparative Studies

The new breaker failure rates result in improved station reliability characteristics. In particular, the new bus failure rate is equal to 0.114 (1/year) compared with 0.131 (1/year) in the base case. The repetition of a BES reliability studies with this new information yields the revised LOLC values. The difference between new results and the ones in the base case is shown in Fig. 23.

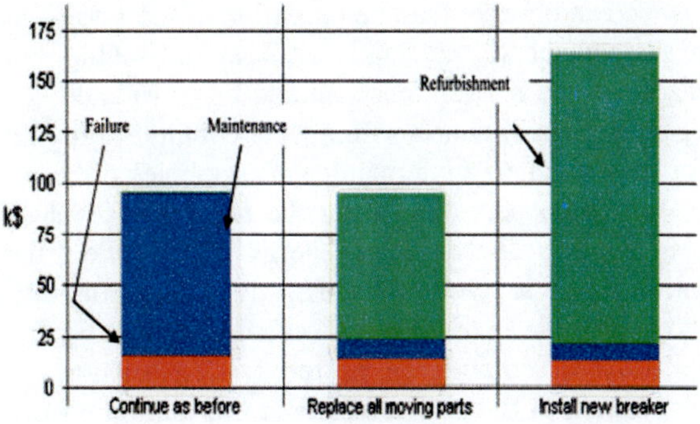

Fig. 22 Cost diagram for various actions with a 3-year delay. The components of each bar are as follows (from the *bottom*): failure cost, maintenance cost, refurbishment action cost

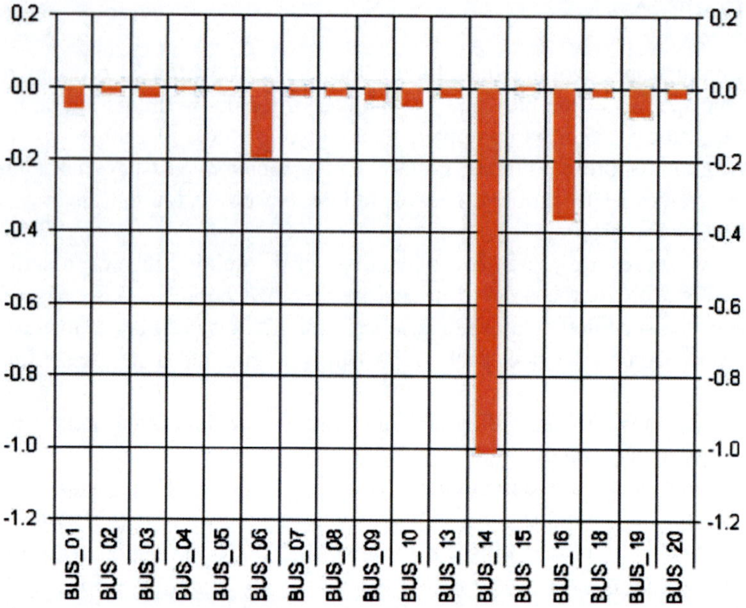

Fig. 23 The decrease in $M of the LOLC values with SF6 breakers installed at the substation represented by bus 14 compared to the base case

Table 9 Comparison of two maintenance alternatives

Index	Present maintenance policy	Replace breakers at bus 14
LOLC (M$/year)	1,391	1,390
EENS (GWh/year)	363.4	362.9
Frequency of interruptions to customers at Bus_14 (1/year)	32.6	32.5
Average duration of interruptions at Bus_14 (h)	61.6	61.6
PV of Expected total cost per breaker at Bus_14 (k$/10 years)	95	170
Average life of a breaker (years)	40	60

We can observe that the reduction of the loss of load cost over a 1 year period reaches about $1 M. This is a substantial saving, but before the final decision is made, all the relevant costs should be considered. Table 9 shows the comparison of the results obtained with all the computer programs of the computer platform for the two alternative maintenance policies discussed above.

We can observe that from the reliability point of view, both alternatives are very similar. The expected energy not supplied, the frequency and duration of interruptions are very close. However, the economic considerations would favor the alternative involving breaker replacement.

The cost comparison of the two alternatives shows that even though the installation of the new breakers at this station will result in an additional cost of about $300,000 over the 10-year planning horizon, the savings in the cost of load interruptions are much greater, reaching about $1,000,000 per year. In addition, the average life of the replacement breaker is about 50% greater than the old one.

6 Conclusions

In this review, a survey is offered of the various maintenance methods available to operators. The methods range from the simplest, "follow the manual"-types to detailed probabilistic approaches. To get most out of maintenance, one would have to select a mathematical model where optimization is possible—optimization for highest reliability or lowest operating costs. There can be little doubt that such probabilistic models would be the best tools for identifying policies that provide the highest cost savings.

Another choice of which operators are becoming more and more aware is to apply a maintenance policy based on no rigid schedule but on the "as needed" principle. This can be implemented with or without mathematical models; example for the latter is the RCM approach. RCM, steadily gaining in popularity, is based on an analysis of failure causes and past performance, and helps to decide where to put the next dollar budgeted for maintenance. The method is good for comparing policies, but not for true optimization.

This chapter discusses models that can be applied for finding an optimal maintenance policy for power equipment. The emphasis is placed on an optimization formulation of a maintenance policy model based on semi-Markov processes. The model allows the analysis of the influence of the maintenance policies, defined by the durations of repairs, the effects and the costs of various repair actions on the remaining life and the lifetime utilization costs of the equipment. Different types of possible Markov model optimizations were discussed and a simulated annealing algorithm was introduced. This algorithm was found to be very well suited to the solution of the maintenance optimization problem.

The operation of the proposed model was demonstrated on a numerical example for high-voltage circuit breakers with a significant improvement of the three important parameters defining a maintenance policy: the remaining life of equipment, the total utilization cost and the unavailability.

In today's competitive environment, cost optimization is becoming even more important. This is particularly true for transmission and distribution equipment where the maintenance choices described in this chapter fully apply. As for generating units, the situation is somewhat different.

In the past, the practice was to centrally plan and coordinate the maintenance of generators within a given jurisdiction. Maintenance was done during low-load seasons and the timing was influenced by such considerations as system risk and production cost. In the deregulated scenario, maintenance may not be centrally planned or even coordinated. Generator owners may tend to keep the units running when the market clearing price of electric energy is high, and perform maintenance only when the market price is low. Even then, they may wish to sell energy to another jurisdiction where the periods of high load (and high market price) are different from those near the unit's location. Therefore, the decision when to maintain a generator will be heavily influenced by profit incentives and the optimal cost of maintenance and repair would be assessed in this context. But even then, some of the approaches and programs discussed in this review would retain their relevance.

The consequences of alternative maintenance actions can be analyzed from three different points of view. Engineers might be mostly interested in the effect of the asset sustainment policy on the asset condition and the probability of failure of the equipment. The condition of the asset can be visualized in a form of a life curve. Development of such curves constitutes a significant part of the analysis. Markov models can be used for this purpose. Solving the equipment Markov model brings information about the probability of failure during the specified time horizon. This probability is, in turn, used to determine the expected cost of the equipment maintenance and failure during this time. The financial information is presented in terms of present values taking into account the anticipated inflation rate and corporate discount rate. The system effects are analyzed with the help of two different approaches: an area and bulk electric system reliability programs, the last one also takes into account the customer interruption costs.

The proposed approach was illustrated in a study to analyze the effect of changes in breaker maintenance policy on the performance of the equipment itself and the reliability of a small area around the substation where the breakers are installed as well the reliability of the entire bulk electric system.

The main feature of this approach is the requirement for a seamless transition between various computational modes. The information transferred can be as basic as the equipment failure rates under various maintenance scenarios, as happens between BES and small area analysis programs or between small area and component analysis modules, or as complex as the complete maintenance strategy and life curves transfer between the computational modules. This innovative design will allow a comprehensive analysis of asset sustainment alternatives in a way that was not possible until now.

The ability to combine engineering and financial information coupled with the ease of use of the computer platform has proven to be an important asset for the re-investment decision-making process.

Acknowledgments The author would like to thank Dr. John Endrenyi with whom he has collaborated for many years and whose contributions appear in several parts of this chapter.

References

1. IEEE Task Force (Endrenyi J, Chair), (2001) The present status of maintenance strategies and the impact of maintenance on reliability. IEEE Trans Power Syst 16(4):638–646
2. Endrenyi J, Anders GJ (2006) Aging, maintenance and reliability. IEEE Power Energy Mag 4(3):59–66

3. Bertling L (2002) Reliability centered maintenance for electric power distribution systems. PhD Dissertation. Royal Institute of Technology, Stockholm
4. Endrenyi J, Anders GJ, Kalinowski B, Bertling L (2004) Comparison of methods for evaluating the effects of maintenance on component and system reliability. In: PMAPS'2004 conference in Ames, Iowa, September 14–17
5. Allan RN, Adraktas AN (1982) Terminal effects of protection system failures in composite system reliability evaluation. IEEE Trans Power Apparatus Syst 101:4557–4562
6. Allan RN, Billinton R, Shahidehpour SM, Singh C (1988) Bibliography on the application of probability methods in power system reliability evaluation, 1982–1987. IEEE Trans Power Syst 3(4):1555–1564
7. Allan RN, Billinton R, Briephol A, Grigg MCH (1994) Bibliography on the application of probability methods in power system reliability evaluation, 1987–1991. IEEE Trans Power Syst 9(1):41–49
8. Allan RN, Billinton R, Briephol A, Grigg MCH (1999) Bibliography on the application of probability methods in power system reliability evaluation, 1992–1996. IEEE Trans Power Syst 14(1):51–57
9. Billinton R (2001) Bibliography on the application of probability methods in power system reliability evaluation, 1996–1999. IEEE Trans Power Syst 16(4):595–602
10. Billinton R, Medicherla TKP (1981) Station originated multiple outages in reliability analysis of a composite generation and transmission system. IEEE Trans Power Apparatus Syst 100:3870–3878
11. Billinton R, Tatla J (1983) Composite generation and transmission system adequacy evaluation including protection system failure modes. IEEE Trans Power Apparatus Syst 102(6):1823–1830
12. Billinton R, Vohra PK (1987) Station initiated outages in composite system adequacy evaluation. IEE Proc C 134(1):10–16
13. Singh C, Patton AD (1980) Models and concepts for power system reliability evaluation including protection-system failures. Int J Electric Power Energy Syst 2:161–168
14. Allan RN, Ochoa JR (1988) Modeling and assessment of station originated outages for composite system reliability evaluation. IEEE Trans Power Syst 3(1):158–165
15. Moubray J (1992) Reliability-centered maintenance. Industrial Press Inc., New York
16. Anders GJ (1990) Probability concepts in electric power systems. Wiley, New York
17. Endrenyi J, Anders GJ, Leite da Silva AM (1998) Probabilistic evaluation of the effect of maintenance on reliability—an application. IEEE Trans Power Syst 13(2):576–583
18. Ross S (1995) Stochastic processes. Wiley, New York
19. Jardine AKS (1973) Maintenance replacement and reliability. Pitman Publishing, London
20. Stopczyk M, Sakowicz B, Anders GJ (2008) Application of a semi-Markov model and a simulated annealing algorithm for the selection of an optimal maintenance policy for power equipment. Int J Reliab Saf 2(1/2):129–145
21. Anders GJ, Leite da Silva A (2000) Cost related reliability measure for power system equipment. IEEE Trans Power Syst 15(2):654–660
22. Løken E, Botterud A, Holen AT (2006) Decision analysis and uncertainties in planning local energy systems. In: PMAPS 2006, Stockholm, Sweden, June 2006
23. Anders GJ, Hamoud G, Leite da Silva AM, Manso L (2003) Optimal outage scheduling—example of application to a large power system. Int J Electr Power Energy Syst 25(8):607–614
24. Leite Da Silva AM, Anders GJ (2003) Composite reliability evaluation for large scale power systems. In: Proceedings of 2003 power technology conference in Bologna, Italy, June 2003
25. Leite Da Silva AM, Manso LAF, Mello JCO, Billinton R (2000) Pseudo-chronological simulation for composite reliability analysis with time varying loads. IEEE Trans Power Syst 15(1):73–80
26. Kumar S, Billinton R (1989) Adequacy evaluation of a small area in a large composite power network. IEEE Trans Power Syst 4(2):551–558

27. Endrenyi J, Maenhaut PC, Payne LE (1973) Reliability evaluation of transmission systems with switching after faults—approximations and a computer program. IEEE Trans Power Apparatus Syst 92:1863
28. Hamoud G, El-Nahas I (2003) Assessment of customer supply reliability in performance-based contracts. IEEE Trans Power Syst 18(4):1587–1593
29. Endrenyi J (1978) Reliability modeling in electric power systems. Wiley, Chichester
30. Billinton R, Allan RN (1983) Reliability evaluation of engineering systems. Plenum Press, London
31. IEEE RTS Task Force of APM Subcommittee (1999) IEEE reliability test system—1996. IEEE Trans Power Syst. 14(3):1010–1020
32. EPRI (1989) Customer demand for service reliability, RP2810
33. Anders GJ, Yung C, Endrenyi J (2001) Risk-based planner for asset management. IEEE Comput Appl Power 14(4):20–26

Mathematical and Physical Properties of Reliability Models in View of their Application to Modern Power System Components

Elio Chiodo and Giovanni Mazzanti

Abbreviations and list of symbols

ALT	Accelerated life test
a.s.	Almost surely
BS	Birnbaum–Saunders (distribution)
cdf	Cumulative distribution function
CLT	Central limit theorem
CV	Coefficient of variation
CRF	Conditional reliability function
DHR	Decreasing hazard rate
DRA	Direct reliability assessment
D[Y]	Standard deviation of the RV Y
E[Y]	Expectation of the RV Y
EV	Extreme value (distribution)
$F(x)$	Generic cdf
$f(x)$	Generic pdf
$f(x)$, $F(x)$	pdf and cdf of stress
$g(y)$, $G(y)$	pdf and cdf of strength

Abbreviations and list of symbols: The singular and plural of names are always spelled the same. "Log" always denotes natural logarithm. Random variables are denoted by uppercase letters.

E. Chiodo (✉)
Department of Electrical Engineering, University of Naples, "Federico II"
Via Claudio 21, 80125 Naples, Italy
e-mail: chiodo@unina.it

G. Mazzanti
Department of Electrical Engineering, University of Bologna, Viale Risorgimento 2, 40136 Bologna, Italy
e-mail: giovanni.mazzanti@mail.ing.unibo.it

G. Anders and A. Vaccaro (eds.), *Innovations in Power Systems Reliability*, Springer Series in Reliability Engineering, DOI: 10.1007/978-0-85729-088-5_3,

$G(r, \phi)$	Gamma distribution with parameters (r, ϕ)
$H(\ )$	Cumulative hrf
hrf	Hazard rate function
HRM	Hyperbolic reliability model
HV	High voltage
IDHR	First increasing, then decreasing hazard rate
IG	Inverse Gaussian (distribution)
IHR	Increasing hazard rate
IID	Independent and identically distributed (random variables)
IPM	Inverse power model
IRA	Indirect reliability assessment
IW	Inverse Weibull (distribution)
LL	Log-logistic (distribution)
LN	Lognormal (distribution)
LT	Lifetime
MRL	Mean residual life
MV	Medium voltage
$N(\alpha, \beta)$	Normal (Gaussian) random variable with mean α and standard deviation β
pdf	Probability density function
$r(s)$	Mean residual life function at age s
RF	Reliability function
$R(t)$	Reliability function at mission time t
$R(t\|s)$	Conditional reliability function at mission time t, after age s
RV	Random variable
SD, σ	Standard deviation
s-independent	Statistically independent
SP	Stochastic process
SS	Stress-strength
Var, σ^2	Variance
$W(t)$	Wear process at time t acting on a device
$W(a, b)$	Weibull model with RF: $R(x) = \exp(-ax^b)$
$W'(\alpha, \beta)$	Alternative form of Weibull model, with RF: $R(x) = \exp[-(x/\alpha)^\beta]$
$\delta(\cdot)$	Dirac delta function
X, $X(t)$	Stress (RV or SP)
Y, $Y(t)$	Strength (RV or SP)
$\Gamma(\ )$	Euler–Gamma function
$\Gamma(,)$	Incomplete gamma function
μ	Mean value (expectation)
$\Phi(z)$	Standard normal cdf
$\varphi(z)$	Standard normal pdf

1 Introduction

1.1 A Premise

In this introduction, particularly from Sect. 1.3 on, reference will be made to some mathematical probability theory which is thought to be useful for the methodological development of the sequel. Since only some particular aspects are highlighted here, it is assumed the reader being acquainted with the analytical definitions and basic properties in the theory of random variables and stochastic processes, with particular reference to the *reliability function* and related functions, such as the *hazard rate function*, the *mean residual life* (MRL) *function*, the *conditional reliability function* (CRF), which will be briefly recalled below. Suggested books in reliability theory are,[1] e.g., [3, 6, 10, 11, 16, 21, 30, 83, 88, 97, 103, 113, 123, 124, 146, 151, 152], for the above basics and also further details. It must be highlighted that the literature in this field is huge and fast growing in the large number of applicative fields, so only some of the most representative papers or books will be cited in the chapter. Some of the references are not mentioned in the text.

The need to develop new methodologies for reliability estimation of technological products is becoming increasingly important in every field of engineering. This is due to the combination of two main reasons: one is the fast extension of liberalization (which is discussed below with reference to the power systems); the other is the ever-increasing level of technological innovation, which brings about higher and higher reliability values for components, thus implying scarcity of failure data. Also, as pointed out in [68], the tendency to very short product-development times and tightened budgets imply that reliability tests must be conducted with severe time constraints, so that frequently no failures occur during such tests. These aspects are widely recognized in literature in almost all technological fields, from electrical [107] to marine [22] engineering. Therefore, the opportunity arises to adopt models and methodologies allowing the most efficient reliability estimation, based not only on the experimental data, but also on technological and physical information usually available to the engineer or analyst, as highlighted in some recent literature [42, 75, 76, 81, 158]. This information is related to wear and stresses acting on the device, e.g., overvoltages or short-circuits in the case of electrical components—and/or its typical models of aging, as expressed mathematically, e.g., by the time behavior of its hazard rate function: as a trivial example of the latter aspect, one may consider that most electronic components are scarcely affected by aging, being their failure largely "accidental", so that a constant hazard rate function could be reasonably expected.

---

[1] The references at the end of the chapter are listed alphabetically, due to the length of the Bibliography, for an easier search by the reader. They are referred to in the chapter by their number.

This leads to the widespread adoption of an Exponential reliability model in such field, even in the absence of many data to support it on a statistical basis.

These two aspects (physical evolution of wear and mathematical models of aging) are of course closely related with each other, but in a way that may sometimes appear divergent from—if not contrary to—"intuition". For instance, this chapter points out that many lifetime distributions deduced by wear models evolving as a random increasing function of time (such as the Gaussian processes leading to the Inverse Gaussian or the Birnbaum–Saunders distribution) are characterized by a decreasing hazard rate for large values of service times; instead, an increasing hazard rate function with time could be expected on purely intuitive grounds (as erroneously reported in some books). This is a property which is seldom discussed in the relevant literature, but it has found experimental evidence, e.g., in the field of electrical insulation [93, 148], and has been debated theoretically [93, 146]. Nevertheless, it is still misunderstood in some recent texts. This fact is not surprising, since many examples are found in probability theory that lead to results appearing, at first sight, illogical or paradoxical even to academics, sometimes[2] nonetheless, it highlights the need to acquire further insight into aging properties, wear mechanisms and the relations between them.

The use of "prior"[3] technological and physical information leads—as discussed throughout the chapter—to the so-called "indirect reliability assessment" (IRA). Such approach is so denoted in that it infers lifetime characteristics from the properties of the stochastic process (SP) describing the wear affecting the device, rather than using statistical fitting, which may result poor due to the limited number of data.

Such methodology based upon "prior information" may be perhaps paralleled to Bayesian estimation methodology in reliability, which—after a first popular systematic treatise in [113]—is fast developing [146, 147]. This is just an aspect of what is happening in all the fields related to inference, and caused Press in a recent important book affirm that "many believe that a paradigm shift has been taking place in the way scientific inference is carried out, away from what is sometimes referred to as classical, or frequentist, statistical inference… (toward Bayesian inference)" [135].

In a different setting, indeed, the Bayesian approach—as well known [14, 57, 113, 132, 135, 138, 146, 147]—typically uses prior information for assigning prior distributions to unknown parameters. Often, indeed, such kind of prior information is deduced by technological information, so that Bayesian estimation has found many applications in recent reliability studies, also for its proved big efficiency—with respect to "classical" statistical inference, which is

[2] A significant citation from C.S. Pierce comes to mind: "Probability is the only branch of mathematics in which good mathematicians frequently get results which are entirely wrong".;

[3] Here, the term "prior" means, loosely, something not closely related to observed data, but coming from pieces of information "outside the data", in analogy with Bayesian estimation terminology.

mainly based upon data [139]—when only very few experimental data are accessible. Of course—apart from the different framework (probabilistic modeling for IRA, statistical inference for Bayesian estimation) a basic difference between the two approaches lies in the so-called (and often criticized) "subjectivity" of Bayesian statistics, which is indeed a key aspect of such methodology, often taken to its extreme limits from its more influential adherents, such as the big names of De Finetti [61] and Lindley [108]. The information which constitutes the foundation of IRA is instead generally to be considered as "objective". This is of course not the place to discuss epistemological problems, as if there exists a clear cut difference between "subjectivity" and "objectivity" (which is questionable in our opinion), but it has to be remarked that a branch of Bayesian statistics is recently making some effort toward so-called objectivity, as can be deduced even from the title of [135]: "Subjective and objective Bayesian statistics".

We agree with [30] when stating that, although the subjective meaning of probability is considered unorthodox in many empirical fields, "it is useful in reliability studies since quantification of the degree of belief is essential in using all available information when experimental data are scarce" (p. 232).

The above interpretations and discussions may be of not so much interest in view of practical engineering applications: indeed, in many applied engineering studies, Bayesian estimation is adopted for "ad hoc" reasons, without the necessity of adhering to its philosophy (even if this "non-adherence" may be questionable on theoretical grounds). As clearly reported in [71]—which is a key paper for the application of technological information to prior assessment (for the Weibull parameters, in that case)—"...when only very few (e.g., 3–5) experimental data are accessible... the controversy about whether to use Bayes or classical methods is surmounted since classical estimators, like maximum likelihood, give estimates that often appear unlikely on the basis of technical knowledge of the engineers."

Concerning IRA in itself, although its roots can be traced back in time to the classic book of Miner [128] or to [82], such studies were mainly theoretical, and continued to be such in the 1970s, as can be seen in papers such as the fundamental [77], which—although revealing itself as a seminal paper in the future—appeared indeed in a purely mathematical review ("*Annals of Probability*"), not typically read by many engineers. The first engineering applications came from the mid-1980s on, with fundamental papers such as [59, 60, 62, 70]. In particular, the original work of Erto in this field will be referenced to in the chapter, since it spans the last two decades (from the above-cited [70] to the very recent [75]) in researches leading to new physical motivations—and finding new properties—for various reliability models, such as the rarely adopted Inverse Weibull of the above papers or the new Hyperbolic Model [76]. By the way, none of these two models, which can be both derived by postulating a wear process, has an increasing hazard rate function $h(t)$ (the Inverse Weibull has a first increasing, then decreasing hrf; the Hyperbolic Model has a decreasing hrf), in accordance with what above discussed. It may not be a chance that, as above hinted at, the same author used an analogous technological approach in various applications of Bayesian inference to reliability, e.g., [69, 71, 72, 73].

Finally, among the vast bibliography it is obvious to quote [68], which apparently introduced the term "IRA". Recently, such methodology has reached high levels of mathematical sophistication by the use of advanced properties of SP (see, e.g., [81, 105, 153, 154], or some contributions in a recent book edited by Erto [74]). These approaches have even brought about new studies in theoretical mathematics and other related fields ([47], see also Sect. 3.3.12).

In the specific field of electrical engineering applications, apart quoting the same references as above, the advanced state of the art in the field of electrical insulation has to be highlighted, as witnessed, say, by [119] and in more recent times by [127]. Many more specific contributions in this field will be referred to in the appropriate Sect. 4.

In this respect, the present chapter is primarily devoted to reliability assessment of modern power system components, using some of the methods proposed by the authors of this chapter in previous papers [37, 39, 40, 42]. Indeed, the present chapter takes its origin from a previous paper published in 2008 by the same authors [42]. In addition, more models are here discussed, some of which recently developed, such as the "hyperbolic reliability model" (HRM) [76]; also some brand-new results about known models, e.g., the Inverse Weibull model [75], are hinted at. Moreover, in the chapter it is shown how popular reliability models, like Gamma, Weibull, etc., can be obtained in a straightforward way, that turns out to be particularly useful for engineering applications.

It is remarked that the present chapter is only devoted to reliability modeling of single components. So, the problem of reliability evaluation of the power system as a whole is deliberately not dealt with: excellent books such as, e.g., the ones by Billinton and Allan [16, 17] and Billinton et al. [18] are available on the subject, not to mention seminal paper such as [19]. In some cases, some of the results of the present chapter may be used perhaps to discuss the adequacy, when analyzing the whole power system, of the use of the Exponential model for any single component, a model widely—if not uniquely—adopted, for evident reasons of simplicity.

1.2 Outline of the Chapter

As discussed in [42] and mentioned above, the new deregulated market of electrical energy and the restrictions imposed by economic constraints to production times and preventive maintenance programs have stressed the need for a careful reliability analysis of electrical components. This emphasizes, on one hand, the problem of an economically optimized design of components in view of the stress levels that are expected in-service and, on the other hand, the issue of a careful conditional reliability or quantiles (percentiles) evaluation, in view of optimizing maintenance procedures. However, as previously mentioned, the deregulation also involves a greater uncertainty or lack of data in system operation and management, due to both the use of highly reliable components (characterized by a considerable level of technological innovation, and, consequently, by a high degree of reliability and large costs) and the expected very fast variability of network configurations. This makes the development of timely and economically adequate maintenance procedures more difficult.

In this framework, the aging failures of system components are a major concern and a driving factor in system planning of many utilities. Indeed, more and more system components are approaching their end-of-life stage, hence aging failures should definitely be included in power system reliability evaluation in order to avoid a severe underestimation of the system risk, as shown in [107], where ad hoc methods to incorporate aging failures in power system reliability evaluation are presented.

For all these reasons, the choice, selection, or estimation of adequate probabilistic models for the assessment of the residual (or "conditional") reliability of electrical components is the first, and often the most critical part, of any statistical investigation devoted to system reliability analysis.

The "classical" models for component reliability estimation are based on direct statistical fitting of component failure data coming from the field. This is generally accomplished in two stages:

- first, a model is selected on the basis of "goodness of fit" statistical tests, such as Kolmogorov–Smirnov test, Chi-square test, etc. [139];
- then, its parameters are estimated by well known methods such as the "Maximum likelihood" (ML) one. This so-called "direct reliability assessment" (DRA) is commonly used at the maintenance stage of devices that are already in-service. Nevertheless, such fitting may result poor due to the limited number of data for modern components, as previously mentioned.

From this respect, a help could come from the knowledge acquired over the years about the physical processes that are responsible for the degradation of materials that compose the aging electrical devices. In particular, since insulation is often the weakest part of an electrical device—particularly in medium voltage (MV) and high voltage (HV) systems—phenomenological and physical aging and life models of electrical insulation (that can be found in the vast scientific literature about this subject) can be used for achieving a so-called IRA [68]. IRA can be an effective tool for reliability evaluation of a large series of components, such as transformers, cables, motors, capacitors, etc.; it may be noteworthy that such approach is rooted on classic studies on the physics of failure, as the ones of Dasgupta and Pecht [60], conceived outside the electrical engineering field.

This is the perspective from which the problem of the selection of adequate probabilistic models for reliability assessment of power system components is analyzed in this chapter.

First, in Sect. 2, the most adopted reliability models in the literature about electrical components are synthetically reviewed and the classical DRA, i.e., reliability assessment via statistical fitting directly from in-service failure data of components, is illustrated, that is commonly used at the maintenance stage of components already at work. The properties of these models, as well as their practical consequences, are discussed, thereby arguing that direct fitting of failure data may result poor or uncertain due to the above discussed limitations. Thus, the selection—or the correct identification—of a suitable probabilistic model for power system component reliability in the field of high-reliability devices and large mission times should be better supported by probabilistic information that leads to reasonable modeling, as those coming from the study of the phenomenology and the physics of aging in the already-mentioned example of MV and HV components. The opportunity of using such kind of information for these components, pointed out in recent literature [37, 38, 59, 68], moves the treatment from "direct" to "indirect" reliability estimation. For this purpose, in Sect. 3, the main stochastic models for IRA are discussed, denoted as "stochastic wear models"—which include the "Degradation", "Stress-Strength" and "Shocks" failure models—showing how they can originate particular reliability models, thereby giving further support to the adoption of a given model (e.g., the Weibull one), beyond the simple DRA, which can be not only unsatisfactory, but even misleading.

Following the same approach with reference to electrical devices, in Sect. 4, reference models developed over the years for the estimation of insulation time-to-failure (life) and aging are illustrated, that are based mainly on experimental results coming from laboratory tests carried out on specimens. When inserted in a proper probabilistic framework, they give rise to "physical reliability models", that are usually employed for a preliminary characterization and comparison of the various materials candidate for the realization of the insulation of electrical components, as well as for the design of the insulating systems of such components. However, they can provide useful guidelines also for reliability estimation from in-service failure data and this closes the "loop" between direct and indirect reliability estimation of electrical components.

Up to this point, the discussion has a prevailing methodological aspect, referring to a vast bibliography for the numerical applications to electrical devices. Then, in the final Sect. 5—in order to better highlight the possible pitfalls brought about by DRA from an applicative and numerical point of view—it is shown, by means of numerical and graphical examples referred to typical insulation data—that seemingly similar models can possess very different lifetime percentiles, CRFs and hazard rate functions. Thus, the power system engineer must be aware of the "mathematical" consequences of the selected models, particularly in view of their aging properties. As already stated in the introduction, availability of repairable components is not considered here, only for space limitations. It is dealt with in many of the books referred above, and is thoroughly discussed in [4]. However, it is obvious that the assessment of a reliability model for a unit considered as a non-repairable unit is also the crucial starting point for its availability assessment when it is instead repairable. Moreover, anyone of the reliability features here discussed with reference to lifetimes have a natural correspondent for the RV "times to repair".

1.3 Reliability Function and Other Measures of Aging: Hazard Rate Function, CRF, MRL Function

Starting from the above considerations, the chapter tackles the overall problem of the assessment of the reliability function (RF) of a given component, considered as a non-repairable unit. It is assumed that the reader is familiar with the basic concepts of probability and statistics, such as properties of the cumulative distribution function, probability density function (pdf), and other relevant concepts and definitions, such as the moments and the quantiles of a random variable (RV), and also of the basic facts of estimation theory [7, 25, 134, 139, 141].

Denoting by T the non-negative RV "time to failure", or "lifetime" (LT), of the component and by $F(t)$ its cumulative distribution function (cdf), the RF is defined as follows:

$$R(t) = P(T > t) = 1 - F(t) \tag{1}$$

being $P(A)$ the probability of the generic random event A. The above RF is sometimes denoted also as "survival function" in literature.

The RV T is taken as continuous and distributed according to a pdf $f(t)$ such that:

$$f(t) = \mathrm{d}F/\mathrm{d}t = -\mathrm{d}R/\mathrm{d}t \tag{2}$$

and

$$R(t) = \int_t^{\infty} f(u)\mathrm{d}u. \tag{3}$$

It is remarked that the RF $R(t)$ and the cdf $F(t)$ are relevant to a time interval, and not to the end point t of the interval, as the notation seems to imply: e.g., $R(t)$ is the probability that the device operates successfully in the *whole* interval $(0, t)$.

The expectation of the LT, denoted as MTTF (mean time to failure) can be obtained, provided that the integral exists, by:

$$E[T] = \int_0^{+\infty} R(t)\mathrm{d}t. \tag{4}$$

Basic facts about aging that are sometimes misunderstood (as discussed) are presented here, without claiming to be exhaustive (again, the reader should consult [4, 16, 21, 30, 146], and other fundamental works mentioned above).

Reliability theory is, from a purely mathematical point of view, a sort of applied probability theory devoted to the study of positive RV. However, it possesses some peculiar functions and parameters which are defined ad hoc to describe RV representing times (in particular LT, but not only), and which do not have, in practice, counterparts in other branches of probability theory.

The most popular and the most used (and sometimes even abused and misused, as we shall see) to describe the aging of the devices is the hazard rate function (hrf). Differently from the RF and the cdf, which are, as remarked above, relevant to a time interval, the hrf $h(s)$ is relevant to the instant $s \geq 0$, to be intended as the "age" of the device; it represents, in a sense, the "instantaneous failure rate" at a given point in time. Formally, if the LT, T, possesses a pfd $f(t)$—as will be tacitly assumed throughout the chapter—the hrf is defined, at any time $t \geq 0$ for which $R(t) \neq 0$, as:

$$h(t) = \frac{f(t)}{R(t)} = -\frac{\mathrm{d}}{\mathrm{d}t}[\log(R(t))]. \tag{5}$$

The "physical" meaning of hazard rate function $h(t)$, as well as the origin of its name, lies in the following property, which is easily seen to be equivalent to the above definition:

$$\begin{aligned} h(x) &= \lim_{\Delta x \to 0^+} \frac{P\{(x < T \leq x + \Delta x) | (T > x)\}}{\Delta x} \\ &= \lim_{\Delta x \to 0^+} \frac{F(x + \Delta x) - F(x)}{\Delta x \cdot R(x)}. \end{aligned} \tag{6}$$

So, as $\Delta x \to 0^+$, the product $h(x)\Delta x$ equals the *conditional* probability that the failure occurs in the interval $(x, x + \Delta x)$, given that the device has survived until age x; i.e., such product may be interpreted as the *instantaneous* failure (conditional) probability for a device of age x.

From the above definitions, and the obvious condition $R(0^+) = 1$, it is possible to deduce the following integral relation which allows to express the interval RF, $R(t)$, in terms of the instantaneous hrf:

$$R(t) = \exp\left(-\int_0^t h(\xi)\mathrm{d}\xi\right), \quad t > 0. \tag{7}$$

Therefore, any reliability model is fully specified once either its pdf, or its RF or its hrf is given, as each pair of these three quantities is directly deducible from the remaining: e.g., starting from $h(t)$, whose parametric form can be sometimes derived from its physical meaning, the RF is obtained from (7). Then, the pdf can be attained from the following relationship:

$$f(t) = h(t)R(t). \tag{8}$$

Further, also the cdf, MTTF, etc., are easily obtainable, e.g., this is the most direct way to introduce the Weibull model (by far the most adopted in applied reliability studies) characterized by the following hrf, RF, pdf as functions of time $x > 0$, with positive parameters (a, b):

$$h(x) = abx^{b-1} \tag{9a}$$

$$R(x) = \exp\left(-ax^{b}\right) \tag{9b}$$

$$f(x) = abx^{b-1}\exp\left(-ax^{b}\right). \tag{9c}$$

In the sequel, the above model will be denoted by the symbol $W(a, b)$; it covers the popular Exponential model when $b = 1$. As well known and discussed in any textbook on the discipline, the behavior of the hrf in time may provide insight as to what is causing the failures. Indeed, a decreasing hrf suggests "infant mortality" or "wear-in", i.e., defective items fail early because of frailty, production defects, etc., and the overall hrf decreases over time as they fall out of the population. A constant hrf rate (which is peculiar of the Exponential model alone) suggests that the device fails, irrespectively of its age, because of random "accidents". An increasing hrf rate suggests that the device is subject to "wear-out", so that it is more and more likely to fail as time goes on. Experimentally, these three kinds of behavior (decreasing, constant and then increasing hrf) are, for many (but not all) products, observed to occur in succession during the whole product life, describing the so-called "bathtub curve" of the hrf. This is a very popular curve which is widely discussed in any book on the matter (see [146]), and in some detailed analytical papers, such as the article by Glaser [84].

In order to better understand the meaning of the above possible behaviors of the hrf, and considering that the hrf is not a probability (and neither a conditional probability),[4] it is perhaps preferable to introduce the "conditional reliability function" (CRF), which is a function of two-time variables defined as follows[5]:

$$R(t|s) = P\{T > s + t | T > s\} = \frac{R(t+s)}{R(s)} (t, s \geq 0). \tag{10}$$

The above CRF, denoted as the *CRF* for a mission time t of a device of age s, equals the *conditional* probability that, the device having survived until age s, it will survive at least until time $(s + t)$, i.e., its age will be increased at least of t time units after age s. This is why it could be also denoted as "residual reliability function".[6] It is obvious that the above CRF, with respect to the time argument t, must behave as a RF, e.g., it must satisfy (at any age s): $R(0^{+}|s) = 1$,

[4] It should be clear, by the way, that the hrf must be positive, but not necessarily less than 1: it can even diverge, as happens for models possessing a pdf which vanishes at some finite point in time, as the Uniform model. It has little to do with a pdf, too: e.g., its integral over the whole interval $(0, \infty)$ must be ∞, since $R(\infty) = 0$, etc.

[5] The notation $R(t|s)$ is purely symbolic, being used for suggesting the conditional aspect of the RF, and should not be confused with the *conditional* probability $P(A|B)$, the main difference being that (r, s) are deterministic numbers, while (A, B) are random events.

[6] This latter would probably be a better name: here we use the term "conditional" instead of residual since it is more adopted in literature.

$R(\infty|s) = 0$, and be decreasing with t. Its behavior with age s might appear less obvious. For instance, one could naively expect that, since age should weaken every object (an indubitable fact), $R(t|s)$ is a decreasing function of s. However, it is not always so; indeed, such reasoning would ignore the fact that the CRF is a *conditional* probability, and that such conditioning may significantly change our information: e.g., sometimes, knowing that a device has survived until age s may render us more confident in its "future" survival than we could be without that information, so that the device *appears* to "strengthen" with age, and the CRF may increase with s. For instance, the CRF $R(t|s)$ increases with s—which is closely related, as remarked, to a decreasing hrf—during the abovementioned "infant mortality" period in the early life of a product, or in "accelerated life tests" (ALT), or when the LT is generated from mixtures ([10], p. 55; such property is also recalled here at the end of Sect. 3). The idiomatic expression "the device appears to strengthen with age" should be interpreted cautiously, in terms of "change of information" rather than effective strengthening of the object, a fact which of course seldom occurs in practice.[7] The statement, if not correctly interpreted (with due emphasis on the word "appears"), may seem to conflict with the obvious property that the RF *must* always decrease with time. This concept will be discussed again later.

In many cases, devices weaken (again, in a "conditional" way) with age: this is generally thought to be typical, e.g., of mechanical devices subjected to increasing wear as they work, and in practice for *all devices* (also electronic ones)—if they should be left to operate indefinitely—when their age is large enough. However, some already hinted at examples of aging related to wear (as will be also discussed in Sect. 2) should render us careful also with this observation. There is no doubt, instead, for what concerns the human beings and living organisms: their hrf $h(s)$ increases and their CFR $R(t|s)$ decreases with age s. Soon it will be recalled that the two properties are indeed equivalent.

Another measure of aging is the *MRL*. It appears to be very useful, although not so popular (strangely, since it has a clearer physical meaning than the hrf). The MRL, $r(s)$, is a function of time (age) s representing the expected residual lifetime of a device that has reached age s. So, it is a "conditional expectation", i.e., the mean value of the "residual" LT at age s—namely, the difference $(T - s)$—conditional to the event $(T > s)$. So, $r(s)$ is defined as:

$$r(s) = E[(T - s)|T > s] \tag{11}$$

and it is computable in terms of the CRF as follows:

---

[7] When discussing aging and hrf and CRF properties, in the authoritative [10] it is observed that "certain materials increase in strength as they are work-hardened" (p. 55). This may be true, but it is unlikely that it holds for very long time intervals: wear-out should ultimately prevail for any device, corresponding to a CRF $R(t|s)$ decreasing with s, for s large enough. Anyway, it is possible that in practice the device is maintained or retired before wearing-out, so that the ultimate, decreasing part of the CRF is not observed.

$$r(s) = \int_0^\infty R(t|s)\mathrm{d}t = \int_0^\infty \frac{R(t+s)}{R(s)}\mathrm{d}t. \quad (12)$$

It appears that $r(s)$ is increasing, constant or decreasing with age s in the same way as the CRF. Also, it can be easily shown that $r(s)$ uniquely specifies the reliability model, in that RF, pdf and hrf can be uniquely expressed in terms of $r(s)$ [30].

1.4 On the Relation Between Hazard Rate and CRF

Previously, some discussions on the physical meaning of the hrf behavior in time, deriving from the observation of the "bathtub curve", have been intuitively explained in terms of the CRF, $R(t|s)$, which possesses an easier interpretation being a probability (differently from the hrf). The CRF behavior versus age s is indeed univocally related to the hrf behavior. Using the above relations between the RF and the hrf, it is not difficult to show, for any LT distribution for which the hrf is defined, the validity of the following equivalences (see, e.g., [10, 30]), whose statements are assumed to hold for each value of $s > 0$, and for any given mission time t, which is to be intended as a constant in the right-hand side of the equivalences:

1. hrf $h(s)$ increasing with $s \Longleftrightarrow$ CRF $R(t|s)$ decreasing with s (at any given time t).
2. hrf $h(s)$ decreasing with $s \Longleftrightarrow$ CRF $R(t|s)$ increasing with s ("").
3. hrf $h(s)$ constant with $s \Longleftrightarrow$ CRF $R(t|s)$ constant with s ("").

As well known, property (c) uniquely characterizes the Exponential model, and assesses its being "memoryless". Property (a) explains why the hrf increases for devices subjected to wear; property (b) explains why the hrf decreases for devices subjected to "infant mortality" (or for devices strengthening with age). From the above properties, the abovementioned bathtub curve, as well as any hrf behavior, can be easily interpreted in terms of CRF properties, an approach which, surprisingly, is seldom found in the relevant applied literature: in most cases the CRF is not even mentioned, or finds much less space than the hrf when introducing the study of aging. For brevity, it is instead reasonable, having interpreted the hrf behavior with the help of the CRF, to use then the hrf $h(t)$ alone. Indeed, it is a function of a single time variable (just as the MRL), it is more easy to be represented graphically, and various procedures for its statistical estimation have been devised. So, also here in the sequel, only the hrf expression, if available, will be reported when discussing the various models. Nonetheless, a look at the CRF curves of the various models—as done in Sect. 5 (see Figs. 6, 7)—could be very helpful for a better understanding of their aging properties.

For the abovementioned properties of the MRL: in case (a) the MRL $r(s)$ is decreasing with s; in case (c) the MRL $r(s)$ is constant with s (so, it is equal to the MTTF; again, such independence from s uniquely characterizes the Exponential model); in case (b) the MRL $r(s)$ is increasing with s. The converse implications are not trivial, and require further assumptions [30].

At this point, it is useful, as done in most books, to operate a classification of reliability models in terms of aging, based upon hrf or CRF properties. First, we notice that we used above the symbol s to denote "age" (a past time), the symbol t to denote a mission time (a future time), so that the symbol s has the same meaning in the hrf $h(s)$ and in the CFR $R(t|s)$; although using the same symbol t to denote time in $h(t)$, $R(t)$ and $R(t|s)$ (as done in most books) would not be an error, nevertheless it may easily induce the reader into confusion for what concerns the relation among these quantities. As long as only the mathematical properties of the hrf are of interest, in Sect. 2 and in the sequel, we shall use $h(t)$ as it is customary in literature.

1.5 A Classification of Reliability Models: IHR, IDR, IDHR Models

A reliability model (or, for brevity, the LT described by such model) is defined as:

- "increasing hazard rate model" (IHR), if its hrf $h(s)$ is an increasing *function* of $s > 0$ (over the whole domain of the hrf, generally $0 < s < \infty$); example: Weibull $W(a, b)$ with $b > 1$.
- "decreasing hazard rate model" (DHR), if its hrf $h(s)$ is a decreasing *function* of $s > 0$ (*idem*); example: Weibull $W(a, b)$ with $b < 1$.
- "increasing, then decreasing hazard rate model" (IDHR), if its hrf $h(s)$ firstly increases with $s > 0$, from $s = 0$ to a given point s^*, then it *decreases* with s; examples: Lognormal and Inverse Weibull models.

The above models are recalled in more detail in Sect. 2. In case of IDHR models, also the denominations: "unimodal hrf", or "reverse bathtub-shaped hrf" are used. As reported in [93], and also illustrated for some of the basic models here presented, the LT is often represented by such models for the situations where the failure is mainly caused by fatigue.

In view of the above-recalled relations between hrf and CRF, the above classification may be equivalently formulated in terms of the CRF and some authors (e.g., [10, 30]) prefer this latter formulation, because it is more general, as it does not require the existence of a pdf (such mathematical details are omitted in this chapter, as all models here considered possess a pdf).

There are a lot of more possible classifications, e.g., in terms of "Average hazard rate", MRL, etc. [10, 30]. We only emphasize here that, as above remarked, a IHR (DHR) model is characterized by a decreasing (increasing) MRL function of time.

Of course, those reported in the above classification are not the only possible behaviors of the hrf, but they are the most useful for the description of aging for the simple analytical models which will be considered from Sect. 2 onwards; these are indeed "models", so that they must not be expected to express the true hrf of the device for its whole life, but only to represent a reasonable approximation to it, for the actual time interval during which the device operates. As an example, it should be clear that the Exponential model cannot be truly valid in real world: it is only an ideal model, in a sense it is "the most ideal" of all models, since no device can be completely memoryless : as remarked above, wear-out—from a certain point in time—must occur for every device. But, if wear appears very late with respect to the time interval (say, 20 years) for which the device is used (before its withdrawal, e.g., for technological innovation), so that the hrf can be considered roughly constant in that interval, the Exponential model can provide, also in view of its simplicity, a good approximation to the "true", unknown, model.

One of the hrf behaviors not considered in the above classification is the famous "bathtub curve", which is, in fact, an experimental curve, that matches none of the models here considered in Sects. 2 and 3; it is in fact seldom found in operating devices, if they are, as generally happens, subjected to "burn-in". Moreover, it has been already pointed out that in practice there is no need of a model which describes the device reliability for all its "theoretical" life. Finally, models with bathtub-shaped hrf can be built analytically, but are either rather complex or difficult to estimate. For instance, two models capable of representing such behavior, reported in [103, p. 47] and practically almost never used, are reported—as functions of age $x > 0$—here:

$$h(x) = \frac{\alpha}{x+\beta} + \gamma x, \quad \alpha, \beta, \gamma > 0 \tag{13}$$

$$h(x) = \frac{\beta x^{(\beta-1)}}{\alpha^{\beta}} \exp\left[\left(\frac{x}{\alpha}\right)^{\beta}\right], \quad \alpha, \beta > 0. \tag{14}$$

1.6 Final Remarks: Some Popular Misconceptions in Applied Reliability Studies

A few final remarks on some common misconceptions, or "pitfalls", related to erroneous interpretations of the hrf are deemed to be useful here, since often also authoritative bibliography seems to "slip" on this concept. Two of such pitfalls are discussed here, which are both related in a sense to the existence of models with a decreasing hrf, which, as opportunely pointed out in [146, p. 69], appears to be a subtle concept, perhaps only fully understandable (we agree with Singpurwalla) from a subjective (or even "psychological") probabilistic reasoning, even if also a sound "objective" explanation is available [141]. A recent paper on some common misconceptions about the modeling of repairable components (a problem not dealt here), with reference to power system applications, is [160]. In this paper, some basic points already in raised in the fundamental book [4] are summarized and further discussed.

1.6.1 The Pitfall of the "Average Hazard Rate" for Two-state (or Multi-State) Reliability Models

Let a device be potentially subjected, in a given time interval, to (only) one of two operating conditions, say "normal condition" (NC) or "adverse condition" (AC), depending on chance. This may be the case of a overhead transmission or distribution line subjected to normal or adverse weather conditions.[8] Let each of the two conditions correspond to a different value of an assumed-as-constant hazard rate, and let:

$$\nu = \text{hazard rate value in NC}; \ \alpha = \text{hazard rate value in AC}; \tag{15a}$$

$$p = \text{probability of NC}; \ q = 1 - p = \text{probability of AC}. \tag{15b}$$

A value $\alpha > \nu$ is of course expected (often, the ratio α/ν can be very high in practice, e.g., 50, for overhead lines), hence this relationship will be assumed to hold. Of course, any situation of a "binary" hrf can be dealt with in such a way, by calling "normal condition" ("adverse condition") the one with the lower (higher) hazard rate value.

Under these hypotheses, the following "average value" assignment of the hrf (still assumed to be constant) to the device is often found (especially in power system studies, a recent example being [90]):

$$h = p\nu + q\alpha, \tag{16}$$

i.e., a weighted average of the two values ν and α. So, the RF model would become:

$$R(t) = \exp(-ht). \tag{17}$$

The above hrf is denoted here as the "average hazard rate". Equation 16 seems to be very reasonable, since the random events "NC" and "AC" are incompatible and exhaustive, and it is indeed used also by some utilities. It is in fact wrong, as well as (17), if the problem is dealt with by the basic tools of elementary probability. Although it may seem paradoxical, the true result is that the overall HR is a

[8] The extension of the following reasoning to three-state or multi-state models is straightforward (e.g., a three-state model occurs in power distribution studies when also "extremely adverse" weather conditions, or similar, are considered [156].

decreasing function of time, of which the value h in (16) is only the initial value [it can be shown that (16) may be—but only in some cases—a good approximation for the hrf for very short time intervals, but the problem is that it is always presented as a true value].

The point is that the total probability theorem [which is evidently the basis of (15)] cannot be applied to the hrf, which is not a probability! (see also the following Sect. 1.6.2). It can be instead be applied, e.g., to the RF $R(t) = P(T > t)$, so that its value is not given by (16), but by:

$$R(t) = p \cdot \exp(-vt) + q \cdot \exp(-\alpha t), \tag{18}$$

which is quite different.[9] By derivation of the above RF with respect to time t, using (5), it is easy to get the right expression of the hrf, which is readily shown to be not a constant at all. Indeed, the hrf is a decreasing function of time. In Sect. 3.3.11 (devoted to "mixture models") also this apparent paradox will be discussed, namely that—even though the individual hrf are constant over time—the "overall" model has a decreasing hrf. The paradox can be fully justified by subjective or Bayesian reasoning, and some very interesting papers or books, such as [10, 136, 146], discuss it. Here, we only observe, as in [141], that—if one should not know the true condition under which the device operates—the larger the observed lifetime t of the device, the more likely it is that the item is subjected to NC (rather than to AC), i.e., the conditions corresponding to a lower hazard rate value. Thus, the older the device, the less likely it is to fail, so that the above "mixture" of constant hazard rates gives rise to a DHR model. A brilliant way to show this, as in [141], is using Bayes' theorem for obtaining the following conditional probability, for any given time t:

$$\begin{aligned} P(\mathrm{NC}|T > t) &= P[(\mathrm{NC}) \cap (T > t)]/P(T > t) \\ &= p \cdot \exp(-vt)/[p \cdot \exp(-vt) + \mathrm{q} \cdot \exp(-\alpha t)], \end{aligned} \tag{19}$$

which is indeed, assuming $\alpha > v$, increasing in t [maybe the easiest way to show this is considering the reciprocal of (19), which is clearly decreasing in t]. Of course, $P(\mathrm{AC}|T > t)$, the conditional probability of the complementary event, is—for the same reason—decreasing in t.

It should be remarked that using (16) means using an overestimation of the hrf, this implying an underestimation of system performances.

From *a* practical point of view, such discussion shows that the development of time-varying models for the hrf is highly opportune, as those proposed by Wang and Billinton in [156] for incorporating the effects of weather conditions and restoration resources in reliability evaluation of distribution systems.

For the purpose of the present chapter, however, the main point to be highlighted above is *not* the right expression or behavior of the hrf, but the mistake one

---

[9] Mistaking (18) for (17) is in practice equivalent to mistaking—as to the computation of the expectation of a function ϕ of a RV X– the expectation $E[\phi(X)]$ with $\phi(E[X])$, which is a trivial error, if ϕ is not a linear function.

can make by treating the hrf as it were a probability. The same can be pointed out as far as the following topic is concerned.

1.6.2 Does a DHR Model Imply a "Strengthening" of the Device?

From one of the above relations, here reported again:

$$R(t) = \exp\left(-\int_0^t h(\xi)\mathrm{d}\xi\right), \quad t > 0, \tag{20}$$

it should be clear that a DHR model in no way implies a "strengthening" of the device under study, as already discussed above. The RF $R(t)$ always decreases with time, and this is assured by its very definition, or—looking at the above integral relation between RF and hrf—by the hrf being positive, with no regard to its behavior in time. So, it should be remarked that phrases like "the reliability of a DHR system improves with age", which are sometimes reported also in books, are at least ambiguous: in the function $R(t)$ only one time-argument appears, which cannot be arbitrarily deemed to be a "mission time" or an "age" as one likes. The truth is that $R(t)$ decreases with t, so it can never improve with "age" t. For what above discussed, a right way—maybe the only way—to express the peculiarity of a DHR model is that, for such model: "the conditional reliability $R(t|s)$ of a DHR system improves with age s, for any given mission time t". This "conditional" aspect (which is present also in the hrf itself) is sometimes forgotten, e.g., when it is stated, as in some books, that "the hrf expresses the probability that the device fails after reaching age t". It has to be noticed that, for approaching such meaning, the hrf—which, by the way, has a dimension of (1/time)—should be at least multiplied by a time interval Δt, *and even so it cannot be claimed to be a probability* (for instance, no-one can assure that this product is less than 1, also in view of the fact that the hrf can be infinite). Nor the hrf can be resembled to a pdf: as recalled in a note, its integral over the whole interval $(0, \infty)$ must be ∞, not 1 as a pdf.

Returning to DHR models (which are the ones—but not the only ones—capable of possessing an infinite hrf, as we shall see in a few lines) one could also think that a DHR (or a constant hrf) model is characterized by a RF having a slower decreasing attitude in time, with respect to a IHR model and assuming as fixed some parameters (e.g., MTTF and/or median), so that it is "better" than a IHR model in some way. Even if it can seem trivial, it may be not useless to remark that this is generally false, or it may be true only for some large enough mission times. Still from the above relation, one can only deduce that, for what concerns the comparison of 2 hrf, if $h_1(x) > h_2(x)$ in a whole interval $(0 < x < c)$, then—for the corresponding RF—$R_1(t) < R_2(t)$ for every t in $(0 < t < c)$. So, it should come as no surprise that a IHR model as a Weibull one with $b > 1$, e.g., $h(x) = ax$ $(a > 0)$, is largely "better" than a constant hrf model, since for the first we have $h(0) = 0$

and a smaller hrf values in a whole interval containing the time origin. In other words, a memoryless model is by no means a "good" or desirable model in the early age of the device, since "chance failures" are of course no better than no (or "very unlikely") failures. Even more pronounced, of course, would be a comparison between the above IHR model and a DHR model such as a Weibull one with $b < 1$ [e.g., $h(x) = a/\sqrt{x}$ $(a > 0)$], to which an infinite value of the hazard rate and a rapidly decreasing RF corresponds in early times. The IHR model $h(x) = ax$ is much better in this early period (which may also be the most crucial in terms of warranties), and possibly also better on the whole (depending, e.g., on the time interval for which the device will be used in practice).

2 "Direct" Reliability Assessment: A Review of Reliability Models

2.1 A Premise on Reliability Models, with Hints at Electrical Applications

A significant set of reliability models, which should cover almost all kinds of practical applications, are briefly reviewed in the present section, in alphabetical order.

For the purpose of the present approach, the way by which these (or other) models can be deduced from wear processes is important. This aspect is omitted here, but is tackled in the next section, so that a few of these models will be met again there.

As defined above, DRA concerns reliability analysis of components on the basis of failure data coming from devices in-service. Performing a DRA requires that the most adequate probability distribution for the reliability analysis (to be chosen from a family of commonly employed distributions for such components) is selected on the basis of data fitting, previous experience, literature or expert judgment, or better on the basis of a combination of all these aspects. Such distribution should both exhibit a good fitting to the data (proven by a proper statistical fitting test) and possess a relatively simple form, with no more than two or three parameters to be estimated from data. Only two-parameter models will be considered here, since they are by far the most adopted, together with the single parameter Exponential model, this latter being so popular that it is here considered only as a particular case of other models, namely the Gamma, Weibull and the less known HRM. The most adopted reliability models—in particular, for electrical components—are by far the *Gamma, Normal, Lognormal* (LN), and *Weibull* models. However, also some other LT distributions are worth being considered, such as the *Inverse Gaussian* (IG) distribution, the Inverse Weibull (IW) distribution, the *Birnbaum–Saunders* (BS) distribution, the *Log-logistic* (LL) distribution, and more. They have found recently some significant application for electrical components reliability, so they are briefly reviewed here, too. A particular and very significant case is that of the Weibull model: it is by far the most adopted in the field of electrical insulation, as illustrated with some detail in Sects. 3 and 4. Such model, together with some related "physical law" of aging—e.g., the *Inverse power model* (IPM)—has kept proving over the years one the most adequate for the statistical fitting of insulation lifetime data, and the fact that it possesses some physical background or motivation is a very desirable property in the present discussion.

Neither the statistical fitting nor the parameter estimation is addressed in this chapter. Relevant references well cover these topics: exhaustive treatises on the subject, in its most general form, are popular books such as [25, 96, 98, 149]. Books specifically devoted to (classical) statistical estimations for LT models are, e.g., [55, 97, 103, 124, 130]. For what concerns Bayesian inference, the key reference is [113], but the approach is also significantly present in many other books, e.g., in [30, 124, 146, 152]. Moreover, the reader should be aware that most of the references on the presented models (appearing in the reference list reported at the end of this chapter) often present also estimation methods for the model parameters. Only a final small hint at estimation, very interesting for our purposes: it is well worth highlighting that the use of Weibull model for insulation applications has stimulated many peculiar estimation methods, both in classic [52] and Bayesian statistics [39]. An excursus on basic reliability models and their key features (excluding deduction from wear models) follows from the next Sect. 2.2 to the end of present Sect. 2. For all the models here presented, the possible derivations from wear processes are reviewed in Sect. 3. Further details on the mathematical features here briefly reviewed can be found in the books cited at the beginning of this chapter, and in the numerous references at the end of this chapter; in particular, aside from the monographic books on the single models (e.g., [34] on the Inverse Gaussian distribution, or [53] on the Lognormal distribution), very detailed accounts on all the models are present in the authoritative volumes of [96], while a brief but complete review is reported in [124], where also many graphs, here omitted for the sake of brevity, are reported illustrating the cdf, pdf, hrf, etc., of the various models.

2.2 Birnbaum–Saunders Model

The Birnbaum–Saunders (BS) model was introduced by Birnbaum and Saunders in 1969 [20], in relation to fatigue-affected lifetimes. It has the following cdf and pdf for $t > 0$:

$$F(t;\alpha,\beta) = \Phi\left\{\frac{1}{\alpha}\left[\left(\frac{t}{\beta}\right)^{\frac{1}{2}} - \left(\frac{\beta}{t}\right)^{\frac{1}{2}}\right]\right\} \tag{21}$$

$$f(t;\alpha,\beta)=\frac{1}{2\sqrt{2\pi}\alpha\beta}\left[\left(\frac{\beta}{t}\right)^{\frac{1}{2}}+\left(\frac{\beta}{t}\right)^{\frac{3}{2}}\right]\exp\left[-\frac{1}{2\alpha^2}\left(\frac{t}{\beta}+\frac{\beta}{t}-2\right)\right]. \quad (22)$$

The parameter (α, β) are positive. The hrf has no simpler form than the ratio:

$$h(t)=f(t)/(1-F(t)). \quad (23)$$

It is defined for all $t>0$ (as happens for all models here considered, unless otherwise stated).

In a recent paper [102], it has been shown analytically that the hrf of the BS distribution is always an IDHR—or "unimodal hrf"—model for all values of the shape parameters, and the change point of the hrf can be determined as a solution of a non-linear equation. These authors have provided an approximation to this change point, and also proposed different methods for estimating the change point. After this change point, the hrf approaches a positive limit as $t\to\infty$, similar to the IG model, whose resemblance with the BS model is illustrated also later. Mean and variance of the BS model are:

$$\mu=\beta\left(0.5\alpha^2+1\right) \quad (24)$$

$$\sigma^2=\beta^2\alpha^2\left[(5/4)\alpha^2+1\right]. \quad (25)$$

2.3 Gamma Model

The Gamma $G(r,\phi)$ model is one of the most popular in applied probability, and is characterized by the following pdf:

$$f(t;r,\phi)=\frac{\phi^r t^{(r-1)}}{\Gamma(r)}\exp(-\phi t), \quad t>0 \quad (26)$$

were $\Gamma(x)$ is the Euler–Gamma special function, ϕ and r are positive constants representing the shape and scale parameters, respectively. The cdf is expressed through the incomplete Gamma function $\Gamma(x, y)$:

$$F(t;r,\phi)=\Gamma(r,\phi)/\Gamma(r). \quad (27)$$

In its simplest formulation, denoted as "Erlang model", the Gamma model describes a positive RV obtained by the sum of r Exponential independent and identically distributed RV with parameter (hrf) ϕ.

For what concerns the hrf, which is not expressible analytically, it can be shown [103] that, if $r>1$ (the most frequent case), the Gamma model implies a hrf which, starting from zero in $t=0$, increases with time, approaching the positive limit ϕ as $t \to \infty$. If $r < 1$, the hrf diverges as $t \to 0^+$, then decreases with time, approaching the same limit ϕ as $t \to \infty$.

Mean and variance of the Gamma model are:

$$\mu = r/\phi \tag{28}$$

$$\sigma^2 = r/\phi^2. \tag{29}$$

A limit case is the Exponential one, obtained when $r = 1$, which has the constant hrf $h(t) = \phi$.

A hint at transformed Gamma RV, i.e., “Inverse Gamma” and “Generalized Gamma” RV, is given in this section.

2.4 Gaussian Model

The Gaussian or Normal model plays a fundamental role in statistical analyses because many distributions are well approximated, in view of the CLT, by the Normal probability distribution. The Normal pdf has the following expression:

$$f(t;\mu,\sigma) = \frac{1}{\sigma\sqrt{2\pi}}\exp\left[-\frac{1}{2\sigma^2}(t-\mu)^2\right], \tag{30}$$

where $-\infty < t < \infty$, $-\infty < \mu < \infty$, $\sigma > 0$. As can be seen, the Normal pdf is characterized by two parameters, μ (the mean of the pdf, a real number) and σ (the SD). Its main properties (bell-shaped form, symmetric around the mean μ) are well known.

The Normal pdf and cfd can be conveniently expressed as a function of the standard Normal pdf, $\varphi(z)$, and cdf, $\Phi(z)$, that correspond to a Normal RV with zero mean and unit variance, which are thus defined as follows:

$$\Phi(z) = \int_{-\infty}^{z} \frac{1}{\sqrt{2\pi}}\exp\left(-\frac{u^2}{2}\right)du; \quad \phi(z) = \mathrm{d}\Phi(z)/\mathrm{d}z \tag{31}$$

so that pdf and cdf of the Gaussian model are:

$$f(t;\mu,\sigma) = \frac{\varphi((t-\mu)/\sigma)}{\sigma}; \quad F(t;\mu,\sigma) = \Phi((t-\mu)/\sigma). \tag{32}$$

It can be shown that the hrf $h(t)$ is an increasing function of time, roughly increasing linearly as t diverges. The model is not theoretically adequate, of course, for lifetimes, being defined also for $t < 0$. However, it is sometimes adopted, provided that the probability of attaining negative values is negligible (this happens in practice if $\mu > 3\sigma$). However, principally due to its scarce flexibility, discussed elsewhere in the chapter, it is almost never a good candidate as a reliability model, while it is sometimes used for "repair times" in availability studies, but not so frequently as the LN and Gamma models.

2.5 Gumbel Model

The Gumbel model here illustrated can be obtained—like the Weibull one, but under different hypotheses—as the minimum of a large (ideally infinite) set of RV, so that its pdf is also denoted as the "smallest extreme value" pdf, or also "type 1-extreme value" pdf [28, 80, 131]. The Gumbel model has the following cdf:

$$F(t;\chi,\delta) = 1 - \exp\{-\exp[(t-\chi)/\delta]\} \ (-\infty < t < +\infty). \tag{33}$$

The two parameters (χ, δ) are, respectively, real and positive. It can be seen that, theoretically, the argument of the cdf may be negative, as in the Gaussian case. By an adequate choice of χ and δ, the probability of negative values may be rendered practically zero. However, also a truncated form of the Gumbel pdf exists, restricted to positive argument values, denoted as "Gompertz model" [103]), which is also used in LT applications. The mean and SD of the model are

$$\mu = \chi - \gamma\delta \tag{34}$$

$$\sigma = \left(\pi/\sqrt{6}\right)\delta \tag{35}$$

being γ the Euler constant (0.5772...). The RF and pdf are easily evaluated from the above cdf, and the hrf is an increasing exponential function of time:

$$h(t;\chi,\delta) = (1/\delta)\exp[(t-\chi)/\delta] \ (-\infty < t < +\infty). \tag{36}$$

This model is widely used—especially in its truncated form—for devices largely affected by wear with increasing age (such as mechanical products, and also human beings).

Apart from lifetimes, the Gumbel model finds—for intuitive reasons—application also as a model for RV representing material "strength". Indeed, it has been used since decades as a possible alternative to the (more adopted) Weibull model, for characterizing electrical strength of insulators [87].

Also a "largest extreme value"—or "double exponential"—model exists. It is used for characterizing the maximum of a large set of RV, and has the following cdf:

$$F(t;\chi,\delta) = \exp\{-\exp[-(t-\chi)/\delta]\}(-\infty < t < +\infty). \tag{37}$$

It is very popular in engineering applications but it is seldom used for lifetimes [131, p. 40].

2.6 HRM Distribution

A new reliability model, the so-called HRM was introduced by Erto and Palumbo in 2005 [76], who showed that many failure mechanisms can produce mortality laws of Hyperbolic type: the "Deterioration", "Stress-Strength", and "Shocks" failure models are some of the above models which can lead to a HRM, as shown in the following section. The same authors illustrated also some applicative examples, with a noteworthy electrical application. Actually, the model was not completely unknown before, e.g., Lawless [103] briefly mentions it as a "generalized Pareto model". However, its properties were not fully analyzed, nor it appears to have ever been applied, before 2005.

For the purpose of the present section, a decreasing hazard rate function, approaching a value greater than zero, is the distinctive characteristic of the model. The HRM indeed takes its name from the hyperbolic form of its hrf, which is quite peculiar in the wide range of all the existing models, and can be expressed as follows (76):

$$h(t) = r + \frac{a}{t+1}, \quad a > 0,\ r > 0, \tag{38}$$

which is strictly decreasing from the early maximum value $(a + r)$, to the asymptotic minimum r, with a being the limit decrement. So, as to the hrf properties, this model shows some analogies with the LN and IW ones, except that the maximum of the hrf is attained at $t = 0$ (while, in the LN and IW models, the maximum of the hrf is attained at some mission time $t > 0$). From Eq. 1, the cumulative hrf, RF, cdf, and pdf are easily derived as:

$$H(t) = \int_0^t h(t)\mathrm{d}t = rt + a\ln(t+1), \tag{39}$$

$$R(t) = \mathrm{e}^{-H(t)} = \frac{\exp(-rt)}{(t+1)^a}, \tag{40}$$

$$F(t) = 1 - R(t) = 1 - \frac{\exp(-rt)}{(t+1)^a}, \tag{41}$$

$$f(t) = h(t)R(t) = \left[r + \frac{a}{t+1}\right]\left[\frac{\exp(-rt)}{(t+1)^a}\right]. \tag{42}$$

It is apparent from (38) that, for t increasing infinitely, the Hyperbolic Model reduces to the Exponential model with constant hrf: $h(t) = r$ (and, so, MTTF = $1/r$).

Erto and Palumbo [76] deduce all the non-trivial statistical properties (mean, variance, etc.) of the HRM by means of the Moment generating function $\Phi_T\{\cdot\}$ of the RV T (lifetime), i.e.:

$$\Phi_T\{x\} = E[\exp(xT)]$$
$$= \int_0^\infty \exp(xt)\left(r + \frac{a}{t+1}\right)\frac{\exp(-rt)}{(t+1)^a}\mathrm{d}t. \quad (43)$$

Denoting by $\Gamma(\cdot,\cdot)$ the incomplete Gamma function, after some manipulations the following expressions for the mean $E[T]$ and variance $\mathrm{Var}[T]$ are obtained:

$$E[T] = \Phi_T'(0) = \frac{1}{r}[r^a \exp(r)\Gamma(-a+1,r)], \quad (44)$$

$$\mathrm{Var}[T] = E[T^2] - E^2[T]$$
$$= \frac{1}{r^2}\{2r^a\exp(r)\Gamma(-a+1,r) - 2ar^a\exp(r)$$
$$\times \Gamma(-a+1,r) + 2r[1 - r^a\exp(r)\Gamma(-a+1,r)]$$
$$- [r^a\exp(r)\Gamma(-a+1,r)]^2\}, \quad (45)$$

where

$$E[T^2] = \Phi_T''(0)$$
$$= \frac{1}{r^2}\{2r^a\exp(r)\Gamma(-a+1,r) - 2ar^a\exp(r)$$
$$\times \Gamma(-a+1,r) + 2r[1 - r^a\exp(r)\Gamma(-a+1,r)]\}. \quad (46)$$

Interesting properties concerning the MRL are also illustrated in [76]; e.g., it is proven to be an increasing function of time, toward the maximum asymptotic value $1/r$ (and this is in agreement with known theoretical relations between hrf and MRL).

2.7 Inverse Gaussian Distribution

The Inverse Gaussian (IG) distribution [34], although not very popular in the field of power systems, has found many applications in theoretical reliability literature for those situations in which the LT distribution is greatly affected by early failures due to the so-called "infant mortality". The Inverse Gaussian model belongs to the IDHR family and is very similar to the LN distribution. In practice, they are in most cases undistinguishable on the basis of field data, so that it is important to understand the kind of aging process which may give rise to the IG or the LN distribution.

The IG distribution has been introduced as the first passage time of a Wiener process [34], as will be recalled in the next section. Its pdf is given by:

$$f(t;\mu,\lambda) = \sqrt{\frac{\lambda}{2\pi t^3}}\exp\left[-\frac{\lambda}{2\mu^2 t}(t-\mu)^2\right] \quad (47)$$

being t, μ, $\lambda > 0$. Using the above recalled Gaussian cdf $\Phi(x)$, the RF and hrf are given by the following expressions:

$$R(t;\mu,\lambda) = \Phi\left[\sqrt{\left(\frac{\lambda}{t}\right)}\left(1-\frac{t}{\mu}\right)\right] - \exp\left[\frac{2\lambda}{\mu}\right]\Phi\left[-\sqrt{\frac{\lambda}{t}}\left(1+\frac{t}{\mu}\right)\right] \quad (48)$$

$$h(t;\mu,\lambda) = \frac{\sqrt{\frac{\lambda}{2\pi t^3}}\exp\left[-\frac{\lambda(t-\mu)^2}{2\mu^2 t}\right]}{\Phi\left[\sqrt{\frac{\lambda}{t}}\left(1-\frac{t}{\mu}\right)\right] - \exp\left[\frac{2\lambda}{\mu}\right]\Phi\left[-\sqrt{\frac{\lambda}{t}}\left(1+\frac{t}{\mu}\right)\right]}. \quad (49)$$

Although not easily, it can be seen that the hrf first increases, reaching its maximum at a time t^* which is not analytically expressible, then approaches the positive limit $\lambda/(2\mu^2)$ as $t \to \infty$. This is a small difference with respect to the LN and LL reliability models, whose hrf goes to zero as t diverges, and a similarity with the BS model, which is indeed resembles very closely the IG model, also in its derivation (see Sect. 3).

Mean and variance are:

$$E[T] = \mu \quad (50)$$

$$\sigma^2 = \mu^3/\phi. \quad (51)$$

2.8 Inverse Weibull Distribution

The Inverse Weibull (IW) model was deduced—although often named in a different way, i.e., Frechet model—as a model for the asymptotic distribution of the maximum value from a succession of independent RV [28]. Subsequently, it was proposed with the present name when it was obtained as the distribution of the inverse (reciprocal) of a Weibull RV. Most of its properties, in particular those of its hrf, were first deduced by Erto [70], and are being developed in a forthcoming paper [75]. In [70], also the identification of the IW model within a "Stress-Strength" (SS) model has been illustrated (see Sect. 3 for details).

The pdf of a IW RV, with parameters σ and β is:

$$f(t,\alpha,\beta) = \alpha\beta(\alpha t)^{-(\beta+1)}\exp\left[-(\alpha t)^{-\beta}\right], \quad (52)$$

where $t \geq 0$, α, $\beta > 0$.

The reliability function and the hazard rate function are:

$$R(t;\alpha,\beta) = 1 - \exp\left[-(\alpha t)^{-\beta}\right] \quad (53)$$

$$h(t,\alpha,\beta) = \frac{\alpha\beta(\alpha t)^{-(\beta+1)}\exp\left[-(\alpha t)^{-\beta}\right]}{1-\exp\left[-(\alpha t)^{-\beta}\right]}. \quad (54)$$

Also such function is of the "IDHR" family; the peak value of the hazard rate of a IW model is obtained at a mission time value belonging to an interval whose extreme points are: $T_m = [\beta/(\beta+1)]^{1/\beta}/\alpha$ (the mode of the IW distribution) and $T_n = \beta^{1/\beta}/\alpha$; the hrf is infinitesimal when $t \to \infty$. The mean (which exists only if $\beta > 1$) and the variance (which exists only if $\beta > 2$) are, setting $\theta = 1/\alpha$:

$$E[X] = \mu = \theta\,\Gamma(1-1/\beta) = \theta\,\Gamma(1\,+1/\beta) \quad (55)$$

$$\mathrm{Var}[X] = \theta^2\Gamma(1-2/\beta) - \mu^2. \quad (56)$$

2.9 Log-Logistic (LL) Distribution

The log-logistic (LL) distribution was adopted by the authors in insulation reliability studies [37, 40]. A recent application of the LL model in ALT is shown in [148], which also refers to [37, 39] for insulation reliability applications. This model is named after the fact that it characterizes a RV: $X = \exp(T)$, where T has a logistic distribution, whose cdf is:

$$F(t;\alpha,\beta) = \frac{1}{1+\exp\left[-\frac{(t-\alpha)}{\beta}\right]} \quad (57)$$

with: $-\infty < \alpha < \infty$, $\beta > 0$, $-\infty < t < \infty$.

Thus, since the variable $T = \log(X)$ is a logistic RV, then X is a so-called log-logistic RV, which is characterized by the following cdf and pdf, in which the parameters $\lambda > 0$ and $b > 0$ are functions of (α, β) above:

$$F(x) = \frac{(\lambda x)^b}{\left[1+(\lambda x)^b\right]},\ x > 0 \quad (58)$$

$$f(x) = \frac{b\lambda^b t^{b-1}}{\left[1+(\lambda t)^b\right]^2}. \quad (59)$$

It is often convenient to use, instead of the scale parameter λ, the parameter $c = 1/\lambda$, which is the median of X, so that:

$$F(x) = \frac{(x/c)^b}{\left[1 + (x/c)^b\right]}, \; x > 0. \tag{60}$$

It is indeed apparent that $F(c) = 0.5$, no matter the value of b. Although the LL model received some attention in survival data analysis since 1983 in a paper by Bennett [13], it is neither frequently used nor well-known in literature—apart from the popular Cox and Oakes' monograph [51]. In [37], the authors also discuss the similarity between the LL and the Weibull model, apart their hrf. In the LL model, the hazard rate function $h(x)$ is:

$$h(x) = \frac{b\lambda^b x^{b-1}}{[1 + (\lambda x)]^b}, \tag{61}$$

which is always decreasing with x if $b \leq 1$; first increasing, then decreasing with time if $b > 1$. In particular, in the latter case $h(x)$ starts from $h(0) = 0$, then reaches its maximum at $x^* = (1/\lambda)(b - 1)^{1/b}$, then $h(x)$ goes to zero as x diverges.

It must be pointed out that also another, more popular model features these properties of the hrf function, i.e., the Lognormal (LN) model. In fact, the LL distribution—as also discussed in the above references [37, 40, 51]—shares many properties with the LN distribution. The LL model is simpler analytically than the LN one, but appears to be more difficult to estimate, while methods—particularly the Maximum Likelihood (ML) one—for assessing the LN model are well established. The mean value (which only exists if $b > 1$) and the standard deviation SD (which—as the variance—only exists when $b > 2$) are given by:

$$E(X) = \frac{c\pi}{[b\sin(c\pi)]}; \; [X] = E[X]\mathrm{CV}[X], \tag{62}$$

where CV is the coefficient of variation, that in this case has the following expression if $b > 2$:

$$\mathrm{CV}[X] = \sqrt{[(b/\pi)\tan(\pi/b) - 1]}. \tag{63}$$

It can be shown (see [51] for some graphical illustration) that the Skewness coefficient of the LL model is positive and always larger than the corresponding Weibull one, possessing the same CV value. Thus, the LL model possesses generally larger "tails" than the Weibull one with the same central parameters and this may lead to underestimate the upper quantiles of the lifetime if a Weibull model is fitted to data generated in fact from a LL model (this can happen, as shown in [37]).

Such distribution may take its origin from a "Gamma mixture" of a Weibull RV, as will be shown in Sect. 3.

2.10 Lognormal Distribution

The Lognormal (LN) model [53] has become more and more popular in last years, also in reliability applications. The LN pdf with parameters (ξ, δ) and argument t is given by:

$$f(t;\xi,\delta)=\frac{1}{\delta t\sqrt{2\pi}}\exp\left[-\frac{1}{2\delta^2}(\ln(t)-\xi)^2\right], \tag{64}$$

where $0\leq t<\infty$, $-\infty<\xi<\infty$, $\delta>0$, being $\xi=E[\ln(T)]$ and $\delta^2=\mathrm{Var}[\ln(T)]$.

Denoting, as above, by $\Phi(z)$ and $\varphi(z)$ the standard Normal pdf and cdf, respectively, the LN RF and hrf are as follows, respectively:

$$R(t;\xi,\delta)=1-\Phi\left(\frac{\ln(t)-\xi}{\delta}\right) \tag{65}$$

$$h(t;\xi,\delta)=\frac{\varphi\left(\frac{\ln t-\xi}{\delta}\right)}{t\delta-t\delta\Phi\left(\frac{\ln t-\xi}{\delta}\right)}. \tag{66}$$

The behavior of the hrf is not easy to analyze, and was sometimes mistaken in literature, so that specific papers were devoted to it (e.g., [150]). However, it is a IDHR model: the hrf at first increases from zero, then decreases toward zero [53]. Differently from the LL model, the "change point" of the hrf cannot be evaluated analytically (as in the BS model).

The mean and the SD are given by:

$$\mu=\exp(\xi+\delta^2/2) \tag{67}$$

$$\sigma=\mu\{\exp(\delta^2)-1\}^{\frac{1}{2}}. \tag{68}$$

2.11 Weibull Distribution (Featuring also the Exponential Model)

The Weibull model (in particular, the two-parameter Weibull model) is quite popular, probably the most popular model in reliability applications—since its birth, in 1939, for application in mechanical engineering (e.g., fatigue life of steel). Its popularity is due to two basic features: (1) its flexibility (e.g., the Gamma, Normal and the Lognormal models can be satisfactorily approximated, under many respects, by a suitable Weibull pdf; the hrf may be increasing, decreasing or constant); (2) the fact that the Weibull belongs (as it was proved in 1945 by Gnedenko) to the family of extreme-values distributions, being able to represent the failure mechanisms of "chain-like" systems that fail when the weakest link is broken [28].

The Weibull model, in the form denoted as $W(a, b)$, being a and b positive parameters, has the following hrf, RF, pdf:

$$h(x) = abx^{b-1}\exp(-ax^b) \tag{69}$$

$$R(x) = \exp(-ax^b) \tag{70}$$

$$f(x) = abx^{b-1}\exp(-ax^b). \tag{71}$$

Also an alternative parameterization, denoted as $W'(\theta, \beta)$, is often used, in which:

$$\theta = 1/a^{\kappa}; \quad \kappa = 1/b; \quad \beta = b \tag{72}$$

so that the RF is expressed by:

$$R(x) = \exp\left[-(x/\theta)^{\beta}\right]. \tag{73}$$

This latter formulation is the most adopted for expressing the mean and the variance:

$$E[X] = \theta\Gamma(1 + 1/\beta) \tag{74}$$

$$\mathrm{Var}[X] = \theta^2\,\Gamma(1 + 2/\beta) - \mu^2. \tag{75}$$

As well known, the Exponential model is a particular case of the Weibull one. Physical motivations for both of them will be discussed later.

Finally, the following relationship holds between Weibull and abovementioned Gumbel model:

$$Y = \log(X) \tag{76}$$

in which Y is a Gumbel RV and X a Weibull RV. This relationship is often useful for parameter estimation.

2.12 Caveats About Using "Popular" Reliability Models

We close this section by noting that often only some simple analytical and/or statistical considerations about probabilistic distributions are needed to select a proper reliability model, or at least to exclude some of them from subsequent analyses.

For instance, the use, and sometimes the abuse, of the "classical" Gaussian and Weibull models is typical in power systems literature. For instance, in [107] these two models are employed in order to represent HV cables LT data, with mean of 45 years and SD of 15 years. For such case-study, the Gaussian and Weibull models imply both RF and hrf which are very similar, and also extreme percentiles can be shown to be fairly close (see also Sect. 5).

However, the adoption of a Gaussian model is at least questionable for no less than three, very simple, reasons which are—rather surprisingly—often neglected in literature:

1. A Gaussian random variable may always assume (even if with small probability, provided that the mean value is larger than three times the standard deviation) negative values, and this fact makes such model theoretically not suitable to describe LT values.
2. The Gaussian model is not flexible (its pdf can have only one shape, the well known so-called "bell-shaped" one).
3. The Gaussian model has always a monotone hrf; it is indeed a IHR model, regardless of the parameter values (this is another aspect of the lack of flexibility of the model).

On the contrary, the choice of the Weibull model has some good theoretical reasons supporting it for application to a LT distribution, e.g., the "Extreme Value" theory, while also the Weibull model has a monotonic hrf; in particular, as well known, such model belongs to one of the three families: IHR (if $\beta > 1$), DHR (if $\beta < 1$), or constant hrf ($\beta = 1$). This kind of property may result unsatisfactory for the purpose of describing the component reliability over large LT intervals, as nowadays requested within the "life extension" programs of deregulated electric market.

In authors' opinion, such kind of motivations shows clearly that it is very useful to identify the reliability model on the basis of both *theoretical* and "physical" reasons. The next section, about the so-called "physical reliability models", illustrates the most adopted kind of physical motivations behind the identification of a reliability model. Experience in power systems operation shows indeed that, in many cases, failures are associated with "stresses", e.g., rated voltage and temperature (that are steady) as well as overvoltages, fault currents, temperature and mechanical stresses, etc. (that can occur randomly during component lifetime). Fortunately, probabilistic aging and life models about endurance ("strength") of electrical components to stresses, are available, often from "accelerated tests". By this way, it is possible to take advantage of available data on the physical processes of stress and/or strength, according to what has been called an "indirect" assessment of item's reliability—as discussed in [42].

A hint at "Inverse Gamma" and "Generalized Gamma" models

Since they are referred to in Sect. 3, and sometimes (not often) used in literature, only some hints at two kinds of "transformed" Gamma RV, i.e., the "Inverse Gamma" and "Generalized Gamma" models [96, 103] are given here. Their pdf are not difficult to express by means of the well known rule of transformations [134] and the reader may consult the references for more details.

Let X be a Gamma $G(r, \phi)$ RV, then:

$$Y = 1/X \tag{77}$$

is a so-called "Inverse Gamma" RV. Now, letting k be a positive parameter, the RV T defined as

$$T = X^k \tag{78}$$

has a "Generalized Gamma" pdf.

The Inverse Gamma model has the following pdf, with argument y:

$$f(y; r, \phi) = \frac{\phi^r}{y^{(r+1)}\Gamma(r)} \exp(-\phi/y), \quad y > 0. \tag{79}$$

The Generalized Gamma model has the following pdf, with argument t:

$$f(t; r, \phi) = \frac{b\phi^r t^{(rb-1)}}{\Gamma(r)} \exp(-\phi t^b), \quad t > 0, \tag{80}$$

in which $b = 1/k$. Such model may be very useful in some applications, for instance in selecting a proper model from data, since it implies the Gamma (obviously, for $k = 1$) and the Weibull (for $r = 1$) as particular cases; moreover, also the LN model is well approximated if r is large enough.

3 Identification of Probabilistic Life Models from Stochastic Process of Wear

3.1 *Outline of the Section: Inferring Probabilistic Life Models From the Stochastic Process of Wear*

This section and the successive ones are devoted to the IRA, i.e., to the lifetime model assessment deduced or inferred from the knowledge of the probabilistic laws of the stochastic processes of degradation and stresses which unavoidably affect any device. Often, also the device "strength", i.e., the maximum stress amplitude that the device is able to withstand before failing, is a RV or, in general, a SP, due to the unavoidable randomness intrinsic in its aging, because of uncontrollable variations from item to item in the manufacturing processes, to randomness of environmental conditions, etc. (these aspects will be dealt with more detail in relation to specific applications of the next section). The combination of stress and strength is generically denoted under the name of "wear" in the following. The distinction between "continuous" and "discontinuous" wear or failure processes, although it may be useful sometimes (the "stress" or "shock" processes should be framed into the latter, according to some publications), is not maintained here, also because it is very difficult to define and distinguish clearly the two kinds of processes, which, in fact, are superimposed in practice. As already discussed, by means of the probabilistic knowledge of wear the model assessment can be performed by means of lifetime data analysis, as in DRA. DRA should be, of course, always performed, but it cannot be at all discriminatory among several "similar" models (e.g., the LN and the Gamma one) when only a few data (say, less than 20) are available, as often occur in practice. Needless to say, the IRA alone cannot be claimed to be the solution to the problem of model assessment as well, a problem that is always very critical within the reliability analysis of modern technology products, for which no definitive solution can be easily found.

Indeed, it must be highlighted that IRA requires that the wear process be known by prior information and/or be somewhat measurable, directly or indirectly (e.g., by measuring its effects on lifetime reduction of similar devices).

This is not always the case in practical applications, of course, so that, in general, an adequate and feasible lifetime model assessment should be better performed by a reasonable combination of both direct and IRA. Often this will not bring to definite conclusions, but sometimes even the exclusion of some models in favor of a restricted choice can be a useful result: this happens, for example, when the allowable models are in practice very similar (as often happens, e.g., for the LN and the IG one: see the final example of Sect. 5), so that "mistaking" one for the other does not bring about remarkable errors (never forgetting the obvious principle that the "only true" model does not exist).

However, the above possible limitations of IRA might be overcome for what concerns the application of the chapter—illustrated in the following sections—which are devoted mainly to electrical components (and, in particular, to electrical insulation): indeed in this field the above requirement of the "measurability" of stress and wear is mostly satisfied, also with the help of extensive experimental surveys conducted by means of ALT. After many decades of experience, such tests allowed to validate well-established models that relate lifetime and applied stress (voltage, temperature, etc.), such as the popular IPM [35, 51, 97]. In the next Sect. 3.2, general *Stress-Strength* (SS) models [99] are reviewed. While describing such models, some reliability distributions (as those already reviewed in previous Sect. 2) are directly obtained. Then, in Sect. 3.3, a complete list of all the models of Sect. 2 together with their possible generative mechanisms based upon degradation is illustrated: the list is by no means meant to be exhaustive, but it only serves as a reference and for illustrating a methodology. Some of the "dynamic" models here reported have been deduced following the same approach as in a recent book by Singpurwalla [146]. Recent accounts of SS and fatigue damage models, particularly devoted to mechanical engineering applications (which were the origin of such models) can be found in [22, 29, 158]. In [90], a generalized Stress-Strength model is considered with reference to stochastic loading and strength aging degradation in a more general way with respect to those dealt with here.

For the sake of brevity and simplicity, only a hint is made here at more complex wear models such as those based on the advanced theory of SP, such as Wiener diffusion processes [157], Gamma processes [1, 22, 45, 145, 153–155], Markov and semi-Markov models of deterioration [44, 46]: most of such models were derived in the framework of Structural Reliability, and are still seldom adopted for electrical devices. Advanced models for comparing DRA and IRA are exposed with emphasis to the statistical estimation point of view in [110, 111], which assume a measurable degradation process, also taking into account the measurement error: it is shown that IRA is, under the assumed degradation model, more efficient for estimating extreme quantiles of the LT distribution. In the same field, a new formulation of degradation modeling with random coefficients models has been recently proposed by [81], using the so-called *Bernstein distribution* in a sensor-based prognostics framework, with the purpose of predicting residual life distributions.

3.2 Stress-Strength Models

A classical family of models for "physical" reliability evaluation alternative to probabilistic life models is made of the so-called probabilistic "Stress-Strength" (SS) models, based on the characterization of the stochastic process describing the wear caused by random stresses [94, 99].

3.2.1 Static and "Quasi Static" Stress-Strength Models

Let us denote by X and Y the two random variables (RV):

- X: the "Stress".
- Y: the "Strength".

For instance, in the application to components insulation (see following section), the random variable X ("Stress") is the peak value of stress (voltage surge); the RV Y ("Strength") is the insulation electric strength. It is apparent that both Strength and Stress are, in general, affected by randomness.

Then, in its simplest, "static" form, the SS model is based on the following expression of the reliability function (RF), i.e.:

$$R = P(X < Y). \tag{81}$$

The model is "static" in that the mission time t does not explicitly appear and only RV (X and Y) are used instead of SP, as would be more appropriate (see Sect. 3.2.3). This means that Strength and Stress are assumed as constant, although random, in the time interval to which the RF is referred. So, the pdf of X and Y are assumed as time-independent, or the mission time pre-determined; the more realistic "dynamic" version of SS models is discussed later.

Denoting with $f(y)$ ($F(y)$) the pdf (cdf) of Y, and with $g(x)$ ($G(x)$) the pdf (cdf) of X, the RF of the device is given—under the reasonable hypothesis that the RV X and Y are statistically independent—by:

$$R = \int_0^\infty g(x)P(X<Y|X=x)\mathrm{d}x = \int_0^\infty g(x)(1-F(x))\mathrm{d}x. \tag{82}$$

In the above equations, time does not appear (at least in explicit form). Various models derived from (82) are illustrated in the above and related references. The analytical solution of (82) exists in a few cases, among which the "Weibull case" for X and Y is illustrated as an example here below.

3.2.2 Example: A Weibull Stress-Strength Model Leading to a Log-Logistic Distribution

We use here as "quasi static SS model", i.e., a simple dynamic generalization of a static SS model, obtained by letting some parameter vary with time. Let X and Y be two Weibull RV with equal shape parameter β, and with scale parameters θ for the Strength X, and α for the Stress Y, i.e., let the cdf of X and Y be given, respectively, by:

$$G(x) = 1 - \exp\left[-(x/\theta)^\beta\right]; \quad F(y) = 1 - \exp\left[-(y/\alpha)^\beta\right]. \tag{83a}$$

As for time dependence, it is reasonable to consider- as proposed in [37]—the following "Inverse power" characterization of the Strength scale parameter α with time t, in which k and m are positive constants:

$$\alpha = \alpha(t) = k/t^m. \tag{83b}$$

Indeed, since the expectation of Y is proportional to α—it is recalled that $\mu = \alpha\Gamma(1+1/\beta)$—relationship (83b) implies that Y decreases with time t as a power function of t, a popular model in LT analyses, which will be met throughout the chapter. Then, after easy computations shown in [37, 41], the following log-logistic (LL) model [51, 96] is obtained:

$$R(t) = 1\Big/\left[1+(\lambda t)^b\right], \tag{84}$$

where $b = m\beta$; $\lambda = (\theta/k)^{1/m}$.

The LL model belongs to the IDHR or (less frequently) to the DHR family of reliability models, depending on the value of the shape parameter b. Indeed, its hazard rate function $h(t)$ has the following expression:

$$h(t) = b\lambda^b t^{b-1}\Big/\left[1+(\lambda t)^b\right], \tag{85}$$

which is always decreasing with time if $b \leq 1$; first increasing, then decreasing with time if $b > 1$ (such properties were already discussed in more detail in Sect. 2.8, after Eq. 61).

In fact, the LL distribution—as also discussed in the above references—shares many properties, such as being IDHR, with the LN distribution. Although appearing against intuition, the IDHR (or DHR) property has been sometimes observed, as already recalled, for some electrical components, and has often been motivated in theoretical reliability literature in relation with random heterogeneity of materials or subjective probabilistic reasoning [10, 79, 109, 146].

3.2.3 Dynamic "Stress-Strength" Models

The static model is, of course, of limited application, for at least two reasons implying a time variation of the pdf of stress and/or strength:

1. The stress X is always best described by a *stochastic process* in time [134, 151], $X(t)$, since many random variables (fault time occurrence, duration, amplitude, location, etc.), most of which are time-dependent, are involved in its definition.
2. The strength $Y(t)$ is generally decreasing in time due to aging effects. Being a SP, the fact that $Y(t)$ is decreasing in time must be defined on a probabilistic basis, as discussed below.

For a typical example of item (1), one can imagine the stress process as constituted by a succession of random "shocks" events which occur at random times: $T_1, T_2, \ldots, T_n$. This is denoted as a "Shock type" stress, which is the most common one in the case of electrical systems (examples: overvoltages, short circuit currents).

However, although quite common, the above "Shock type" stress process is not the most general, as there always exists, in practice, an "ordinary" stress which is continuous in time (i.e., caused by weather conditions, or also by nominal voltage, etc.), upon which the Shock type Stress is "superimposed". A general view of dynamic Stress-Strength Models allowing for the description of Stress or Strength processes by means of continuous SP is given in the following.

Let us define the following stochastic processes:

- $X(t)$ = "Stress" Process.
- $Y(t)$ = "Strength" Process.

Then the RF over the interval $(0, t)$ is given by:

$$R(t) = P[X(s) < Y(s), \quad \forall \, s \, \text{in} (0, t)]. \tag{86}$$

Accordingly, the LT of the component—i.e., the RV here denoted by T—is given by the first time instant at which the Strength is greater than the Stress:

$$T = \inf\{t : t > 0, \, X(t) > Y(t)\}. \tag{87}$$

Some simple examples of Stress and Strength processes are illustrated below. For instance, being of course the Strength process $Y(t)$ closely related to the aging of the device (possibly due also to the wear cumulated up to time t because of all previous shocks), it may be considered as a continuous process which is generally decreasing in time, in a stochastic sense; i.e., denoting by (*a*, *b*) generic time instants:

$$a<b \to P[Y(a) > y] > P[Y(b) > y], \quad \forall y > 0. \tag{88}$$

Equivalently, by introducing the cdf of $Y(t)$:

$$F(y;t) = P[Y(t) < y)], \quad y > 0; \, t > 0. \tag{89}$$

Equation 88 may be written as follows:

$$a<b \to F(y;a) < F(y;b), \quad \forall y > 0. \tag{90a}$$

Sometimes, the milder condition may be imposed that $Y(t)$ is decreasing in the mean value sense, i.e.:

$$a<b \to E[Y(a)] < E[Y(b)]. \tag{90b}$$

It is easy to show that (90a) implies (90b), but the converse is not generally true except in some cases, as in the following example Sect. 3.2.4.

A stronger condition of decreasing Stress may be the a.s. (almost sure, i.e., with probability = 1) one, i.e.:

$$a<b \to Y(a) > Y(b), \text{ a.s.} \tag{90c}$$

Of course, this implies both (90a) and (90b).

3.2.4 Example: A Weibull Strength Model with IPM Time Variation

Based on already recalled extreme-value theory, the strength Y of a material can be characterized by a Weibull distribution. Moreover, let us assume that the time dependence of strength is contained in the scale parameter α, so that the time-dependent cdf of the Stress is expressed by:

$$F(y,t) = 1 - \exp\left[-\left(\frac{y}{\alpha(t)}\right)^{\beta}\right], \quad y > 0, \; t > 0. \tag{91}$$

Moreover, let $\alpha = \alpha(t)$ be decreasing in time, e.g.: $\alpha(t) = k/t^m$, as in the IPM model. In this case, it is easy to see that both (90a) and (90b), hold; the first is immediate, the second comes from the mean value expression:

$$E[Y(t)] = \alpha(t)\Gamma(1 + 1/bgr;). \tag{92}$$

In the above Weibull example, the "a.s. decreasing Y" property is not assured.

Similar properties—"mutatis mutandis"—may be adopted for the Stress process $X(t)$ which may be considered as a continuous process, generally increasing (in a stochastic sense) in time.

3.2.5 Dynamic Stress-Strength Models with "Shock Type" Stress: A Cumulative-Damage Model

As previously pointed out, stresses are often caused by repeated "shocks" (this is indeed the case of overvoltages, fault currents, etc.), whose succession constitutes a typical example of a stochastic process. Their effect may be cumulated or not, depending on the kind of component (e.g., in the case of an insulation, this may also depend on whether it is self-healing or not). So, let us consider a stress process as constituted by a succession of random "shocks" events which occur at random times: T_1, T_2, …, T_n,… This SP, denoted as a "Shock type" stress, can be considered as a "point process" [134, 151] of random variables Z_j ($j = 1$, …, n,…) occurring at the random instants T_j ($k = 1$, …, n, …): i.e., the RV Z_j represents the stress amplitude associated with the shock event occurring at time T_j. In practice, the stress "process", viewed as a continuous function of time t, $W(t)$, is always zero except for "spikes" (of negligible duration) with amplitude Z_k at times T_k. Moreover, also the number of events occurring in a given interval (0, t) is random. So, let us denote by $N(t)$ the following stochastic process: $N(t)$ = number of stresses occurring in the interval (0, t).

Due to the fact that the stresses (overvoltages) are purely accidental, a reasonable hypothesis is that the process $N(t)$ can be described by a (homogeneous) Poisson process [43, 134, 140], so that its probability distribution is expressed by:

$$p(k,t) \equiv P[N(t) = k] = \frac{(\phi t)^k}{k!} \exp(-\phi t) \quad k = 0, 1, \ldots, \infty, \tag{93}$$

where ϕ is the mean frequency of occurrence of the event (i.e., the mean number of shocks per unit time). The above Poisson model has always found many applications for describing the fault process in the case of power systems [3, 5], and here—for brevity—it will be the only one considered. Extension of SS theory to non-homogeneous Poisson processes is dealt with, e.g., in [89]. Let us suppose, as a typical case which finds many applications in literature (see, e.g., [2] for an application to HV circuit breakers), a "cumulative wear process" and denoting by Z_K the stress amplitude at time T_K, the total wear acting at the end of the interval (0, t) on the component is given by the SP:

$$W(t) = \sum_{k=0}^{N(t)} Z_k, \quad \text{if } N(t) > 0 \tag{94a}$$

$$W(t) = 0, \quad \text{if } N(t) = 0. \tag{94b}$$

Of course, Z_K is generally a RV, since its value cannot be predicted. So, the wear process $W(t)$ is characterized as a "Compound Poisson process" [134].

Let us suppose, for the moment, that the strength Y is not a RV, but it is a constant y (time-independent). Since $W(t)$ is an (almost surely) increasing function of time—provided that the RV Z_K are non-negative, as reasonable—and the fault occurs as soon as the total wear $W(t)$ is greater than the strength y, the device reliability function is, in term of the wear cdf:

$$R(t) = P(W(t) < y) = F_w(y, t), \quad t > 0. \tag{95}$$

The probability distribution—and so the RF of (95)—of the process $W(t)$ of (93), (94) may be deduced as follows [134, 140], assuming as reasonable that $N(t)$ and the Z_K are s-independent: let Q_n be the whole damage conditioned to the occurrence of a deterministic number, n, of stresses:

$$Q_n = \sum_{k=1}^{n} Z_k. \tag{96}$$

Denoting by $F_{Q_n}(w)$ the cdf of Q_n, the cdf of $W(t)$, i.e., the probability of the event $W(t) < w$, is given—according to the total probability theorem—by:

$$F_W(w,t) = \mathrm{e}^{-\phi t} + \sum_{n=1}^{\infty} F_{Q_n}(w) p(n,t). \tag{97}$$

It is remarked that the above function depends both on wear magnitude, w, and on the time instant t. Both w and t are positive. Only in some cases, the pdf of $W(t)$ can be expressed in analytical way (if not in a closed form), and this happens only if it is assumed that the RV Z_k are identically distributed and independent of the process $N(t)$; for example, if the Z_K RV are exponential, then $W(t)$ has a Bessel distribution. However, if the variables Z_k are independent, $W(t)$ is a stochastic process with independent increments; as t increases—according the Central Limit Theorem—it approaches a Gaussian process [134]; the process mean and variance can be obtained as follows, assuming—as above said—that the Z_K RV are independent, with equal mean and variance:

$$E[Z_K] = \mu_z, \ \forall k \quad V[Z_K] = \sigma_z^2, \ \forall k. \tag{98}$$

Then, it is easy to show that the mean value of $W(t)$ at time t is equal to:

$$E[W(t)] = \mu_z \phi t \tag{99}$$

and the variance and the auto-covariance function C_W [134] of the process $W(t)$ are given—denoting by (t, t_1, t_2) generic time instants—by, respectively:

$$\mathrm{Var}[W(t)] = \left(\mu_z^2 + \sigma_z^2\right)\phi t \tag{100}$$

$$C_W(t_1, t_2) = \varphi\left(\mu_z^2 + \sigma_z^2\right)\min(t_1, t_2). \tag{101}$$

For the purpose of the RF evaluation, the fact that $W(t)$ approaches—according to the Central Limit Theorem—a Gaussian process implies that the LT distribution may be represented, at least approximately, by a "Birnbaum–Saunders" distribution for the time to failure T, as will be highlighted in Sect. 3.3. Such model, as recalled, is again IDHR. A more complete discussion of aging properties, for such cumulative models deriving from shocks, can be found in the classical book [10], characterized by a superb level of mathematics.

The extension of (95) to the case when the strength Y is a RV, with (time-independent) pdf $g(y)$, is straightforward, still using the total probability theorem:

$$R(t) = P(W(t) < Y) = \int_{o}^{\infty} F_w(y,t)g(y)\mathrm{d}y. \tag{102}$$

In this case, the LT distribution is—under the above hypotheses—again IDHR, being so the time-dependent integrand of (102).

3.2.6 Dynamic Stress-Strength Models with "Shock type" Stress: A memoryless Dynamic Stress-Strength Model

With reference to the general expression (81) of the RF over the *interval* (0, t), let us assume that:

- $X(t)$ is a SP which can be described as a "Shock type" stress.
- The shocks occurring at the time instants T_j.

The device fails only because of the occurrence of a stress, i.e., at the time $t = T_j$ when stress amplitude is greater than Strength $Y(t) = Y(T_j)$; of course, such failure time is a RV.

It is observed that, in order that the device does not fail in the *whole* interval (0, t), then every Stress within the given interval must be smaller than the relevant Strength, i.e., $(X_j < Y_j)$ must be verified for every index $j = 1, \ldots, N(t)$, where $X_k = X(T_k)$, and $Y_k = X(T_k)$, T_k being the RV "time of k-th stress occurrence".

The RF can be obtained first by conditioning on the event $E_n = [N(t) = n]$:

$$R(t|\,E_n) = P[(X_1 < Y_1) \cap (X_2 < Y_2) \cap \cdots \cap (X_n < Y_n)|E_n\,]. \tag{103}$$

Note that $R(t|E_n)$ is indicated as R_n in what follows.

Then, once the functions R_n have been computed, the RF $R(t)$ can be obtained applying the total probability theorem as a function of the R_ns and of the distribution of the point process $N(t)$:

$$R(t) = \sum_{n=0}^{\infty} R_n(t)p(n,t), \quad t > 0, \tag{104}$$

where $p(n, t) = P[N(t) = n]$. In the following, the above introduced Poisson law will be used, but the methodology is not dependent on such assumption.

In the model here hypothesized, both stress and stress are time-independent. In this case, assuming also that the RV X_j ($j = 1,\ldots n,\ldots$) and Y_j ($j = 1,\ldots n,\ldots$) are statistically independent of each other and of $N(t)$, then:

$$R_n = R(t|E_n) = P[(X_1 < Y_1) \cap (X_2 < Y_2) \cap \cdots \cap (X_n < Y_n)|E_n)] = \prod_{k=1}^{n} P(X_k < Y_k). \tag{105}$$

So letting $\prod_{k=1}^{n} P(X_k < Y_k) = r_n$, the RF is given by:

$$R(t) = \sum_{n=0}^{\infty} r_n(t)p(n,t), \quad t > 0. \tag{106}$$

A simple case where the RF is analytically computable is when:

- the X_j are IID with common cdf $G(x) = F_X(x) = P(X_j < x), \quad \forall j = 1, 2, \ldots n, \ldots$ (independent of time) and pdf $g(x)$;
- the Y_j are IID with common cdf $F(y) = F_y(y) = P(Y_j < y), \quad \forall j = 1, 2, \ldots n, \ldots$, (independent of time), and pdf $f(y)$;

Then:

$$r_n = r^n, \tag{107}$$

where using the same approach as in (82), and denoting by X a generic one of the X_j RV (and the same for Y and Y_j):

$$r = P(X < Y) = \int_0^{\infty} g(x)(1 - F(x))\mathrm{d}x. \tag{108}$$

Under the above hypotheses, the wear process can be defined as a "memory-less" one, since wear at age t does not depend on previously occurred shocks. It is possible that such model applies to self-healing insulating materials.

According to (106), by resorting to well known properties of series expansion of the exponential function appearing in the assumed Poisson law $p(n, t)$, the result is straightforward:

$$R(t) = \exp[-\phi t(1 - r)] = \exp(-\phi q t) \tag{109}$$

having defined q as the elementary failure probability $q = 1 - r = P(X > Y)$ = probability that the generic stress X_j is greater than the generic strength Y_j.

In the case that the Stress is a constant, y, then:

$$q = 1 - F_X(y) \rightarrow R(t) = \exp[-\phi t(1 - F_X(y))]. \tag{110}$$

The above RF is clearly an Exponential one, i.e., it may be expressed as:

$$R(t) = \exp(-\lambda t) \tag{111}$$

with parameter λ(hazard rate) $= \phi q =$ (mean stress occurrence) $\times$ (elementary failure probability).

As an example, a particular case of interest is that in which the generic Strength Xj has—as already used above—a Weibull cdf:

$$F_X(x) = 1 - \exp(-ax^b), \quad \text{for } x > 0 \tag{112}$$

in this case, the RF of (110) has the following expression:

$$R(t) = \exp[-(\phi t)\exp(-ay^b)] \tag{113}$$

or alternatively, by setting $\tau = 1/a^{1/b}$:

$$R(t) = \exp\left[-(\phi t)\exp\left(-(y/\tau)^b\right)\right]. \tag{114}$$

The above RF may be expressed as a function of the mean value m of the Strength X, given by $m = E[X] = \tau\Gamma(1 + 1/b)$.

DRA and IRA are compared for this model, also in view of estimation, in [38].

In conclusion, it is highlighted that the illustrated SS model is, in this case, characterized by a constant hazard rate. This fact—together with the IDHR property of many models previously examined (LN, IG, IW, LL)—may appear at a first sight conflicting with intuition, again, since one could expect that the presence of stresses, whose number surely increases with time, implies an increasing hazard rate. This confirms that a careful analysis of the hypotheses which yield the reliability model may lead to non-trivial conclusions, difficult to be anticipated by pure intuition. By the way, by means of the above SS model a further (and seldom reported in literature) justification for the LT being an Exponential random variable is obtained, which is to be added to those leading to the Weibull or the Gamma model with shape parameter 1, examined in the following.

3.3 *Identification of Main Probabilistic Lifetime Models by IRA*

In this section, the same list of reliability models of Sect. 2 is considered, with their generative mechanisms from wear processes. It is recalled that the identification of the Exponential and the log-logistic model have been already deduced previously (Sect. 3.2), so that they will be considered again here with different motivations; also the Birnbaum–Saunders distribution has been hinted at before (see the comment following Eq. 101).

3.3.1 Birnbaum–Saunders Model

The BS model [62, 63, 96]—sometimes denoted as "fatigue life" model—was originally derived on the basis of a "discrete" stress process, accounting for accumulating cracks on a material, which can cause its failure when a given "critical dimension", y, is overcome. Let the material be subjected to repeated cycles of a common stress, the single stress amplitudes being s-independent Gaussian RV; so, using an approach similar to that used in Sect. 3.2.5, assuming a "cumulative wear process", and denoting by Z_K the stress amplitude at k-th cycle, the total stress after n cycles is of course

$$W_n = \sum_{k=0}^{n} Z_k, \quad n = 1, 2, \ldots, \infty. \tag{115}$$

Then, the original BS model was obtained by observing that the "discrete" failure time N (i.e., the number of cycles after which the failure occurs) has the following cdf (defined for discrete values of n in the set of natural numbers):

$$P(N \leq n) = P(W_n > y) \tag{116}$$

(it can be easily deduced, e.g., that the random events: $(N > n)$ and $(W_n < y)$ are equivalent; then, considering the complementary events, the above relation is obtained).

Assuming that the Z_K RV are IID Gaussian RV, with $N(a, b)$ distribution ($a > 0$ for obvious reasons), or that they are in a high number so that the central limit theorem holds, then W_n is a $N(an, b\sqrt{n})$ RV, so that:

$$P(N \leq n) = 1 - P(W_n \leq y) = 1 - \Phi\left[(y - an)/\mathrm{b}\sqrt{\mathrm{n}}\right] = \Phi\left[(an - y)/b\sqrt{n}\right], \tag{117}$$

where the known property of the standard Normal cdf: $\Phi(-x) = 1 - \Phi(x)$ has been used. It is remarked that the symbol n above denotes the "time" argument of the cdf. Now, in the original deduction of the model, discrete time n was "transformed" into continuous time t (this is somewhat incorrect, but can be roughly justified by letting the cracks occurring at a constant rate r in time, so that $n = rt$), then the following cdf can be easily obtained—with an understandable meaning of the positive constants (α, β):

$$F_T(t; \alpha, \beta) = \Phi\left\{\frac{1}{\alpha}\left[\left(\frac{t}{\beta}\right)^{\frac{1}{2}} - \left(\frac{\beta}{t}\right)^{\frac{1}{2}}\right]\right\} \tag{118}$$

which is indeed a BS cdf.

The limits of the above derivation of the BS model were highlighted, e.g., by [15], where it is correctly observed that a Gaussian RV can assume—even if with small probability, by an adequate choice of the relevant constants—values less than 0: so, the above stress model does not guarantee that W_n is an increasing function of time n (indeed, a "negative-amplitude" crack can occur every now and then), thereby resulting in a non-realistic model for accumulated stress. In [15], the authors, by comparing the BS with the similar IG model (to be dealt with afterwards), find the IG model superior in that it is free of the above limitations and directly formulated in continuous time.

However, as also reported in [96], Desmond [62] derived a more general form of the BS model relaxing the hypothesis of Gaussian crack amplitude and of s-independence. He showed that the BS pdf can be obtained by a mixture of two appropriate IG pdf, too.

However, the BS model remains less attractive and less adopted than the IG one, mainly because well-established estimation methods for the IG exist, while only some "ad hoc" methods are available for the BS model.

3.3.2 Gamma Model (featuring also the Exponential model)

The Gamma $G(r, \phi)$ pdf is given by:

$$f(t; r, \phi) = \frac{\phi^r t^{(r-1)}}{\Gamma(r)} \exp(-\phi t), \quad t > 0. \tag{119}$$

Two main methods of deducing the Gamma model from wear processes are presented here below at points (a) and (b), followed by the particular Exponential model at point (c). Two (minor) transformed forms generated by the model, denoted "Inverse Gamma model" and "Generalized Inverse Gamma model", are only hinted at for brevity afterwards, when dealing with the LN model.

1. It is well known [96] that a Gamma RV $G(r, \phi)$, if r is a positive integer, can be obtained as the sum of r Exponential statistically independent and identically distributed RV with parameter ϕ, which is also the value of their common, constant, hrf. Although this motivation for its use appears to be rarely adopted (apart from the particular case of an "Exponential unit" with $r - 1$ stand-by redundant identical units), it can be advocated when the LT of the devices passes through a series of s-independent "stages", each one lasting a time interval which is distributed according the same Exponential RV. This may be the case, e.g., when the device, starting from an initial "good" state, reaches the failed state after a few "partial failure" states. In practice, however, it is difficult to imagine a situation in which all the stages have the same pdf (thus the same value of parameter ϕ) and are s-independent. On the other hand, if the s-independence subsists, and the single-stage RV are Exponential RV with different values of the parameter ϕ, the pdf of their sum has a closed form expression [96], which can be approximated in many cases by a Gamma pdf, even when the number of stages, r, is unknown (as realistic). In the general case of unknown single pdf, with r high enough (say, $r > 5$), the CLT may be a valid reason for approximating the LT by a Gaussian RV. Indeed, it is true that the Gamma model $G(r, \phi)$ is closely approximated by the Gaussian one for r high enough.
2. Although the following motivation appears to be rarely, if ever, highlighted in literature (maybe for its triviality), the Gamma model may also be deduced by means of a "cumulative wear process", i.e., the SP denoted as $W(t)$ in Eq. 94a, 94b,

as briefly accounted in the following. Let us suppose that the stress amplitudes Z_K at time T_K—appearing in (94a)—are not RV, but possess the same constant (deterministic) value. With no loss of generality—by choosing a proper stress amplitude unit—it may be supposed that $Z_K = 1$ for each k. Thus, the SP $W(t)$ simply equals the Poisson process $N(t)$. Then, assuming—as in (95)—that the strength Y has also a constant value $y > 0$, the device reliability function is given—from (95), using the Poisson distribution of (93), and denoting by r the largest integer value smaller than y—by:

$$R(t) = P(N(t) < y) = \sum_{k=0}^{r} \frac{(\phi t)^k}{k!} \exp(-\phi t). \tag{120}$$

This is a well known formulation—denoted as "Erlang"—of the Gamma RF [96]. It is easy to see indeed that also in this case, the time to failure is the sum of r Exponential independent and identically distributed RV with parameter (hrf) ϕ, where ϕ and r are positive constants representing the shape and scale parameters, respectively.
In view of relaxing some of the hypotheses leading to the above model, it is interesting to remark that the Gamma model has proven to be a satisfactory approximation to the true model even in the case of random stress (Z_K): this has been shown numerically, in some cases, in [2], but cannot be, at present, claimed to be always true.

3. (*Exponential model*) The above points also apply to the Exponential RV, when the particular case $r = 1$ is considered. Moreover, at least two additional deductions of the Exponential model, which appear more fundamental, must be considered. The first is the classic one of the "memoryless" property, already recalled, by which such model is the only one allowable for devices whose failure is only due to accidents, with no regard to age. Indeed, the Exponential model is the only one possessing a constant hrf, and so a CRF $R(t|s)$ independent from age s. The second deduction has been obtained as a consequence of the particular Dynamic Stress-Strength Models with "Shock type" Stress of Sect. 3.2.6. Moreover, the Exponential model can be obtained as a particular case of the Weibull model with shape parameter 1, which will be considered afterwards.

3.3.3 Gaussian Model

Despite its high popularity, it is difficult to justify the adoption of the Gaussian model as a lifetime model, first of all because it is not restricted to positive values. In addition, such model can be theoretically obtained only—by virtue of the CLT—when the LT can be expressed as the sum of many s-independent RV. This is why the peculiar cases already illustrated for the Gamma model $G(r, \phi)$—with r high enough—can be often approximated also by the Gaussian one, but actually there are only drawbacks in doing so, for theoretical and practical (statistical) reasons already presented, i.e., the non-positivity, the inadequacy in representing skewed pdf (which are by far the most part of LT pdf occurring in the real world), and, in general, the scarce flexibility of the model, also with reference to the hrf (which can assume only a roughly linear form in time). It is also to be remarked that other, more flexible models—such as the Gamma $G(r, \phi)$, the LN [LN(α, β)], and the Weibull $W(a, b)$—can also closely approximate symmetrical pdf, by an opportune choice of their shape parameter values (the first if $r > 10$, the second if $\beta \to 0$, the third if $b \approx 3.6$). The CLT motivations can support the use of the Gaussian model as a repair time model in availability studies that are outside the scope of this chapter. Obviously, the Gaussian model fully maintains its well-established supremacy—also in reliability applications—for what concerns inference studies (e.g., for obtaining the distribution of statistical estimators, confidence intervals, etc.).

3.3.4 Gumbel Model

It is recalled that the Gumbel model is the "smallest extreme value" model, and this constitutes also a clear motivation for its use, even if it might be in many cases better advocated for systems rather than for components. From a statistical modeling point of view, the fact that the hrf can have only one shape (the increasing exponential one) is a limit of the model; at the same time, this may make it useful for devices largely affected by wear with increasing age (such as mechanical products; for the same reasons, also human being LT are often characterized by this or similar models).

Instead of LT values, a physical reason for adopting the Gumbel model can be found in the case of strength values. This may be the case when dealing with dielectric strength of electrical insulation; indeed, the value of breakdown voltage of a large-size insulation system may be considered as the minimum between the values of breakdown voltage of smaller elements. Very significant applications in this field, also comparing the Weibull and the Gumbel models (that can be used both as a *smallest extreme value model*) are in [66, 87], the first also containing physical properties which can justify the models.

3.3.5 "Hyperbolic Reliability Model"

The HRM is characterized by the hrf:

$$h(t) = r + \frac{a}{t+1}, \quad a > 0, r > 0. \tag{121}$$

Erto and Palumbo [76] showed that the HRM identification can be performed observing al least three mechanisms of failure leading to this law of mortality, which they shown to be: (1) a "Deterioration" mechanism of failure; (2) a "Stress-

Strength" mechanism of failure; (3) a "Shocks" mechanism of failure. For sake of brevity, referring to the above paper for the first and second, only the third mechanism is illustrated here. Let a device be subjected to a random succession of shocks, which can potentially cause system failure; let the shocks occur following, as discussed previously, a Poisson law:

$$\Pr\{N_s = n_s\} = \frac{(\delta t)^{n_s}}{n_s!}\exp(-\delta t), \quad \delta > 0. \tag{122}$$

Now, if the survival probability for each shock depends on the run time (but not on the number of previously suffered shocks) following the law:

$$S_v(t) = \rho - \frac{\tau}{t+1}, \quad 0<\rho<1,\ \tau > 0, \tag{123}$$

then the probability of failure in a time interval Δt is:

$$F(t+\Delta t) - F(t) = R(t)\sum_{i=1}^{\infty}\left\{\frac{(\delta\Delta t)^i}{i!} \times \exp(-\delta\Delta t)\left[1 - S_v^i(t)\right]\right\} \tag{124}$$

from which

$$h(t) = \lim_{\Delta t\to 0}\frac{F(t+\Delta t) - F(t)}{R(t)\Delta t} = \delta[1 - S_v(t)], \tag{125}$$

and substituting $S_v(t)$ with its time-dependent function

$$h(t) = \delta(1-\rho) + \frac{\delta\tau}{t+1}. \tag{126}$$

Renaming the two products

$$\delta(1-\rho) = r, \quad \delta\tau = a, \tag{127}$$

the hrf in Eq. 121 is obtained.

As previously hinted at, an electrical application—relevant to the times-to-breakdown of an insulating fluid, working at constant voltage equal to 32 kV—was successfully illustrated by means of the HRM in [76].

3.3.6 Inverse Gaussian Model

Also, the IG model can be obtained from a "Stress-Strength" (SS) model arising from a Wiener Stress process and a deterministic Strength. Indeed, as hinted at previously, the IG distribution [34] has been introduced as the first passage time of a Wiener process [134], i.e., a Gaussian Stochastic Process with independent increments. Let us hypothesize that the SP *W*(*t*) describing wear is a Wiener process with "drift" $\mu > 0$ and "diffusion constant" $v > 0$. Then, *W*(*t*) satisfies the differential equation:

$$-\mathrm{d}W/\mathrm{d}t = \mu + G(t), \tag{128}$$

where μ is a positive constant and *G*(*t*) is a Gaussian SP with the following mean and covariance functions [134]:

$$E[G(t)] = 0,\ \mathrm{Cov}[G(t)G(t-s)] = v\delta(s) \quad (v > 0). \tag{129}$$

If the wear process is defined by a stress characterized by the above Wiener process and a deterministic strength *b*, the associated LT RV is defined by the first instant, *T*, in which the SP *W*(*t*) crosses the barrier *b*, i.e.:

$$T = \inf\{t : t > 0,\ W(t) = b\}. \tag{130}$$

As shown in [34], the pdf of *T*—by a proper choice of the parameters μ and λ (both having the dimensions of time), and letting, with no loss of generality, $b = 1$—is given by:

$$f(t;\mu,\lambda) = \sqrt{\frac{\lambda}{2\pi t^3}}\exp\left[-\frac{\lambda}{2\mu^2 t}(t-\mu)^2\right], \quad t,\mu,\lambda > 0 \tag{131}$$

i.e., the above-mentioned IG pdf as in Sect. 2.

It has been already recalled in Sect. 2 that the IG and the BS models are very similar, and this may be also justified by the above similar derivations from Gaussian wear processes; it is again highlighted that they are both IDHR.

3.3.7 Inverse Weibull Model

The reliability function of the IW model is:

$$R(t;\alpha,\beta) = 1 - \exp\left[-(\alpha t)^{-\beta}\right]. \tag{132}$$

Apart from deriving the abovementioned peculiar properties of aging of this IDHR model, Erto [70] also showed that it can be originated by reasonable Stress-Strength models and deduced at least two possible "physical" derivation of the IW model, shown in the following.

1. (*Stress is a Weibull RV and strength is a deterministic function of time*) Let the distribution of the RV stress, *X*—assumed as time-independent—be a RV distributed according a Weibull $W'(u, v)$ law, thereby having cdf:

$$F_X(x) = P(X \le x) = 1 - \exp\{-[x/u]^v\} \quad (b > 0). \tag{133}$$

Let the strength $y(t)$ be a deterministic decreasing function of time, described by a process with an "inverse power" aging law as:

$$y = y(t) = k/t^h, \quad h, k > 0 \tag{134}$$

(a lowercase letter is used for strength y since it is not a RV here; h and k are deterministic constants, even if unknown in practice); since X has the above $W'(u, v)$ cdf, the RF at time t is:

$$\begin{aligned} R(t) = P[y(t) > X] = F_X[y(t)] &= 1 - \exp\{-[k/(ut^h)]^v\} \\ &= 1 - \exp\left\{-[1/(\alpha t)]^\beta\right\}. \end{aligned} \tag{135}$$

This is indeed an IW model with parameters σ and β both positive, assuming $\beta = hv$ and $\sigma = (u/k)^{1/h}$.

2. Strength is a deterministic constant, Stress is a Weibull SP, increasing in time) In the same framework as above, let the strength y be a deterministic constant, and let the stress $X = X(t)$ be a random function (SP) of time, with Weibull $W(a, b)$ cdf—$F_X(x, t)$—and time-dependent scale parameter $a = a(t)$:

$$F_X(x, t) = P[X(t) \le x] = 1 - \exp\left\{-[x/a(t)]^b\right\}(b > 0). \tag{136}$$

Let us suppose that $a(t)$ is an increasing power function of time:

$$a(t) = kt^m \quad (k, m > 0). \tag{137}$$

This second hypothesis implies, as reasonable, that the mean value of stress increases with time, since, under the Weibull model:

$$E[X(t)] = a(t)\Gamma(1 + 1/b) = kt^m\Gamma(1 + 1/b) \tag{138}$$

with b constant. Thus, the reliability function at time t is:

$$\begin{aligned} R(t) = P[X(t) \le y] = 1 - \exp\left\{-[y/a(t)]^b\right\} &= 1 - \exp\left\{-[y/kt^m]^b\right\} \\ &= 1 - \exp\left\{-[1/(\alpha t)]^\beta\right\} \end{aligned} \tag{139}$$

in which the positive constants, α and β, have a clear meaning as functions of (b, k, m, y). Again, T is an IW RV.

A final remark. Since all the above deductions appear to be coherent with the known, measurable, properties of stress and strength of electrical insulation, in our opinion they should stimulate new applications of this seldom adopted model. It should also be remarked that they can lead to further justification of the already recalled IDHR property observed sometimes, also in data obtained by ALT, for such materials [93, 148].

3.3.8 Log-Logistic (LL) Model (with a Hint at Burr Model)

Two methods of generating the LL model are hinted at here. Indeed, the LL model has been already been deduced—as in [37]—in Sect. 3.2.2 in the framework of a "quasi static" Weibull Stress-Strength model. Moreover, it will be shown in the last part of this section that it can be motivated by a mixture model deriving from a Weibull $W(A, b)$ RV, with random scale A. So, the model is characterized by a conditional RF:

$$R(t|A) = \exp\left(-At^{b}\right) \tag{140}$$

with random scale parameter, A, distributed according an Exponential distribution. As it will be shown when dealing with mixtures (Sect. 3.3.11), if A is an Exponential RV with mean s, then, applying the total probability theorem, the unconditional RF $R(t)$ deduced by the conditional RF above has indeed the following expression, which is clearly seen to be coincident with a LL RF:

$$R(t) = \frac{1}{1 + st^{b}}. \tag{141}$$

Such model can be considered also as a particular case of the "Burr" model (Sect. 3.3.11), and is still IDHR or DHR. Finally, it is observed that the LL model appears to be very similar to the IW model, for which indeed an analogous "Stress-Strength model" motivation has been presented just above. More evident, sometimes impressive, is the analogy with the LN model (including the IDHR property with hrf decreasing toward zero). This analogy arises from the strong similarity between the Gaussian and the Logistic model which give rise to the LN and LL model, respectively, through the exponential function $y = \exp(x)$. In [37], a very good approximation of the LL model was obtained, which worked satisfactorily for the RF, the pdf and the hrf (while it is well known that it is very difficult to approximate all these functions at the same time). This was also shown by many graphs reported in the abovementioned paper, but should be tested with a wider range of parameter values. The above approximation is obtained by the simple method of equating the first and second moment of the logarithms of the LL and LN RV, which of course uniquely determine the parameter values of the two models (it is not practical to equate the simple moments of the two models, since the LL does not possess the first moment if $b < 1$, the second moment if $b < 2$).

In some applications, when both models fitted well data, the LL has been sometimes preferred to the LN for its simpler analytical expression (this is evident especially for the hrf); on the other hand, the LN has no restriction on the parameters for what concerns the existence of its moments, and possesses more desirable properties from the parameter estimation point of view. Indeed, ML and moment estimates are well established and readily available in the LN model, while their deduction is cumbersome for the LL model; for the latter, even the moment estimates can be problematic in view of the above restrictions on the theoretical moments.

3.3.9 Lognormal Model (with Some Reference to the "Inverse Gamma" and "Inverse Generalized Gamma" Model)

Many ways for deducing the LN model on theoretical grounds from wear processes have been illustrated in literature, while—on the "practical" side—many applied papers have shown its great capacity to fit experimental data from very different fields (e.g., speaking of duration data, from the time to first marriage of a person, to the microelectronics lifetime model). In fact, reliability theory is one of the few fields in which the use of the LN model largely surmounts that of the Gaussian model (from which the LN one was derived). Often, the LN model is derived as a proper model for wear itself rather than for lifetimes (as also briefly shown below). Its applications to LT distributions (as witnessed by [53]) are however numerous, due especially to its flexibility: the Lognormal pdf is indeed capable of assuming a large variety of shapes with positive skewness index, which allows for typical large "right tails". In particular, Cox and Oakes [51] show that, among the most popular models, the LN possesses one of the most high "skewness coefficient" values for a given coefficient of variation (CV) value. Furthermore, two properties that appear to be scarcely recalled improve its flexibility: (a) as already hinted at, if the β (shape) coefficient of the LN(α, β) model is small enough (in practice, $\beta < 0.3$) the LN pdf tends to become symmetrical and may satisfactorily approximate even a Gaussian model with the same mean [*this fact can be proved analytically, using the series expansion of* $y = exp(x)$ *for* $x \to 0$]; (b) the CV, v, can assume a wide range of values: in particular—if $\beta = 0.8325$—the value $v = 1$ is obtained, as for the Exponential model, to which the LN appears in this case to be very similar, and often indistinguishable from it (this may explain also its abovementioned applicability to microelectronics).

Also the "decreasing hazard rate" property of the Lognormal distribution for large values of time is a desirable property, for instance in ALT and insulation applications. The hrf properties of the LN (which is, it is recalled, a IDHR model), its high variability and the presence of large right tails are perhaps the main reasons for its being the most applied for repair times, as shown in power systems literature [24, 159]. The above properties all account for the possibility of relatively large times compared to their expected values: thus, the LN assumption may also be justified by a "conservative" approach which seems very appropriate for repair times (and, in general, for "waiting times") when the exact distribution is unknown; such kind of properties, however, are still more evident for the less known and less adopted LL model, which has indeed a higher "skewness coefficient" value for a given CV value, as shown in [51].

All the above reasons and the rapidly growing literature on the model have validated the authoritative forecast of two scholars such as Johnson and Kotz, who in 1970 already stated that "it is quite likely that the LN distribution will be one of the most widely applied in practical statistical work in the near future", as reported in the preface to [53].

Only two significant models which may give origin to the LN model for LT are presented here; by the way, they also can give rise—under different hypotheses—

to the so-called "Inverse Gamma" and "Inverse Generalized Gamma" models, as shown at the end of this section. A LN model for wear can also be the origin of a BS model for the corresponding lifetime distribution, as shown in the end of the present section.

Linear function Stress Process leading to the Lognormal model. In many cases, it may be reasonable to express, or approximate over some time interval, Stress (or Strength) by means of linear (random) functions of time (see [40, p. 103] for the physical deduction of such models, with some examples beyond those here presented, as that of "alpha" pdf, including the effect of temperature as a stress parameter in the *Arrhenius model*).

A linear Stress process may be written as:

$$X(t) = Bt \tag{142}$$

being B, in the general case, a RV, with $B > 0$ almost surely (a.s.), so that the Stress is a.s. increasing in time (for this reason a Gaussian model, e.g., for the RV B is not opportune).

As a simple example of a SS model with linear Stress process and a LN Strength, let us consider the Strength model of last equation with B Lognormal-distributed, and let Y also be Lognormal-distributed, and B and Y be independent. Then, of course, the LT T is such that:

$$BT = Y \rightarrow T = Y/B. \tag{143}$$

Then, since the ratio of two independent LN RV is also a LN RV (this is indeed the same property as that the difference of two independent Gaussian random variables is also Gaussian), T is a LN RV, with parameters which can be obtained very simply.

If the RV Y is deterministic (i.e.: $Y = y$, constant), then T is again a LN RV.

Power Function Stress Process leading to the Lognormal Model. Quite similar results are obtained if stress $X(t)$ is a "power function" of time such as:

$$X(t) = Bt^c. \tag{144}$$

Let again B a Lognormal-distributed RV: also in this case, with a LN strength Y, then the LT is again LN. This can be seen very easily recalling that, if T is a LN RV, then also T^h is a LN RV, whatsoever the real value of the exponent h.

A hint at "Inverse Gamma" and "Inverse Generalized Gamma" Model. The chance is taken here of giving at least a brief mention of two models, which are defined in reliability or survival literature (e.g., in [103]), but are very seldom used. A little diversion from the LN model will be allowed here, only because the two models both appear similar to the LN one, also their deduction being similar. With reference to the linear Stress process of point (1): $X(t) = Bt$ with fixed strength y, let us suppose that B is not LN, but Gamma-distributed (another reasonable hypothesis, since this is an adequate model for positive quantities, and also flexible). Then the LT, which is expressed as above:

$$T = y/B \tag{145}$$

is a so-called “Inverse Gamma” RV, already described in Sect. 2) and used until now mainly as prior pdf in Bayesian estimation, e.g., for the MTTF parameter in the Exponential model [113]. It is curious to notice that, if B is a $G(r,)$ Gamma RV with $r = 0.5$, then T follows a particular form of the *Inverse Gaussian* model.

Finally, let us assume that the stress $X(t)$ is a “power function” of time of the kind:

$$X(t) = Bt^c \tag{146}$$

and that B is again a Gamma RV. Under such hypotheses—with fixed strength y—the LT is given by:

$$T = (y/B)^d \tag{147}$$

with $d = 1/c$, and B^d has a “Generalized Gamma” pdf (already described in Sect. 2). Then T has a so-called “Inverse Generalized Gamma” pdf, which seems to be never used before as a LT pdf. It was used by the authors as a prior pdf in Bayesian estimation for some insulation reliability applications [39], and was also used in relation with heterogeneity studies in statistics with applications to economy (e.g., for the durations of unemployment spells).

3.3.10 Weibull Model (Also Featuring the IPM)

It is not difficult to deduce the Weibull model from wear process considerations, also using its property of being a particular Extreme Value Distribution for Minima. Here, the following simple derivation is proposed, in the framework of Stress-Strength models, using the EV property not for LT itself, but for dielectric Strength (as in [35]). By the way, the same procedure, as it will be seen, may yield a theoretical justification for the well-known IPM, often used especially in ALT on insulation, as discussed in [42, 131]; it is one of the most popular models—among the many proposed since the seminal paper (devoted to survival analysis) by Cox [50]—taking into account the effect of “covariates” (here, the stress, intended as the applied voltage) on the lifetimes.

Let the Strength of the object (e.g., the breakdown voltage for insulation) be a RV Y, and z be the applied constant (in time) stress (e.g., the applied voltage peak value).

The random nature of device strength at time t is to be addressed to the lack of homogeneity of the device material. In order to emphasize this fact in probabilistic terms, let us suppose that the device can be considered (as reasonable for large-size insulation) as a system constituted of a number n of homogeneous elemental components. Denoting as Y_i the strength of the ith elemental component ($i = 1, \ldots, n$), the strength Y of the device can be then expressed as a function of the RVs Yi as it follows (omitting for the moment any relevant time dependence):

$$Y = \min(Y_1, Y_2, \ldots, Y_n) \rightarrow P(Y > z) = P(Y_1 > z, Y_2 > z, \ldots, Y_n > z). \quad (148)$$

Such relationship holds in every fixed time instant t, and implies that the device strength is determined by the strength of its "weakest link", or rather of the least strong elemental component. In what follows, it is supposed that the RVs Y_i are statistically independent; in general, however, they will not be identically distributed. Of course, for such model to be realistic, number n has to be high (ideally, infinite): this allows to get, under very general conditions, a probabilistic characterization of Y resorting to the asymptotic theory of the extreme values [28]. Under general hypotheses, the limit cdf of Y is of the Weibull type, and it can be expressed in the form:

$$F_Y(r) = 1 - \exp(-m \cdot r^k), \quad r \geq 0;\ m > 0,\ k > 0. \quad (149)$$

Parameters m and k are real and positive, and depend on the parameters of the component distributions. Thus, in practice they must be estimated experimentally, since in general the Y_i RVs are not observable. For instance, relationship (149) is verified if the component cdf are of the Gamma, Exponential, Beta, Weibull, Pareto, etc., type, that is for most of the distributions that have a pdf "limited to the left". Since a strength-type RV is intrinsically non-negative (as is in fact the case of dielectric strength), and typically it has a domain (a, ∞), the cdf of the Y_i RVs are of such type, too.

Let us now consider the time dependence of the distribution of the process Y_t. Such dependence will be clearly expressed by the time variation of the parameters m (scale parameter) and/or k (shape parameter) with time. The expectation of the above Weibull model, $W(m, k)$, for the RV Y is expressed as:

$$\mu = \Theta^c \cdot \Gamma(1 + c), \quad \text{where: } \Theta \equiv 1/m, \quad c \equiv 1/k \quad (150)$$

and $\Gamma(x)$ is the Euler-Gamma function.

From (150) it is not difficult to deduce that the mean μ is a decreasing function of m, while it is not a monotonous function of k. Since it has been assumed previously that the strength is decreasing with time, its average value will be decreasing with time as well. The simplest way for accounting of this behavior—on the basis of what previously said about the μ versus m relationship—is to assume that the scale parameter m is an increasing function of time. Moreover, if (149) has to match the properties of a cumulative distribution function, it can be immediately noticed that the strength Y_t has to be infinite in $t = 0$ and zero in $t = \infty$. This gives rise to the following conditions on the time function $m(t)$:

$$m(0) = 0; \quad m(\infty) = \infty. \quad (151)$$

Such conditions are satisfied by the following simple model:

$$m = m(t) = m_0 t^b, \quad m_0 > 0,\ b > 0, \quad (152)$$

where m_0 and b are proper constants to be determined. If such model holds, the time behavior of the mean value of strength in time is expressed by a function of the type:

$$\mu = \mu(t) = \mu_0 / t^p, \quad \mu_0 > 0, \quad p > 0 \tag{153}$$

[in particular, $\mu_0 = (1 + c)/(m_0)^c$; $p = bc$].

By introducing (152) into (149), the distribution function of process Y_t (for $r > 0$ and $t > 0$) is obtained:

$$F_Y(r;t) \equiv P(Y_t \leq r) = 1 - \exp\left[-\left(m_0 t^b r^k\right)\right] \tag{154}$$

being m_0, b, $k > 0$.

Therefore—being z the constant stress applied to the device, the reliability function of the device is obtained, using the above relations:

$$R(t;z) = P(Y_t > z) = 1 - F_Y(z;t) = \exp\left[-\left(m_0 t^b z^k\right)\right]. \tag{155}$$

Equation 155 shows that the reliability function of the device is of the Weibull type; in particular, the RV T follows a Weibull distribution with shape parameter b [defined according to (152)] and scale parameter dependent on stress z, namely: $T \sim W(a, b)$, being:

$$a = a(z) = m_0 z^k, \tag{156}$$

where b, m_0, k are constants >0.

Parameter z is known for hypothesis (constant stress, e.g., applied voltage), whereas constants b, m_0 and k must be evaluated from available experimental data.

The above method leads to the same results if it is assumed that also the (deterministic) stress z varies with time, if this happens by means of a "power function" such as:

$$z(t) = z_0 t^q. \tag{157}$$

The statistical relationship between LT and stress z, expressed by (156) and popular in experimental applications, is the above-mentioned IPM. This denomination comes from the fact that, according to such model, the mean lifetime varies as a (positive) power of the inverse of stress; the same holds also for the percentiles of T. Indeed:

$$E[T] = (1/a)^C \cdot \Gamma(1 + c) = \alpha / z^h \tag{158}$$

being

$$c \equiv 1/b > 0; \; \alpha = (1/m_0)^c \, \Gamma(1 + c) > 0; \; h = ck > 0. \tag{159}$$

An analogous relationship holds for the LT percentiles. Since $h > 0$, the above relationships clearly highlight the decrease of the expected duration with the increase of stress z, as well as relationship (155) points out the reduction of the probability of survival as z increases.

Finally, it must be underlined that the IPM here introduced constitutes a particular "proportional hazard model", as well as an "accelerated time model" [51], both widely applied in LT analysis, and validated by numerous statistical tests.

3.3.11 A Hint at Other Models: Mixture Models, Featuring Burr and LI Models

It is obvious that the LT models are potentially infinite, as the possible ways of constructing them by means of opportune wear models; e.g., even a very simple wear model such as $W(t) = A + B_t$, by choosing among the infinite couples of RV A and B, and allowing for possible randomness of strength, can give rise to an enormous variety of RF.

However, it should kept in mind that the problem of finding the "true" model is insoluble and probably not very interesting from a practical point of view, as long as the resulting models may often not be very different from already established models, with the possible drawback of being characterized by too many parameters to be useful for real applications, in view of the need to estimate those parameters from few data.

So, also for reasons of space, only a peculiar family of models deduced from the combination of two models is here sketched., i.e., the "Mixture models", based upon a random hazard rate.

A large variety of models can be deduced indeed by allowing some parameter of the LT pdf vary randomly among items, accounting for heterogeneity of material or production process, or random variability of environment in which the items operate. Theoretical studies on random hazard rate functions may be found in [85, 143, 146]. A popular model was introduced, among others, by [109], i.e., a model characterized by a "Proportional Hazard Model" [51] with random factor Z accounting for the above randomness, so that for a given value of the RV Z, the random hrf is written as:

$$h(t|Z) = Zh_b(t), \tag{160}$$

in which $h(t|Z)$ denotes a "conditional" hrf, given the positive random factor Z, and $h_b(t)$ is the "baseline", deterministic hrf. In particular, an analytical model is proposed by [109], based on a conditional Weibull hazard rate with shape parameter b:

$$h(t|Z) = Zbt^{b-1}, \quad (t > 0, b > 0). \tag{161}$$

For the random scale parameter characterization, positive RV such as the Lognormal, Gamma and Inverse Gaussian distributions may be considered with reasonable motivations [35, 36]. It is easy to show that, if Z has pdf $g(z)$, the (unconditional) reliability function is given by the so-called [10] "mixture" of the RF according to the pdf $g(z)$:

$$R(t) = \int_0^\infty \exp\left[-zt^b\right] g(z) \mathrm{d}z \tag{162}$$

which is the Laplace Transform of the $g(z)$, evaluated in $s = t^b$. Differentiation of the log of $R(t)$ leads to the unconditional hrf. In the mixture models denoted as "Weibull-Gamma" (meaning that Z has a Gamma distribution) and "Weibull-Inverse Gaussian" (meaning that Z has an Inverse Gaussian distribution) analytical results exist [36]. In particular, let us assume that Z is Gamma-distributed, with pdf:

$$g(z; r, s) = \frac{z^{r-1}}{s^r \Gamma(r)} \exp\left[-\frac{z}{s}\right] \quad z > 0. \tag{163}$$

Then, it is easy to show, applying above relations, that the unconditional reliability function and hazard rate are, respectively, given by:

$$R(t) = \frac{1}{(1 + st^b)^r} \tag{164}$$

$$h(t) = \frac{rsbt^{b-1}}{(1 + st^b)}. \tag{165}$$

The above RF and hrf belong to a particular form of the so-called "Burr model" [96], already hinted at in Sect. 3.3.8. It is noticeable that such model coincides with the already illustrated Log-logistic one when $r = 1$, as anticipated in Sect. 3.3.8.

It is, thus, remarked that in this chapter two peculiar (and completely different) ways have been shown for deducing a LL model from a Weibull model: the first was the SS model of Sect. 3.2 (example 3.2.2), this second is the one of mixture models.

As already discussed, the LL hazard rate function $h(t)$ possesses the following properties:

- if $b \leq 1$, $h(t)$ is decreasing in t, with limit value 0 (DHR model);
- if $b > 1$, $h(t)$ first increases—starting from $h(0) = 0$—then decreases toward 0 (IDHR model).

The particular case $b = 1$ (Exponential-Gamma mixture model) is also denoted as "Lomax" or "Pareto of the second kind" [96, 109].

Noticeably, the above case $b = 1$ shows also a paradox discussed in literature [10, 35, 79, 125, 136, 142], even if often neglected. It was already met, for the discrete case, in Sect. 1.6. The (apparent) paradox is that the random variability of the environment (Z factor) gives rise to a decreasing hrf (DHR), even though the individual hrf is constant over time. The paradox can be fully justified by subjective probability reasoning [10, 136, 146]; it has also been explained in [125] with new motivations.

Still more remarkable, perhaps, is the fact that, if $b > 1$, the individual hrf are increasing (IHR model), while the overall (unconditional) hrf is decreasing for large t (IDHR model).

The above result also shows that DHR models remain, instead, DHR even after any mixture, and also this fact is a general property [10].

The above and many more properties of mixtures of LT have been thoroughly analyzed also in Mathematical Demography [86]. Recently, Singpurwalla (see [146], chapter 7) provided new insight in the study of life distributions derived from the characterization of the hrf as a stochastic process, motivated by the randomness of the dynamic variability of environment.

For completeness, it must be added (even if it is well beyond the topics of the present contribution) that the study of lifetimes influenced by a common random environment provided further methods to explore the statistical dependence between lifetimes of different components of a system [48, 54, 109]. This is a crucial topic which arises naturally in reliability analysis of electrical power systems, since—for obvious "physical" reasons—the components of such systems can seldom be considered as really s-independent.

Finally, returning to the main point, it is noticeable that the LL model may be obtained by a completely different approach from those presented previously.

3.3.12 A Remark on Non-Reliability Applications

Finally, it is noteworthy to remark—not only from an academic point of view—that all the above-mentioned studies on wear process, diffusion processes, Brownian motion, barrier crossings, etc., which were originated by motivations coming from reliability theory, have stimulated new important studies in the theory of stochastic processes, as witnessed by a recent paper by Cinlar [47], which adds significantly to all his previously cited papers and books. Interesting relations between reliability and biology (bio-mathematics) may be found, e.g., in the study of Ricciardi [137] on diffusion processes, which was referred to by Ebrahimi [68] in his key paper on IRA. Ebrahimi used indeed a Lognormal "Ito diffusion processes", typical of biology, as the basis for his methodology. Other noticeable applications in Economy are discussed in [101, 146] among the others.

Law of proportionate effect giving origin to the BS or LN model. A well-known physical model for wear—first analyzed in mechanical engineering in relation to fatigue crack growth process [53, 59]—considers a stress process acting on a device at discrete times $T_{k,}$ producing a succession of "cracks" on the material, which can ultimately yield its failure when the crack size exceeds the device strength. It is assumed a reasonable multiplicative effect—according to a model denoted as "law of proportionate effect"—on the component, such that the "crack size" after the $(k+1)$th stress, W_{k+1}, is proportional to the previous crack size W_k, according to the relation:

$$W_{k+1} = W_k(1 + Z_k), \quad k = 0, 1, \ldots \tag{166}$$

being Z_k a succession of non-negative, independent RV. It is easy to show that, as time (thus index k) diverges, the succession of RV W_k converges to a LN RV, according to the Central Limit Theorem applied to the logarithms of W_k. For instance, assuming for simplicity (but these hypotheses can be relaxed) that the RV $Y_k = \log(1 + Z_k)$ are independent and identically distributed (IID) random variables with common mean μ and common standard deviation σ, then the RV $L_k = \log(W_k/W_0)$ is—if k is large enough—approximately Gaussian, and its mean and variance are, respectively, $k\mu$ and $k\sigma^2$.

So, if the device strength, y, is deterministic, and the failure occurs as soon as the wear process overcomes y, it is easy to see—by introducing the log of both wear and strength—that a BS model is obtained for the LT [103, 112]. A similar model holds if strength is a LN RV (s-independent from stress).

If the failure, under this model, is not due to the "strength overcrossing", but other hypotheses can be assumed for failure mechanism (as in [30], p. 33), then the LT may be instead described itself by a LN model.

4 Probabilistic Life Models for Electrical Insulation

4.1 Basic aspects and main models for electrical insulation reliability

As previously pointed out, the time-to-failure (life)[10] of an insulation, or of a component of which the insulation is a part, is a RV. Thus, it is always associated with the relevant failure probability, namely the probability of failing under the action of applied stresses, or, conversely, with the corresponding reliability, namely the probability of withstanding the applied stresses and surviving. In fact, aging and failure processes are regulated by stochastic laws, as demonstrated by the fact that identical specimens manufactured with the same material, subjected to the same levels of stresses, exhibit different failure times, because of the intrinsic inhomogeneities of the materials, the uncertainties in the manufacturing processes, the imperfect control of the test conditions and so on [127, 144].

In the particular case of electrical breakdown of solid insulation subjected to applied voltage, a huge series of experimental tests have been conducted through

---

[10] The term "life" is used throughout the present Sect. 4 for indicating the generic percentile of the distribution of times-to-failure of an insulation, according to a very common practice in electrical insulation literature since the very early times till now (see, e.g., [58, 65, 118, 127, 144]).

many decades. The results of such tests, combined with sound mathematical reasons related to the "Extreme Value" theory (which became more and more popular since the fifties until our days), and the advances in physical knowledge of the mechanisms of wear and stresses acting on insulation, have shown that the probability distribution that has turned out to be the best for reproducing the relationship between failure probability and life is the above-described Weibull distribution [129, 130]. This model, indeed, seems to realize the already-mentioned desirable combination of DRA and IRA, in a process which is of course dynamic and which, by means of modern literature [41, 87], is gaining further contributions and investigations.

Under the Weibull distribution, the cumulative failure probability (cdf), F, versus time t, is expressed—using for convenience the $W'(\alpha, \beta)$ parametric form—as follows:

$$F(t) = P(T \leq t) = 1 - \exp\left[-(t/\alpha)^{\beta},\right] \tag{167}$$

where α is the scale parameter and β is the shape parameter of the distribution: β is linked to the dispersion of the times-to-breakdown, while α coincides with the breakdown time at 63.2% failure probability, i.e., with the 63.2th percentile of breakdown times. It must be highlighted that both parameters are function of the stresses applied to the insulation, though the dependence of β is usually weaker and is neglected in practice [119, 127]. Hence, by indicating with $S_1, S_2, \ldots, S_N$ the values of the N stresses applied to the insulation (assumed as constant with time) α can be written as $\alpha = \alpha(S_1, S_2, \ldots, S_N)$. As a consequence, equation (167) can be rewritten as follows:

$$F(t; S_1, S_2, \ldots, S_N) = 1 - \exp\left\{-\left[\frac{t}{\alpha(S_1, S_2, \ldots, S_N)}\right]^{\beta}\right\}. \tag{168}$$

From (168), by virtue of the meaning of α, and denoting by t_F the 100 Fth percentile of time-to-breakdown, the so-called "probabilistic life model" of the considered insulating system can be derived, namely a relationship between life, stress levels and failure probability (or, conversely, reliability) [126]. In order to do that, Eq. 168 should be expressed in terms of t_F:

$$t_F(S_1, S_2, \ldots, S_N) = [-\ln(1 - F)]^{1/\beta} \alpha(S_1, S_2, \ldots, S_N) \tag{169}$$

that is the expression of the probabilistic life model implicit in terms of the stresses. It can be noticed that it enables the derivation, for any value of stresses $S_1, S_2, \ldots, S_N$, of the relevant 100 Fth percentile of breakdown time, $t_F(S_1, S_2, \ldots, S_N)$.

Note that, for its statistical significance, the scale parameter $\alpha = \alpha(S_1, S_2, \ldots, S_N)$ of the Weibull distribution is commonly chosen as the reference percentile of the distribution of failure times coming from breakdown tests on electrical insulation. Any other failure time percentile can thus be derived from $\alpha = \alpha(S_1, S_2, \ldots, S_N)$ and β resorting to Eq. 169.

Probabilistic life models are the fundamental tool for carrying out a reliability analysis on the basis of laboratory test results only, before the component is put in-service. Indeed, under the hypothesis that stress levels are constant and fixed, probabilistic life models enable the estimation of key reliability parameters and functions, such as the Mean Time To Failure (MTTF), the hazard function and the reliability function of the insulation, thus of the electrical device of which the insulation is the weakest part (and ultimately of the power system which the device belongs to). In fact, from Eq. 168, the reliability function at mission time t can be evaluated trivially via the following RF $R(t;\ S_1,\ S_2,\ \ldots,\ S_N)$:

$$\begin{aligned} R(t;S_1,S_2,\ldots,S_N) &= 1 - F(t;S_1,S_2,\ldots,S_N) \\ &= \exp\left\{-\left[\frac{t}{\alpha(S_1,S_2,\ldots,S_N)}\right]^{\beta}\right\}. \end{aligned} \tag{170}$$

Thus, failure rate at the same time t can be estimated via the following hazard function:

$$\begin{aligned} h(t;S_1,S_2,\ldots,S_N) &= \frac{\beta}{\alpha(S_1,S_2,\ldots,S_N)}\left[\frac{t}{\alpha(S_1,S_2,\ldots,S_N)}\right]^{\beta-1} \\ &= \frac{\beta t^{\beta-1}}{[\alpha(S_1,S_2,\ldots,S_N)]^{\beta}}. \end{aligned} \tag{171}$$

Actually, in Eqs. 168–171 the functional dependence of the relevant reliability model versus applied stresses is not fully assessed, until the functional dependence of α on S_1, S_2, ..., S_N is not fully assessed; this is needed for the estimation of reliability and related quantities. Such functional dependence can be explained provided that the life model holding for the considered insulation (or component) has been singled out.

4.2 Insulation life models

Life modeling of electrical insulation has the goal of determining the most appropriate mathematical relationship (model) between the time-to-failure (life) of a given insulation and the levels of the various stresses applied to such insulation [119, 127, 144]. Therefore, referring to the 63.2% failure probability a life model in its most general form can be expressed as follows:

$$\alpha(S_1,S_2,\ldots,S_N) = f(S_1,S_2,\ldots,S_N;p_1,p_2,\ldots,p_M), \tag{172}$$

where p_1, p_2, ..., p_M are the model parameters and $f(S_1,\ S_2,\ \ldots,\ S_N;\ p_1,\ p_2,\ \ldots,\ p_M)$ is a proper mathematical function of model parameters and applied stresses. Moreover, the functional dependence of the model on applied stresses and model parameters is generally such that relationship (172) can be recast trivially in the following form:

$$\begin{aligned}\alpha(S_1,S_2,\ldots,S_N) &= f\left(S_{1,0},S_{2,0},\ldots,S_{N,0};p_1,p_2,\ldots,p_M\right)\\ &\quad\times\frac{f(S_1,S_2,\ldots,S_N;p_1,p_2,\ldots,p_M)}{f\left(S_{1,0},S_{2,0},\ldots,S_{N,0};p_1,p_2,\ldots,p_M\right)}\\ &= \alpha_0 f'(S_1,S_2,\ldots,S_N;p_2,\ldots,p_M)\end{aligned} \tag{173}$$

where $\alpha_0 = \alpha(S_{1,0}, S_{2,0}, \ldots, S_{N,0}) = p_1$ is the 63.2th failure time percentile at reference values of stresses $S_{1,0}, S_{2,0}, \ldots, S_{N,0}$ and is sometimes referred to as "scale parameter" of the life model [118]; $f'(S_1, S_2, \ldots, S_N; p_2, \ldots, p_M) = f(S_1, S_2, \ldots, S_N; p_1, p_2, \ldots, p_M)/f(S_{1,0}, S_{2,0}, \ldots, S_{N,0}; p_1, p_2, \ldots, p_M)$ is a dimensionless function that encompasses the whole dependence of the model on applied stresses.

Although the 63.2th percentile is usually chosen as the reference one of the failure-time distribution obtained from breakdown tests on electrical insulation (see Sect. 4.1), other life percentiles than the 63.2th could be considered (see Eq. 169) and this would affect the value of α_0 in (173) [119, 127, 144].

As it can be argued from (172) to (173), a life model valid for a certain insulation provides life estimates for that insulation (or for the component which the insulation belongs to) at selected levels of the applied stresses, on condition that the values of the model parameters are known.

Insulation life models are commonly used first of all for characterizing and comparing the endurance properties of various materials candidate for the realization of the insulation of electrical components. For a given insulating material, the life model parameters are usually derived via laboratory tests performed on small-size specimens, thereby achieving considerable time and cost savings with respect to tests on full-size insulation systems.

Secondly, since insulation is often the weakest part of an electrical device (as highlighted in Sect. 4.1) insulation life models can be employed also for inferring the service life of power components. However, this requires an extrapolation of test results and relevant model parameter values to the full-size insulation system of the considered power component; this introduces a degree of uncertainty in life estimation of the power component itself.

The extrapolation can be performed, e.g., via the statistical "enlargement law" [114–116, 129]. This law provides the relationship between full-size insulation life, t_D (at design values of applied stresses, $S_{1,D}, S_{2,D}, \ldots, S_{N,D}$, and failure probability, P_D) and test-size insulation life, $\alpha(S_{1,D}, S_{2,D}, \ldots, S_{N,D})$ (at the same values of $S_{1,D}, S_{2,D}, \ldots, S_{N,D}$, but at failure probability 63.2%, that is usually the reference probability for test result processing, as pointed out above), namely:

$$t_D = \alpha(S_{1,D},S_{2,D},\ldots,S_{N,D})[\ln(1-P_D)/D]^{1/\beta}, \tag{174}$$

where D is the so-called enlargement factor. As an example, when dealing with power cables, for test minicables of length l_T, conductor radius r_T, outer insulation radius R_T, and power cables of length l_D, conductor radius r_D, outer insulation radius R_D, D can be written as [129]:

$$D = (l_D/l_T)(r_D/r_T)^2\left[1 - (r_D/R_D)^{\beta_E - 2}\right] \Big/ \left[1 - (r_T/R_T)^{\beta_E - 2}\right] \quad (175)$$

where β_E is the shape parameter of the Weibull probability distribution of dielectric strength for both mini cables and full-size cables (β_E is not affected by the scaling process) [129].

On the basis of the enlargement law (174), Eq. 173 can be rewritten as follows:

$$t_D = \alpha_0 f'(S_{1,D}, S_{2,D}, \ldots, S_{N,D}; p_2, \ldots, p_M)\left[-\ln(1 - P_D)/D\right]^{1/\beta}. \quad (176)$$

During the last three decades, the understanding of aging mechanisms for insulating materials subjected to different types of service stresses has grown continuously, leading to significant achievements. Such achievements go from the development of "phenomenological" life models—able to fit failure time data for various stresses (singly or simultaneously applied) and useful for deriving parameters for material evaluation (see e.g., [78, 144])—to "physical" models—that describe different physical–chemical mechanisms responsible for insulation degradation under different types and/or ranges of applied stress (see e.g., [23, 64, 65]). All this information provides a considerable help at the design stage of a full-size insulation, thus of an electrical component on the whole.

The introduction of normative references in IEC and IEEE publications (see e.g., [91, 92]) regarding thermal, electrical, mechanical, environmental and multiple stresses support the maturity and the progresses obtained on this topic. Most of the newly acquired knowledge is associated with the diffusion of polymeric insulation and the requirement of increasing design stresses, in order to reach more compact devices and reduce costs without affecting insulation system reliability, as required by the deregulated electricity market (see above).

In the following part of this section, a few fundamental life models available in the literature and employed for time-to-failure estimation of insulating materials and systems are presented. Focus is made, for the sake of brevity, on electrical and thermal stress, these being the stresses which mostly age and cause failure of the insulation of electrical devices. When such models are inserted in the probabilistic framework outlined at Sect. 4.1, reliability can be evaluated first of all in aprioristic terms with respect to real operating conditions of insulating materials and systems—that usually include time-varying stresses—i.e., on the basis of "rated" or "design" stress levels that are assumed as constant. This kind of indirect reliability evaluation, illustrated in what follows of this section, provides however a fundamental indication for the design of the insulating systems of power components such as cables, capacitors, transformers and motors.

On the other hand, since power devices in their actual service conditions are mostly—if not ever—subjected to time-varying stress levels, the need for a "closer-to-real-world" reliability evaluation for power components arises. For this reason, some recent theoretical developments for IRA under time-varying stresses—relevant to daily load cycles and based on the "good-old" cumulative-damage law of Miner [128]—are illustrated in Sect. 4.3.

4.2.1 Life Modeling Under Electro-Thermal Stress

The problem of insulation life modeling under the combination of electrical and thermal stress (i.e., when both voltage and temperature are applied), referred to as electro-thermal stress, was solved by combining two single stress-life models that hold, respectively, when only either temperature or voltage is applied [78, 144]. It is therefore convenient to review briefly these single stress-life models.

When only a constant temperature (thermal stress) is applied to insulation, electrical breakdown does not occur, of course. Thus, failure is said to take place conventionally when a selected diagnostic property, that has to be monotonous in time and correlated to thermal degradation (e.g., dielectric strength, yield strength, weight, density, etc.), reaches a proper end point, i.e., a fixed limit beyond which the insulation is no more able to perform satisfactorily [92, 144]. The first studies of insulation aging regarded mainly endurance to thermal stress. A fundamental approach was Dakin's theory, dating back to 1948 [58], according to which temperature speeds up the rate of thermally activated degradation reactions (e.g., oxidation, cross-linking, etc.), thereby accelerating the chemical aging of insulation. As a consequence, it can be shown that the logarithm of the time-to-end point, i.e., "thermal life", is inversely proportional to absolute temperature, Θ [92], and that the 63.2th percentile of thermal life, $\alpha(\Theta)$ can be expressed as [127, 144]:

$$\alpha(\Theta) = \alpha_0 \exp[-B(1/\Theta_0 - 1/\Theta)], \tag{177}$$

where Θ_0 is a reference value of absolute temperature, α_0 is the 63.2th percentile of thermal life at such reference temperature, $B = \Delta W/k_B$ is a constant typical of the material, ΔW being the activation energy of the main thermal degradation reaction involved and $k_B = 1.38 \times 10^{-23}$ J/K the Boltzmann constant. Equation 177 is referred to as the Arrhenius model.

For the Arrhenius model it is common practice to choose as the reference temperature a value higher than the operating temperature of the insulation, e.g., the value that corresponds to a mean life of 20,000 h (such value in °C is referred to as the "Temperature Index", TI) [92]. By this way, α_0 is known and fixed a priori, and the Arrhenius model is characterized by two parameters, namely Θ_0 and B. The model is usually represented in the so-called Arrhenius graph, having coordinates log(life) versus $-1/\Theta$, thereby giving rise to a straight line of slope B, that enables the extrapolation from test to service temperatures. In particular, B is very important: an insulation that exhibits a higher value of B (for the same value of α_0 and Θ_0, being Θ_0 higher than the service temperature) features a longer life at temperatures lower than Θ_0, i.e., down to the service temperature of the insulation. Thus, the higher is B, the better is the insulation [127, 144] (see Fig. 1a).

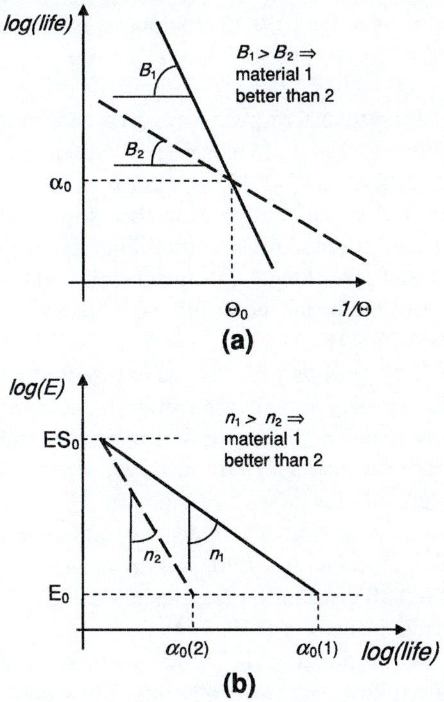

Fig. 1 **a** Arrhenius Model in log(life) versus $-1/\Theta$ coordinates (Arrhenius graph). **b** Inverse Power Model in log(E) versus log(life) coordinates

International Standards have been established for evaluating thermal endurance capabilities of insulating materials via indices such as the above-mentioned TI and the HIC (Halving Interval in Celsius, the temperature difference giving rise to halving of life, starting from the temperature of TI) [92].

When a fast-increasing voltage is applied to an insulation system, electrical breakdown (i.e., the discharge of the whole insulation thickness) takes place as the applied voltage exceeds a value typical of the considered insulation, the breakdown voltage; the relevant electric field is referred to as dielectric strength and depends on several quantities [127, 144].

When a constant (in the rms sense) voltage only is applied to an insulation system, the so-called IPM and Exponential model (EM) are mostly used for expressing the relationship between applied voltage and time to breakdown (electrical life). According to the IPM and the EM, the 63.2th percentile of electrical life, $\alpha(E)$, can be expressed, respectively, as follows:

$$\alpha(E) = \alpha_0 (E/E_0)^{-n} \quad \text{(IPM)} \tag{178}$$

$$\alpha(E) = \alpha_0 \exp(-h(E - E_0)) \quad \text{(EM)} \tag{179}$$

where E is the magnitude of electric field, also referred to as "electrical stress" (proportional to the applied voltage via trivial geometrical factors), E_0 is the value of electric field under which the aging produced by the electrical stress (i.e., the electrical aging) is negligible, α_0 is the value of the 63.2th percentile of time-to-breakdown corresponding to E_0 (i.e., $\alpha(E = E_0) = \alpha_0$).

Equations 178 and 179 provide straight lines in log–log and semi-log coordinate systems, respectively, with slopes $-1/n$ and $-1/h$, if—as usual—E is in ordinate and L in abscissa. Coefficient n (or h) is called Voltage Endurance Coefficient (VEC). The VEC is a fundamental parameter for insulation characterization and design (together with dielectric strength): indeed, for the same values of initial dielectric strength ES_0, an insulation that exhibits a higher value of n features a longer life for electric field values below ES_0. Hence, the larger the VEC, the better the insulation endurance, i.e., its ability to endure electrical stress (see Fig. 1b).

The IPM and the EM have essentially an empirical background, because most of the ALT data can be fitted by straight lines in log–log or semilog plots (the linearization of the stress-life relationship is needed to extract coefficients for material characterization, as well as to derive design field estimation through extrapolation from the results of ALT, carried out at stresses considerably larger than the service one). However, both models can acquire a theoretical background; in particular, the IPM was associated with a statistical approach based on the Weibull distribution (as illustrated in Sect. 3.3.10 above) and was applied to power cable insulation [130, 144].

By combining the Arrhenius thermal model (Eq. 177) and the IPM (Eq. 178), the following "electro-thermal life model" was obtained [144]:

$$\alpha(E,\Theta) = \alpha_0 (E/E_0)^{-(n_0 - bcT)} \exp(-BcT), \tag{180}$$

where $\alpha(E, \Theta)$ is the 63.2th percentile of electro-thermal life, cT is the so-called "conventional thermal stress", defined as $cT = 1/\Theta_0 - 1/\Theta$. Model (180), that features four parameters (i.e., B and n_0 for thermal and electrical endurance, respectively, b for the extent of stress synergism, α_0 as the scale parameter) was fitted satisfactorily to several sets of data, relevant to different materials [119, 127, 144].

From the above considerations—and by comparing Eqs. 172, 173 with 180—it can be concluded that, in the presence of electrical and thermal stresses only, $\alpha(S_1, S_2, \ldots, S_N)$ reduces to $\alpha(E, \Theta)$, that can be expressed through Eq. 180. Hence, by inserting Eq. 180 into Eq. 169 one gets the following probabilistic electro-thermal life model:

$$t_F(E,\Theta) = [-\ln(1-F)]^{1/\beta} \alpha_0 (E/E_0)^{-(n_0 - b\,cT)} \exp(-BcT) \tag{181}$$

while from Eqs. 168, 170, 171, respectively, one obtains the relevant cumulative failure probability, hazard and reliability functions:

$$F(t;E,\Theta) = 1 - \exp\left\{ -\left[\frac{t}{\alpha_0 (E/E_0)^{-(n_0 - b\,cT)} \exp(-BcT)} \right]^{\beta} \right\} \tag{182}$$

$$h(t;E,\Theta) = \frac{\beta t^{\beta-1}}{[\alpha(E,T)]^{\beta}} = \frac{\beta t^{\beta-1}}{\left[\alpha_0(E/E_0)^{-(n_0-bcT)}\exp(-BcT)\right]^{\beta}} \tag{183}$$

$$R(t;E,\Theta) = \exp\left\{-\left[\frac{t}{\alpha_0(E/E_0)^{-(n_0-bcT)}\exp(-BcT)}\right]^{\beta}\right\}. \tag{184}$$

Equation 184 represents a "physical reliability model" that can be used for reliability estimation in the case of a typical solid insulation for MV and HV subjected to electrical and thermal stress. It can be argued that it is a Weibull reliability function, thus it is characterized by the relevant mathematical properties illustrated at previous section.

4.2.2 Life Modeling in Distorted Regime

The electro-thermal life model of Eq. 180 holds for constant temperature and either dc electrical field or sinusoidal field at industrial (or moderately higher) frequency [119, 127, 144]. However, the ever-increasing diffusion of power electronics and non-linear loads in power systems involves a consequent increase in the level of harmonic distortion (both in current and in voltage). Current distortion gives rise to an increase in Joule losses in conducting parts, thereby raising insulation temperature and accelerating thermal degradation. Voltage distortion may increase the peak and/or the rms voltage and the rate of voltage rise with respect to sinusoidal conditions; this can accelerate also the electrical aging of components under distorted voltage [120, 121, 122]. Experimental tests and theoretical studies, carried out on different insulation systems subjected to various distorted current and voltage waveforms, showed that insulation life under distorted regime can be reduced with respect to life at rated sinusoidal voltage and temperature mainly due to the temperature increase produced by harmonic currents in conductors and electrical stress increase due to distorted voltage waveform [32, 122]. The parameters that weigh the severity of the distorted voltage waveform with respect to the nominal sinusoidal voltage are the *peak factor*, K_p, the rms *factor*, K_{rms}, and the *shape factor*, K_f, defined as [32]:

$$K_p = V_p/V_{1p,n} \tag{185}$$

$$K_{\mathrm{rms}} = V/V_{1,n} \tag{186}$$

$$K_f = \sqrt{\sum_{h=1}^{H} h^2\left(V_h/V_{1,n}\right)^2}, \tag{187}$$

where V_p and $V_{1p,n}$ are peak values of distorted voltage and of rated sinusoidal voltage, respectively, V and $V_{1,n}$ are rms values of distorted voltage and of rated sinusoidal voltage, V_h is rms value of the h-order harmonic and H is the maximum order of harmonics occurring in the system. In the nominal sinusoidal regime, $K_p = K_{rms} = K_f = 1$, while in distorted regime K_p, K_{rms} and K_f can exceed unity, thereby causing an acceleration of electrical degradation.

It was shown in [32, 122] that the following electro-thermal life model holds under distorted regime:

$$\alpha_{NS}(E_S, \Theta_S) = \alpha_S(E_S, \Theta_S)\exp(-B\Delta cT_{\text{arm}})K_p^{-n_p}K_f^{-n_f}K_{\text{rms}}^{-n_r}, \tag{188}$$

where $\alpha_{NS}(E_S, \Theta_S)$ is the 63.2th percentile of insulation life in a distorted regime characterized by a rated sinusoidal rms electric field E_S and a rated temperature Θ_S, $\alpha_S(E_S, \Theta_S)$ is the 63.2th percentile of insulation life in rated sinusoidal conditions, n_p, n_f and n_r are exponents that account for the aging acceleration effect of factors K_p, K_f and K_{rms}, respectively, B is the above constant introduced when dealing with the Arrhenius model and ΔcT_{arm} is a quantity depending on the temperature rise due to harmonics, $\Delta\Theta_{arm}$, defined as:

$$\Delta cT_{arm} = 1/\Theta_S - 1/(\Theta_S + \Delta\Theta_{arm}) \tag{189}$$

being $\Theta = \Theta_S + \Delta\Theta_{arm}$ insulation temperature in distorted regime. By means of (188), life in distorted regime can be related directly to life in nominal sinusoidal conditions, pointing out the (possible) life reduction caused by current and voltage harmonics.

Therefore, by assuming that the insulation system works in non-sinusoidal regime and that the life model of Eq. 188 holds, then $\alpha(S_1, S_2, \ldots, S_N)$ reduces to $\alpha_{NS}(E_S, \Theta_S)$, that can be expressed through Eq. 188. Hence, by inserting Eq. 188 into 169 one gets the following probabilistic electro-thermal life model for distorted regime (in the presence of a rated fundamental sinusoidal component of electric field having rms value E_S and of a rated temperature Θ_S):

$$t_{F,NS}(E_S, \Theta_S) = [-\ln(1-F)]^{1/\beta}\alpha_S(E_S, \Theta_S)\exp(-B\Delta cT_{\text{arm}})K_p^{-n_p}K_f^{-n_f}K_{\text{rms}}^{-n_r}. \tag{190}$$

By substituting Eq. 188 into Eqs. 168, 170, 171, respectively, one obtains also the relevant cumulative failure probability, hazard and reliability functions, respectively:

$$F(t; E_S, \Theta_S) = 1 - \exp\left\{-\left[\frac{t}{\alpha_S(E_S, \Theta_S)\exp(-B\Delta cT_{\text{arm}})K_p^{-n_p}K_f^{-n_f}K_{\text{rms}}^{-n_r}}\right]^{\beta}\right\} \tag{191}$$

$$R(t; E_S, \Theta_S) = \exp\left\{-\left[\frac{t}{\alpha_S(E_S, \Theta_S)\exp(-B\Delta cT_{\text{arm}})K_p^{-n_p}K_f^{-n_f}K_{\text{rms}}^{-n_r}}\right]^{\beta}\right\} \tag{192}$$

$$h(t;E_S,\Theta_S)=\frac{\beta}{\alpha_{NS}(E_S,\Theta_S)}\left[\frac{t}{\alpha_{NS}(E_S,\Theta_S)}\right]^{\beta-1}$$
$$=\frac{\beta t^{\beta-1}}{\left[\alpha_S(E_S,\Theta_S)\exp(-B\Delta cT_{\text{arm}})K_p^{-n_p}K_f^{-n_f}K_{\text{rms}}^{-n_r}\right]^{\beta}}. \tag{193}$$

Equation 192 represents a "physical reliability model" that can be used for reliability estimation in the case of a typical solid insulation subjected to distorted voltage. It can be argued that the reliability function follows again the Weibull distribution, as in the sinusoidal case, given that β is constant with applied stresses, as pointed out when dealing with Eq. 167.

In [121], the trend of failure rate versus aging time for a MV power cable feeding a traction system with AC/DC 12-pulse converters, subjected to harmonic voltages characteristic of the converters (i.e., essentially the 11th and 13th) at the limits stated by Standard EN 50160, is reported. It is shown that, although the level of voltage distortion complies with standard limits, nevertheless the reliability of the cable can be severely affected by voltage harmonics. On the other hand, this effect can be more than counterbalanced by a temperature significantly lower than rated temperature, as it is mostly the case for components of the supply system of electrical traction systems subjected to voltage and current harmonics.

4.3 Life Modeling Under Time-Varying Stress: The Case of Load Cycles

Power components in their real service conditions are mostly—if not ever—subjected to time-varying stresses. This involves the need for a "closer-to-real-world" reliability evaluation for power components, i.e., a reliability evaluation that accounts for the time variation of operating stress. Generally speaking, this problem is quite cumbersome.

However, it can be argued that a big deal of electrical devices—e.g., all the components of power transmission and distribution grids—exhibit every day more or less the same rms current and voltage values at the same hours, at least during working days of a given period of the year, under typical operating conditions of the users [118]. Thus, apart from the statistical fluctuations due to the random time-varying nature of the supplied loads and the deterministic fluctuations associated with the weekly and/or seasonal characteristics of the loads, such components are subjected to daily load cycles. Moreover, applied rms voltage is approximately constant with time—apart a generalized voltage increase when load decreases and vice versa; such variations, however, under normal operating conditions are within ±10% of rated voltage of components/systems. Thus, time varying stresses are mostly associated with current variations in the form of daily current/load cycles.

As a consequence, component life (i.e., the generic 100 Fth percentile of time-to-failure, t_F, see footnote at Sect. 4.1) can be divided approximately into K equal intervals. These intervals have all the same duration, $t_{\text{cycle}} = 1$ day, and the same relationship between rms load current, I, and cycling time, $t \in [0, t_{\text{cycle}}]$. Then, t_F can be expressed as a simple function of K, referred to as the number of "cycles-to-failure", namely:

$$t_F = K\, t_{\text{cycle}}. \tag{194}$$

Daily load cycles can be thought of as a sequence of N equally lasting current steps of height I_i (rms current value) and duration $\Delta t_i = t_{\text{cycle}}/N$ ($i = 1, \ldots, N$), as that sketched in Fig. 2 for $N = 6$ (a simplification of a typical load cycle of HVAC cables). Let us refer to such cycles as "stepwise-constant" daily cycles. Stepwise-constant daily cycles can reproduce satisfactorily every daily load cycle, on the condition that a sufficiently high number of steps N is taken. Thus, only daily load cycles of the stepwise-constant kind will be treated here.

Assuming as above that the weakest part of a power device is its insulation, the predominant stresses acting on insulation in-service commonly arise from the *electric field* associated with voltage (electric stress) and the *temperature* associated with Joule losses in conducting elements plus dielectric losses in the insulation (thermal stress). Therefore, in general, the maximum stresses applied to a power device are maximum temperature and electric field in the insulation. In this framework, the life of a power device subjected to load cycles is assumed here to end when its insulation fails because of the degradation caused by the maximum stresses, that act all over its life as a consequence of a fixed stepwise-constant daily load cycle.

As argued above, applied voltage is approximately constant with time. Hence, maximum electric field can be hypothesized as steady and equal to its design value, E_n. On the contrary, maximum temperature varies during each ith interval Δt_i of the load cycle, due to the relevant variation of rms current. The temperature rate-of-change depends on the difference between power losses in the present

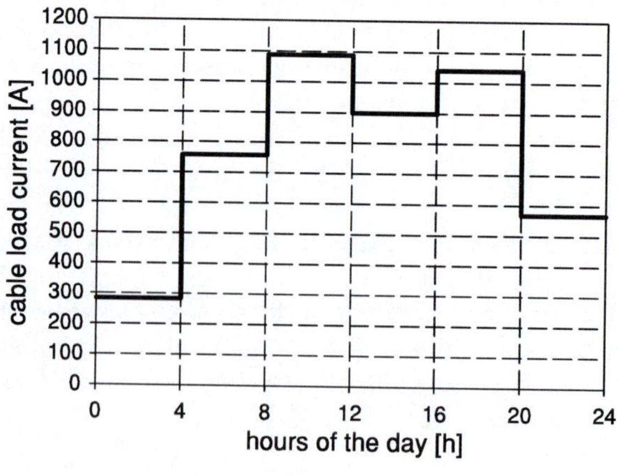

Fig. 2 Stepwise-con stant daily load cycle for a HVAC cable ($N = 6$ equally lasting current steps)

step—proportional to I_i^2—and in the previous step—proportional to I_{i-1}^2—as well as on heat storage and exchange properties of the layers that constitute the insulation and its outer environment. The combination of these effects gives rise to a thermal transient during which temperature varies starting from an initial value $T_{i,0}$ (the temperature at the beginning of Δt_i) and tending toward a steady value, $T_{i,\infty}$ (the regime temperature corresponding to a constant rms current I_i). The transient temperature within each Δt_i can be derived by means of an *ad hoc* transient thermal model (see, e.g., [117, 118] for power cables).

Every step Δt_i of the cycle can be then split into infinitesimal intervals, which range from a generic time t to a subsequent time $t + \mathrm{d}t$. Thus, each infinitesimal interval corresponds to one single value of transient temperature, $T_i(t)$. Hence, the fraction of life lost by component insulation during a given $\mathrm{d}t$ within Δt_i, denoted as dLF, can be written as follows:

$$\mathrm{dLF} = \mathrm{dLF}[E_n, T_i(t)] = \frac{\mathrm{d}t}{t_F[E_n, T_i(t)]}, \tag{195}$$

where $t_F[E_n, T_i(t)]$ is insulation life at constant values of maximum electric field and temperature, E_n and $T_i(t)$, respectively. $t_F[E_n, T_i(t)]$ must be evaluated via an electro-thermal life model valid for the insulation of the examined component [117, 118], e.g., Eq. 180 for an insulation subjected to temperature plus sinusoidal voltage.

According to Miner's cumulative-damage theory [128], the sum of all life fractions lost (referred to as "loss-of-life fractions" from now on) should yield 1 at failure. Therefore, cable life can be estimated by applying (195) to every infinitesimal interval $\mathrm{d}t$ of each step Δt_i of the cycle and setting the sum of all the relevant loss-of-life fractions at failure—in fact an integral—equal to 1. Thus, by defining the loss-of-life fraction relevant to the ith step of the cycle, LF_i, as:

$$\mathrm{LF}_i = \int_0^{\Delta t_i} \mathrm{dLF}[E_n, T_i(t)] = \int_0^{\Delta t_i} \frac{\mathrm{d}t}{t_F[E_n, T_i(t)]} \tag{196}$$

a relationship that contains t_F, i.e., cable life under the considered stepwise-constant load cycle, and K, i.e., the number of cycles-to-failure (see 194) is achieved, namely [128]:

$$\int_0^L \mathrm{dLF} = K \sum_{i=1}^N \left\{ \int_0^{\Delta t_j} \mathrm{dLF}[E_n, T_i(t)] \right\} = K \sum_{i=1}^N \mathrm{LF}_i = 1. \tag{197}$$

Now, K can be attained simply from (197):

$$K = \left[\sum_{i=1}^N \mathrm{LF}_i \right]^{-1} \tag{198}$$

and life t_F can be inferred directly from (194), (198), by computing the various life fractions LF_i ($i = 1, \ldots, N$) via (196). As this latter equation shows, such computation is easy for stepwise-constant daily load cycles, provided that a life model valid for cable insulation is available (see Sect. 4.2.1) and time functions $T_i(t)$ can be calculated [117, 118].

Percent life variation (possibly extension) of the power component under load cycling, i.e., the percent variation of the generic 100 Fth percentile of component time-to-failure under load cycling with respect to rated 100 Fth percentile of time-to-failure at rated voltage and temperature, $t_{F,n}$, can be evaluated by means of the following quantity:

$$\Delta t_{F,100} = 100\left(t_F - t_{F,n}\right)/t_{F,n}. \quad (199)$$

The definition of $\Delta t_{F,100}$ accounts for the sign of the difference between estimated life and rated life. Indeed, such difference is >0 in the case of life extension, <0 in the case of life reduction with respect to rated life.

Note that $\Delta t_{F,100}$ of (199) is based on t_F, which is derived through (198) by considering thermal transients and electric stress plus electro-thermal synergism.

5 Model Selection in View of Aging Properties: A Numerical Illustration

In order that the whole previous discussion does not appear to be purely academic, it seems opportune to illustrate in this final section the advantages which can be brought about by IRA, or else the drawbacks of a mere DRA, by means of practical numerical examples relevant to real life data.

We start from observing that, for an adequate selection of the reliability model, one must be aware that two or more models may often be—or appear to be—similar. This is often the case, e.g., for the LN and Weibull models, as discussed in literature [49, 51, 67, 104], especially as long as one is interested in the estimates of "central" moments or percentiles (e.g., the median value). However, the same is not true for lower LT percentiles, or hazard rate functions (hrf), which are of main practical interest for characterizing the aging properties of the device [103, 124]. Here, some results are presented that were partly illustrated also by the authors of this chapter in [42]. Let us consider for example (as indicated in Table 1) the

Table 1 Values of theoretical 1st, 5th and 50th percentiles, in years, under five different reliability models with the same mean (45 years) and SD (15 years)

Model	Normal	Weibull	Gamma	Lognormal	Inverse Gaussian
Percentile $T_{0.01}$ (years)	10.10	12.65	17.54	20.06	20.05
Percentile $T_{0.05}$ (years)	20.33	20.61	23.48	25.03	25.05
Median = $T_{0.5}$ (years)	45.00	44.92	43.34	42.69	42.65

theoretical value of the 1st and the 5th LT percentile $T_{0.01}$ and $T_{0.05}$—together with the median $T_{0.50}$—for five different reliability models having the same mean (45 years) and SD (15 years), chosen as the typical values reported in [106] for electrical insulation components. Here, it is denoted by T_p a LT value such that: $F(T_p) = p$, being $F(t) = 1 - R(t)$ the cdf of the LT, T. So, for instance, the percentile $t^* = T_{0.01}$ is the life span such that only 1% of the devices fails before t^*.

It can be seen, looking at the values in the rows of Table 1, that very strong differences are obtained for the above percentiles, especially between the above discussed LN (and also IG) and Weibull models, or LN and Normal models. The differences are larger (up to 100%!) for lower percentiles, while they are reasonably small for the median values, as anticipated.

Moreover, very different behaviors of the hrf are obtained, as shown in Fig. 3 for three of the above models (i.e., Gamma, LN and Weibull). Such differences become noticeable particularly after mission times near to the MTTF value (45 years), namely an age that is being approached by many power system components currently in-service.

On the contrary (as noticed in Sect. 2.2) the Gaussian and Weibull models with same mean and SD possess generally extreme percentiles values which are very similar, and the same is true for both RF and hrf. However, they share also the drawbacks discussed in the same Sect. 2.12, among which we mention here again the lack of flexibility with respect to the hrf behavior description.

Also, the LN and IG models are very similar, but this will be discussed later.

On the other hand, due to the popularity of the Weibull model it is interesting to remark two undesirable consequences of the "wrong" assumption of a Weibull model, when in fact a LN model is true:

1. The wrong assumption may cause unnecessary maintenance actions. Indeed, it is known that typical age maintenance programs are opportune if and only if the hrf is increasing in time [9, 83]: on the contrary, often the LN model (which is IDHR) is mistaken for a IHR Weibull one.
2. The wrong assumption implies an improper under-estimation of the RF in the lower tail, as confirmed by the above results on the 1st LT percentile $T_{0.01}$

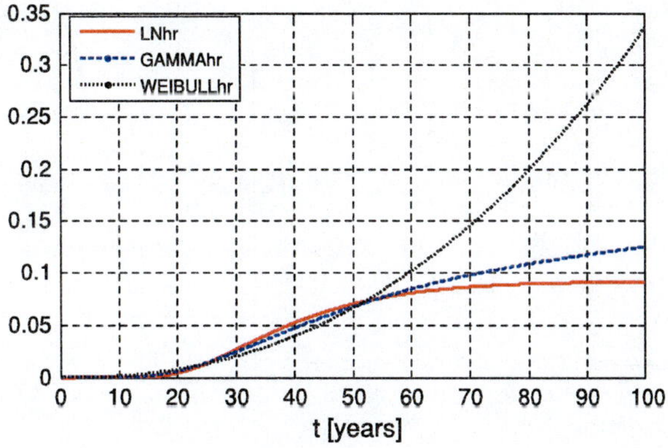

Fig. 3 Hrf curves for the Gamma, LN and Weibull reliability models, with same mean (45 years) and SD (15 years)

(in particular, the Weibull model provides for such parameter an estimate which is about one half of the "true" value). Of course, this under-estimation may bring about unnecessary costs in view of maintenance actions.

As previously pointed out, such results—which are here discussed only in the framework of probabilistic models—imply, in view of the statistical assessment of a reliability model that, in practice, two or more different models may be often undistinguishable, but in the presence of many data. Indeed, as well known from statistical theory, many data are needed to efficiently estimate extreme percentiles, but these are of course rarely available in this field, so that the model identification is an unavoidable, difficult and very critical first step in any reliability analysis. In view of data scarcity, the most useful way to perform such step is (when feasible, as often happens in engineering applications) to use technological and probabilistic information about wear and aging which unavoidably affect every power system component.

On the other hand, fortunately some of the above models are very similar to each other under many respects, especially—among those here considered—the LN model and the IG model, as shown in Figs. 4 and 5, respectively, for the pdf and hrf functions corresponding to the same values of mean (45 years) and SD (15 years) as above. This happens also, often (but not always), as far as the LN and LL models are concerned, as discussed in previous sections and in [37], too.

This kind of property may be exploited when the exact distribution is unknown, but—as sometimes happens in the field of electrical insulation (see Sect. 2)—the researcher knows that the aging mechanism gives rise to a decreasing hazard rate for large mission times. For instance, often the LN distribution is used in such cases because its parameters can be statistically estimated more easily with respect to other models possessing similar characteristics [51].

While, above, differences in the hrf curves for some models were illustrated (see Fig. 3), it is to be remarked that the differences in the CRF curves may be

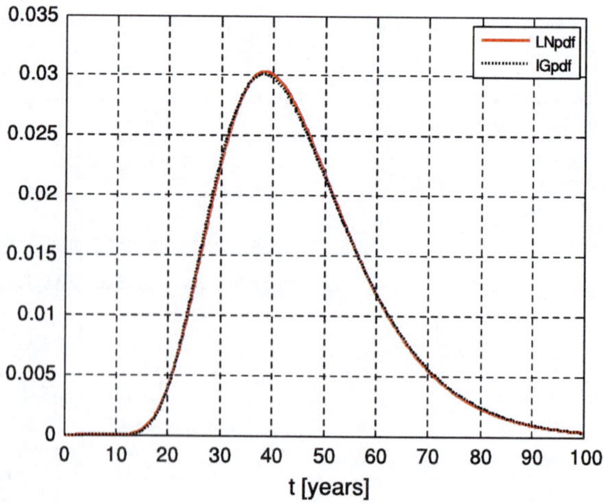

Fig. 4 Pdf of LN model and IG model, with same mean (45 years) and SD (15 years)

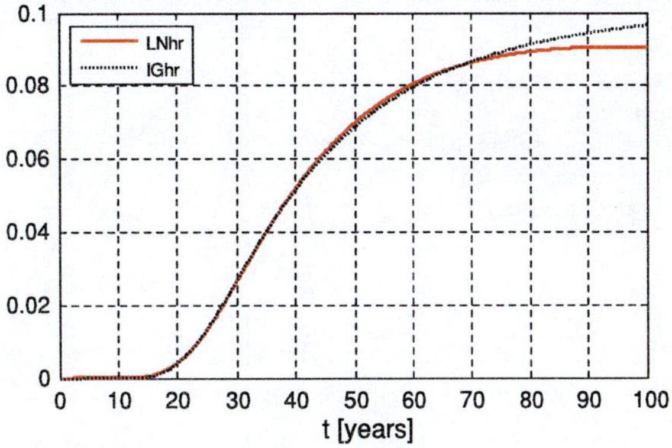

Fig. 5 Hrf of LN model and IG model, with same mean (45 years) and SD (15 years)

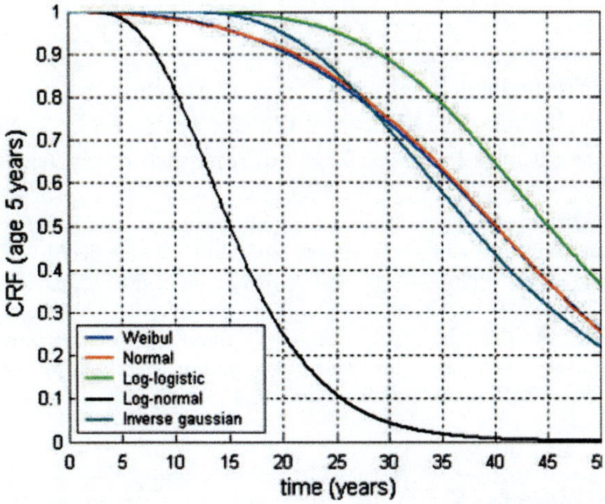

Fig. 6 CRF, relevant to an age of 5 years, for the Weibull, Normal, Log-logistic, Lognormal, Inverse Gaussian distributions with same mean (45 years) and standard deviation (15 years)

even more pronounced. This is shown, e.g., in next Figs. 6 and 7 showing the CRF $R(t|s)$ versus t, relevant, respectively, to an age of $s = 5$ and $s = 35$ years, for the Weibull, normal, Log-logistic, Lognormal, Inverse Gaussian distributions with same mean (45 years) and standard deviation (15 years), in order to appreciate the differences that exist among the different models.

It seems that such comparison is very interesting, also because the CRF is seldom reported, although possessing, in the authors' opinion, a deeper "physical" meaning than the hrf (see Sect. 1.4). Its importance in view of proper maintenance actions is understandable, and this is a key point in the necessity of an adequate reliability model selection.

In the above discussion—as in the whole chapter—the problem of finding statistical estimates of the relevant quantities (here, the percentiles and the hrf) has not been addressed. In practice, the discussed differences between "similar" models may be even emphasized when the data are scarce, as pointed out and illustrated in [104]. In these kind of analyses, characterized by lack of data and

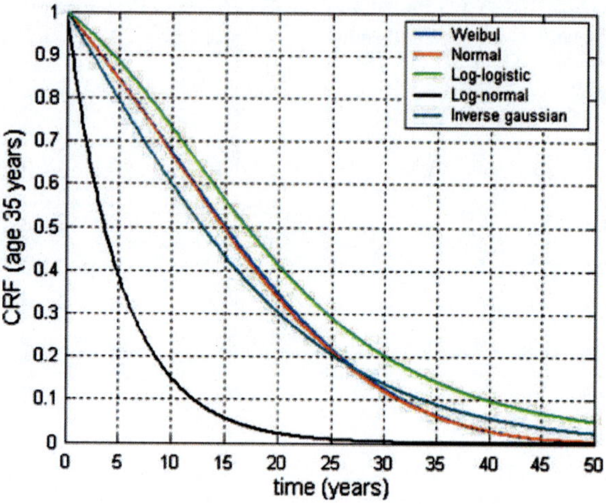

Fig. 7 CRF, relevant to an age of 35 years, for the Weibull, Normal, Log-logistic, Lognormal, Inverse Gaussian distributions with same mean (45 years) and standard deviation (15 years)

prior technological information, the authors believe that the Bayesian estimation methodology [113, 146, 147] can be the most adequate tool, as discussed and shown in [12, 39, 72, 73] with reference to electrical device reliability analyses.

Acknowledgments The authors wish to express their thankful acknowledgments to some friends and colleagues of the University of Naples "Federico II", for their very useful advices and significant contributions to the improvement of the present chapter. In particular, we express our thanks to Profs. Ernesto Conte, Pasquale Erto, Biagio Palumbo. We are also grateful to Profs. Erto and Palumbo and Dr. Giuliana Pallotta for bringing to our attention the main preliminary results of their last paper, while it was still in progress [75]. We also thank Dr. Claudia Battistelli for her help in revising the manuscript.

References

1. Abdel-Hameed A (1975) A gamma wear process. IEEE Trans Reliab 24(2):152–153
2. Allella F, Chiodo E, Pagano M (2002) Dynamic discriminant analysis for predictive maintenance of electrical components subjected to stochastic wear. Int J Comput Math Electr Electron Eng 21(1):98–115
3. Anders GJ (1990) Probability concepts in electric power systems. Wiley, New York
4. Ascher H, Feingold H (1984) Repairable systems reliability. Marcel Dekker, New York
5. Anderson PM, Timko KJ (1982) A probabilistic model of power system disturbances. IEEE Trans Circuits Syst 29(11):789–796
6. Aven T, Jensen U (1999) Stochastic models in reliability. Springer, Berlin
7. Bain LJ, Engelhardt M (1992) Introduction to probability and mathematical statistics. Duxbury Press, Pacific Grove
8. Barlow RE (1985) A Bayes explanation of an apparent failure rate paradox. IEEE Trans Reliab 34:107–108
9. Barlow RE, Proschan F (1965) Mathematical theory of reliability. Wiley, New York
10. Barlow RE, Proschan F (1975) Statistical theory of reliability and life testing. Holt, Rinehart & Winston, New York
11. Barlow RE, Proschan F (1981) Statistical theory of reliability and life testing: probability models. To Begin With, Silver Spring

12. Barrett JS, Green MA (1994) A statistical method for evaluating electrical failures. IEEE Trans Power Deliv 9(3):1524–1530
13. Bennett S (1983) Log-logistic regression model for survival data. Appl Stat 32:165–171
14. Bernardo JM, Smith AFM (2000) Bayesian theory. Wiley, Chichester
15. Bhattacharyya GK, Fries A (1982) Fatigue failure models—Birnbaum–Saunders vs. inverse Gaussian. IEEE Trans Reliab 31:439–441
16. Billinton R, Allan RN (1992) Reliability evaluation of engineering systems: concepts and techniques. Plenum Press, New York
17. Billinton R, Allan RN (1996) Reliability evaluation of power systems. Pitman Publishing Ltd, Plenum Press, New York
18. Billinton R, Allan RN, Salvaderi L (1991) Applied reliability assessment in electric power systems. IEEE Service Center, Piscataway
19. Billinton R, Salvaderi L, McCalley JD, Chao H, Seitz T, Allan RN, Odom J, Fallon C (1997) Reliability issues in today's electric power utility environment. IEEE Trans Power Syst 12(4):1708–1714
20. Birnbaum ZW, Saunders SC (1969) Estimation for a family of life distributions with applications to fatigue. J Appl Probab 6:328–347
21. Birolini A (2004) Reliability engineering: theory and practice. Springer, Berlin
22. Bocchetti D, Giorgio M, Guida M, Pulcini G (2009) A competing risk model for the reliability of cylinder liners in marine diesel engines. Reliab Eng Syst Safe 94:1299–1307
23. Boggs SA (2002) Mechanisms for degradation of TR-XLPE impulse strength during service aging. IEEE Trans Power Deliv 17(2):308–312
24. Brown RE (2002) Electric power distribution reliability, 2nd edn. Marcel Dekker, New York (published in 2008 by CRC Press)
25. Casella G, Berger RL (2002) Statistical inference. Duxbury Press, Pacific Grove
26. Cacciari M, Mazzanti G, Montanari GC (1994) Electric strength measurements and Weibull statistics on thin EPR films. IEEE Trans Dielectr Electr Insulation 1(1):153–159
27. Caramia P, Carpinelli G, Verde P et al (2000) An approach to life estimation of electrical plant components in the presence of harmonic distortion. In: Proceedings of 9th international conference on harmonics and quality of power. Orlando, pp 887–892
28. Castillo E (1988) Extreme value theory in engineering. Academic Press, New York
29. Castillo E, Fernandez-Canteli A (2009) A unified statistical methodology for modeling fatigue damage. Springer, Berlin
30. Catuneanu VM, Mihalache AN (1989) Reliability fundamentals. Elsevier, Amsterdam
31. Cavallini A, Fabiani D, Mazzanti G et al (2002) Life model based on space-charge quantities for HVDC polymeric cables subjected to voltage-polarity inversions. IEEE Trans Dielectr Electr Insulation 9(4):514–523
32. Cavallini A, Mazzanti G, Montanari GC et al (2000) The effect of power system harmonics on cable endurance and reliability. IEEE IAS Annual Meeting, Rome, pp 3172–3179
33. Chang DS, Tang LC (1993) Reliability bounds and critical time for the Birnbaum–Saunders distribution. IEEE Trans Reliab 42:464–469
34. Chhikara RS, Folks JL (1989) The inverse Gaussian distribution. Marcel Dekker, New York
35. Chiodo E (1992) Reliability analyses in the presence of random covariates. PhD Thesis in Computational Statistics and Applications (in Italian). Department of Mathematical Statistics of the University of Napoli "Federico II", Italy
36. Chiodo E, Gagliardi F, Pagano M (2004) Human reliability analyses by random hazard rate approach. Int J Comput Math Electr Electr Eng 23(1):65–78
37. Chiodo E, Mazzanti G (2004) The log-logistic model for reliability characterization of power system components subjected to random stress. In: Proceedings of 2004 SPEEDAM, Capri, pp 239–244
38. Chiodo E, Mazzanti G (2006) Indirect reliability estimation for electric devices via a dynamic stress—strength model. In: Proceedings of 2006 IEEE SPEEDAM, Taormina, pp S31_19–S31_24

39. Chiodo E, Mazzanti G (2006) Bayesian reliability estimation based on a Weibull stress-strength model for aged power system components subjected to voltage surges. IEEE Trans Dielectr Electr Insulation 13(1):146–159
40. Chiodo E, Mazzanti G (2006) A new reliability model for power system components characterized by dynamic stress and strength. In: Proceedings of 2006 IEEE SPEEDAM, Taormina, pp S30_11–S30_16
41. Chiodo E, Mazzanti G (2006) New models for reliability evaluation of power system components subjected to transient overvoltages. In: Proceedings of 2006 IEEE PES General Meeting, Montreal. ISBN 1-4244-0493-2/06/
42. Chiodo E, Mazzanti G (2008) Theoretical and practical aids for the proper selection of reliability models for power system components. Int J Reliab Saf Spec Issue Emerg Technol Methodol Power Syst Reliab Secur Assess 2(1/2):99–128
43. Cinlar E (1975) Introduction to stochastic processes. Prentice-Hall, Englewood Cliffs, New Jersey
44. Cinlar E (1977) Shock and wear models and Markov additive processes. In: Shimi IN, Tsokos CP (eds) The theory and applications of reliability, vol 1. Academic Press, New York, pp 193–214
45. Cinlar E (1980) On a generalization of gamma processes. J Appl Probab 17(2):467–480
46. Cinlar E (1984) Markov and semi-Markov models of deterioration. In: Abdel-Hameed MS, Cinlar E, Quinn E (eds) Reliability theory and models. Academic Press, New York, pp 3–41
47. Cinlar E (2003) Conditional Lévy processes. Comput Math Appl 46:993–997
48. Cinlar E, Shaked S, Shanthikumar JG (1989) On lifetimes influenced by a common environment. Stochastic Process Appl 33(2):347–359
49. Croes K, Manca JV (1988) The Time of guessing your failure time distribution is over. Microelectr Reliab 38:1187–1191
50. Cox DR (1972) Regression models and life tables (with discussion). J Roy Stat Soc B 34:187–220
51. Cox DR, Oakes D (1984) Analysis of survival data. Chapman & Hall, London
52. Cousineau D (2009) Fitting the three-parameter Weibull distribution: review and evaluation of existing and new methods. IEEE Trans Dielectr Electr Insulation 16(1):281–288
53. Crow EL, Shimizu K (1988) Lognormal distributions. Marcel Dekker, New York
54. Currit A, Singpurwalla ND (1988) On the reliability function of a system of components sharing a common environment. J Appl Probab 26:763–771
55. Cohen AC, Whitten BJ (1988) Parameter estimation in reliability and life span models. Marcel Dekker, New York
56. Crowder MJ, Kimber AC et al (1991) Statistical methods for reliability data. Chapman & Hall, London
57. D'Agostini G (2003) Bayesian reasoning in data analysis. World Scientific, Singapore
58. Dakin TW (1948) Electrical insulation deterioration treated as a chemical rate phenomenon. AIEE Trans 67:113–122
59. Dasgupta A, Pecht M (1991) Material failure mechanisms and damage models. IEEE Trans Reliab 40(5):531–536
60. Dasgupta A, Pecht M (1995) Physics of failure: an approach to reliable product development. J Inst Environ Sci 38(5):30–34
61. De Finetti B, Galavotti MC, Hosni H, Mura A (eds) (2008) Philosophical lectures on probability. Springer, Berlin
62. Desmond A (1985) Stochastic models of failure in random environments. Canad J Stat 13:171–183
63. Desmond A (1986) On the relationship between two fatigue-life models. Trans Reliab 35:167–169
64. Dissado LA, Mazzanti G, Montanari GC (2001) Elemental strain and trapped space charge in thermoelectrical aging of insulating materials. Part 1: elemental strain under thermo-electrical-mechanical stress. IEEE Trans Dielectr Electr Insulation 8(6):959–965

65. Dissado LA, Mazzanti G, Montanari GC (2001) Elemental strain and trapped space charge in thermoelectrical aging of insulating materials. Life modeling. IEEE Trans Dielectr Electr Insulation 8(6):966–971
66. Dissado LA (2001) Predicting electrical breakdown in polymeric insulators. IEEE Trans Dielectr Electr Insulation 9(5):860–875
67. Dumonceaux R, Antle CE (1973) Discrimination between the lognormal and Weibull distributions. Technometrics 15:923–926
68. Ebrahimi NB (2003) Indirect assessment of system reliability. IEEE Trans Reliab 52(1):58–62
69. Erto P (1982) New practical Bayes estimators for the 2-parameters Weibull distribution. IEEE Trans Reliab 31(2):194–197
70. Erto P (1989) Genesis, properties and identification of the inverse Weibull survival model. Stat Appl (Italian) 1:117–128
71. Erto P, Giorgio M (1996) Modified practical Bayes estimators. IEEE Trans Reliab 45(1):132–137
72. Erto P, Giorgio M (2002) Assessing high reliability via Bayesian approach and accelerated tests. Reliab Eng Syst Safe 76:301–310
73. Erto P, Giorgio M (2002) Generalised practical Bayes estimators for the reliability and the shape parameter of the Weibull distribution. In: Proceedings of: "PMAPS 2002: Probabilistic Methods Applied to Power Systems", Napoli, Italy
74. Erto P (ed) (2009) Statistics for innovation. Springer, Milan
75. Erto P, Pallotta G, Palumbo B (2010) Generative mechanisms, properties and peculiar applicative examples of the inverse Weibull distribution (submitted)
76. Erto P, Palumbo B (2005) Origins, properties and parameters estimation of the hyperbolic reliability model. IEEE Trans Reliab 54(2):276–281
77. Esary JD, Marshall AW (1973) Shock models and wear processes. Ann Probab 1:627–649
78. Endicott HS, Hatch BD, Sohmer RG (1965) Application of the Eyring model to capacitor aging data. IEEE Trans Comp Parts 12:34–41
79. Follmann DA, Goldberg MS (1988) Distinguishing heterogeneity from decreasing hazard rate. Technometrics 30:389–396
80. Galambos J (1987) The asymptotic theory of extreme order statistics. R. E. Krieger Publishing, Malabar
81. Gebraeel N, Elwany A, Pan J (2009) Residual life predictions in the absence of prior degradation knowledge. IEEE Trans Reliab 58(1):106–117
82. Gertsbakh IB, Kordonskiy KB (1969) Models of failures. Springer, Berlin
83. Gertsbakh JB (1989) Statistical reliability theory. Marcel Dekker, New York
84. Glaser RE (1980) Bathtub and related failure rate characterizations. J Am Stat Assoc 75(371):667–672
85. Harris CM, Singpurwalla ND (1967) Life distributions derived from stochastic hazard function. IEEE Trans Reliab 17(2):70–79
86. Heckman JJ, Singer B (1982) Population heterogeneity in demographic models. In: Land KC, Rogers A (eds) Multidimensional mathematical demography. Academic Press, New York
87. Hirose H (2007) More accurate breakdown voltage estimation for the new step-up test method in the Gumbel distribution model. Eur J Oper Res 177:406–419
88. Hoyland A, Rausand M (2003) System reliability theory. Wiley, New York
89. Huang W, Askin RG (2004) A generalized SSI reliability model considering stochastic loading and strength aging degradation. IEEE Trans Reliab 53(1):77–82
90. He J, Sun Y, Wang P, Cheng L (2009) A hybrid conditions-dependent outage model of a transformer in reliability evaluation. IEEE Trans Power Deliv 24(4):2025–2033
91. IEC 60216-1 (2001) Electrical insulating materials—properties of thermal endurance—Part 1: ageing procedures and evaluation of test results, 5th edn. IEC, Geneva
92. IEC 60505 (2004) Evaluation and qualification of electrical insulation systems, 3rd edn. IEC, Geneva

93. Iyengar S, Patwardhan G (1988) Recent developments in the Inverse Gaussian distribution. In: Krishnaiah R, Rao CR (eds) Handbook of statistics. Elsevier, Amsterdam, pp 479–499
94. Johnson RA (1988) Stress-strength models for reliability. In: Krishnaiah PR, Rao CR (eds) Handbook of statistics 7: quality control and reliability. North Holland, Amsterdam
95. Jiang R, Ji P, Xiao X (2003) Aging property of unimodal failure rate models. Reliab Eng Syst Saf 79(1):113–116
96. Johnson NL, Kotz S, Balakrishnan N (1995) Continuous univariate distributions, vols 1 and 2, 2nd edn. Wiley, New York
97. Kalbfleisch JD, Prentice RL (2002) The statistical analysis of failure time data. Wiley, New York
98. Kendall MG, Stuart A (1987) The advanced theory of statistics. MacMillan, New York
99. Kotz S, Lumelskii Y, Pensky M (2003) The stress-strength model and its generalizations: theory and applications. Imperial College Press, London (distributed by World Scientific, Singapore)
100. Kreuger FH (1992) Industrial high voltage. Delft University Press, Delft
101. Krishnaiah PR, Rao CR (eds) (1988) Handbook of statistics, vol 7: Quality control and reliability. North Holland, Amsterdam
102. Kundu D, Kannan N, Balakrishnan N (2008) On the hazard function of Birnbaum–Saunders distribution and associated inference. Comput Stat Data Anal 52:26–52
103. Lawless JF (1982) Statistical models and methods for lifetime data. Wiley, New York
104. Lawless JF (1983) Statistical models in reliability. Technometrics 25:305–335
105. Lawless JF, Crowder MJ (2004) Covariates and random effects in a gamma process model with application to degradation and failure. Lifetime Data Anal 10:213–227
106. Li W (2002) Incorporating aging failures in power system reliability evaluation. IEEE Trans Power Syst 17(3):918–923
107. Li W (2004) Evaluating mean life of power system equipment with limited end-of-life failure data. IEEE Trans Power Syst 19(1):236–242
108. Lindley DV (2000) The philosophy of statistics. Statistician 49:293–337
109. Lindley DV, Singpurwalla ND (1986) Multivariate distributions for the life lengths of components of a system sharing a common environment. J Appl Probab 23:418–431
110. Lu CJ, Meeker WQ (1993) Using degradation measures to estimate a time-to-failure distribution. Technometrics 35(2):161–174
111. Lu CJ, Meeker WQ, Escobar LA (1996) A comparison of degradation and failure-time analysis methods for estimating a time-to-failure distribution. Stat Sin 6:531–554
112. Mann N, Schafer RE, Singpurwalla ND (1974) Methods for statistical analysis of reliability and life data. Wiley, New York
113. Martz HF, Waller RA (1991) Bayesian reliability analysis. Krieger Publishing, Malabar
114. Marzinotto M, Mazzanti G, Mazzetti C (2006) Comparison of breakdown performances of extruded cables via the enlargement law. In: Proceedings of 2006 conference on electrical insulation and dielectric phenomena, Kansas City, pp 760–763
115. Marzinotto M, Mazzanti G, Mazzetti C (2007) A new approach to the statistical enlargement law for comparing the breakdown performance of power cables—part 1: theory. IEEE Trans Dielectr Electr Insulation 14(5):1232–1241
116. Marzinotto M, Mazzanti G, Mazzetti C (2008) A new approach to the statistical enlargement law for comparing the breakdown performance of power cables—part 2: application. IEEE Trans Dielectr Electr Insulation 15(3):792–799
117. Mazzanti G (2005) Effects of the combination of electro-thermal stress, load cycling and thermal transients on polymer-insulated high voltage ac cable life. In: Proceedings of 2005 IEEE PES general meeting, San Francisco
118. Mazzanti G (2007) Analysis of the combined effects of load cycling, thermal transients and electro-thermal stress on life expectancy of high voltage ac cables. IEEE Trans Power Deliv 22(4):2000–2009
119. Mazzanti G, Montanari GC (1999) Insulation aging models. In: Wiley encyclopedia of electrical and electronic engineering. Wiley, New York, pp 308–319

120. Mazzanti G, Passarelli G (2005) Reliability analysis of power cables feeding electric traction systems. In: Proceedings of 2005 ship propulsion and railway traction systems, Bologna, pp 100–107
121. Mazzanti G, Passarelli G (2006) A probabilistic life model for reliability analysis of power cables feeding electric traction systems. In: Proceedings of 2006 IEEE SPEEDAM, Taormina, pp S30_17–S30_22
122. Mazzanti G, Passarelli G, Russo A, Verde P (2006) The effects of voltage waveform factors on cable life estimation using measured distorted voltages. In: Proceedings of 2006 IEEE PES general meeting, Montreal. ISBN 1-4244-0493-2/06/
123. Modarres M (1993) Reliability and risk analysis. Marcel Dekker, New York
124. Meeker WQ, Escobar LA (1998) Statistical methods for reliability data. Wiley, New York
125. Mi J (1998) A new explanation of decreasing failure rate of a mixture of exponentials. IEEE Trans Reliab 47(4):460–462
126. Montanari GC, Cacciari M (1989) A probabilistic life model for insulating materials showing electrical threshold. IEEE Trans Dielectr Electr Insulation 24(1):127–134
127. Montanari GC, Mazzanti G, Simoni L (2002) Progress in electrothermal life modeling of electrical insulation over the last decades. IEEE Trans Dielectr Electr Insulation 9(5):730–745
128. Miner M (1945) Cumulative damage in fatigue. J Appl Mech 12:159–164
129. Mosch W, Hauschild W (1992) Statistical techniques for HV engineering. Peter Peregrinus, London
130. Nelson W (1982) Applied life data analysis. Wiley, New York
131. Nelson W (1990) Accelerated testing. Wiley, New York
132. O'Hagan A (1994) Kendall's advanced theory of statistics, vol 2B. In: Arnold E (ed) Bayesian Inference, London
133. Okabe S, Tsuboi T, Takami J (2008) Reliability evaluation with Weibull distribution on AC withstand voltage test of substation equipment. IEEE Trans Dielectr Electr Insulation 15(5):1242–1251
134. Papoulis A (2002) Probability, random variables, stochastic processes. Mc Graw Hill, New York
135. Press SJ (2002) Subjective and objective Bayesian statistics: principles, models, and applications, 2nd edn. Wiley, New York
136. Proschan F (1963) Theoretical explanation of observed decreasing failure rate. Technometrics 5:375–383
137. Ricciardi LM (1977) Diffusion processes and related topics in biology. Lecture notes in biomathematics. Springer, Berlin
138. Robert CP (2001) The Bayesian choice. Springer, Berlin
139. Rohatgi VK, Saleh AK (2000) An introduction to probability and statistics, 2nd edn. Wiley, New York
140. Ross SM (1996) Stochastic processes. Wiley, New York
141. Ross SM (2003) Introduction to probability models. A Harcourt Science and Technology Company, San Diego
142. Roy D, Mukherjee SP (1988) Generalized mixtures of exponential distributions. J Appl Probab 25:510–518
143. Shimi IN (1977) System failures and stochastic hazard rate. In: Krishnaiah PR (ed) Applications of statistics. North Holland, New York, pp 497–505
144. Simoni L (1983) Fundamentals of endurance of electrical insulating materials. CLUEB, Bologna
145. Singpurwalla ND (1997) Gamma processes and their generalizations: an overview. In: Cooke R, Mendel M, Vrijling H (eds) Engineering probabilistic design and maintenance for flood protection. Kluwer Academic Publishers, Dordrecht, pp 67–75
146. Singpurwalla ND (2006) Reliability and risk: a Bayesian perspective, Wiley series in probability and statistics. Wiley, New York

147. Spizzichino F (2001) Subjective probability models for lifetimes. Chapman & Hall/CRC Press, Boca Raton
148. Srivastava PW, Shukla R (2008) A log-logistic step-stress model. IEEE Trans Reliab 57(3):431–434
149. Stuart A, Kendall MG, (1994) Kendall's advanced theory of statistics, vol I. In: Arnold E (ed) Distribution theory, 6th edn. Hodder Arnold, London, ISBN: 0340614307, EAN: 9780340614303, pp 704
150. Sweet AL (1990) On the hazard rate of the lognormal distribution. IEEE Trans Reliab 39:325–328
151. Thompson WA (1988) Point process models with applications to safety and reliability. Chapman & Hall, London
152. Ushakov IA, Harrison RA (1994) Handbook of reliability engineering. Wiley, New York
153. Van Noortwijk JM, Pandey MD (2004) A stochastic deterioration process for time-dependent reliability analysis. In: Maes MA, Huyse L (eds) Reliability and optimization of structural systems. Taylor & Francis, London, pp 259–265
154. Van Noortwijk JM, Pandey MD et al (2007) Gamma processes and peaks-over-threshold distributions for time-dependent reliability. Reliab Eng Syst Safe 92(12):1651–1658
155. Van Noortwijk JM (2009) A survey of the application of Gamma processes in maintenance. Reliab Eng Syst Saf 1:2–21
156. Wang P, Billinton R (2002) Reliability cost/worth assessment of distribution systems incorporating time-varying weather conditions and restoration resources. IEEE Trans Power Deliv 17(1):260–265
157. Whitmore G (1995) Estimating degradation by a Wiener diffusion process subject to measurement error. Lifetime Data Anal 1:307–319
158. Xie L, Wang L (2008) Reliability degradation of mechanical components and systems. In: Misra BK (ed) Handbook of performability engineering. Springer, London
159. Zapata CJ, Silva SC, Burbano L (2008) Repair models of power distribution components. In: Proceedings of IEEE/PES transmission and distribution conference and exposition: Latin America, 13–15 Aug. 2008, pp 1–6
160. Zapata CJ, Torres A, Kirschen DS, Rios MA (2009). Some common misconceptions about the modeling of repairable components. In: Proceedings of IEEE Power & Energy Society general meeting, 26–30 July 2009, pp 1–8

Adequacy Assessment of Wind-Integrated Composite Generation and Transmission Systems

Roy Billinton, Yi Gao, Dange Huang and Rajesh Karki

Abbreviations

AR	Auto-regressive
ARMA	Auto-regressive and moving average
BES	Bulk electric system
COPT	Capacity outage probability table
DAFOR	Derated adjusted forced outage rate
DAFORW	Derated adjusted forced outage rate for wind farm
EDLC	Expected duration of load curtailment
EENS	Expected energy not supplied
EFLC	Expected frequency of load curtailment
EFOR	Equivalent forced outage rate
FOR	Forced outage rate
h	Hour
IEAR	Interrupted energy assessment rate
IEEE-RTS	IEEE reliability test system
IEEE-RTSW	IEEE-RTS with wind
LOEE	Loss of energy expectation

R. Billinton, Y. Gao, D. Huang and R. Karki (✉)
Power System Research Group, University of Saskatchewan, 57 Campus Dr, Saskatoon, SK S7N 5A9, Canada
e-mail: rajesh.karki@usask.ca

R. Billinton
e-mail: roy.billinton@usask.ca

Y. Gao
e-mail: yi.gao@usask.ca

D. Huang
e-mail: dange.huang@usask.ca

G. Anders and A. Vaccaro (eds.), *Innovations in Power Systems Reliability*, Springer Series in Reliability Engineering, DOI: 10.1007/978-0-85729-088-5_4,

LOLE	Loss of load expectation
MA	Moving average
MCS	Monte Carlo simulation
MECORE	Monte Carlo and enumeration composite reliability evaluation
MRBTS	Modified RBTS
MRBTSW	Modified RBTS with wind
MW	MegaWatt
MWh	MegaWatt hour
occ	Occurrence
OPF	Optimal power flow
RapHL-II	Reliability analysis program for HL-II
RBTS	Roy Billinton test system
RBTSW	RBTS with wind
WECS	Wind energy conversion system
WTG	Wind turbine generator
yr	Year

1 Introduction

The reliability evaluation of a composite generation and transmission system or bulk electric system (BES) is a relatively complex task that includes detailed modeling of the generation and transmission facilities [1–6]. Composite power system reliability evaluation indicates the ability of the BES to meet the load and energy requirements at the major load points and for the overall system and can be performed using analytical methods or Monte Carlo simulation (MCS). Analytical methods usually involve some form of contingency enumeration and represent the system by mathematical models. The reliability indices are derived from the models using direct numerical analyses and assumptions are frequently required to simplify the overall problem, particularly when operating procedures are included in the studies. This difficulty can usually be reduced or avoided using a MCS approach, in which the reliability indices are obtained by simulating the random behavior of the BES over a suitable period of time. There are two basic MCS techniques that can be applied to BES reliability assessment. These methods are commonly designated as sequential and non-sequential simulation [7, 8]. Both techniques can be used in conventional and wind-integrated BES reliability assessment and are applied in the studies presented in this chapter.

The utilization of the wind to generate electrical energy is increasing rapidly throughout the world. Wind turbine generators (WTGs) can be added and are being added in a wide range of locations in modern electric power systems. These units can be dispersed throughout relatively low voltage networks or in concentrated wind farms located at relatively remote sites with suitable wind regimes. Large wind energy conversion systems (WECSs) include both inland and off-shore facilities and are usually connected to the BES by radial transmission lines. Wind power, however, behaves quite differently than conventional electric power generating facilities due to its intermittent and diffuse nature. The incorporation of WECS in generation and transmission system reliability studies therefore requires distinctive and applicable modeling and data considerations.

Considerable work has been done on the development and application of models and techniques for generating capacity adequacy assessment and is well documented in the published literature [1–6]. Work has also been done on the reliability evaluation of conventional generating systems incorporating WECS [9–13]. These studies do not consider the BES and focus on the ability of the generating facilities including wind power to satisfy the overall system load. The most accurate and comprehensive approach to incorporate wind energy in a generating capacity adequacy evaluation is to use sequential Monte Carlo simulation and suitable time series wind models [13]. Wind energy has also been included in generating capacity adequacy assessment by considering the wind power as negative load and modifying the total system load prior to combining this load with the conventional generating capacity to determine the required risk indices. Wind energy can be incorporated in analytical approaches and in non-sequential or state sampling Monte Carlo simulation methods using multi-state WECS models in the form shown later in this chapter.

A comprehensive reliability analysis of a BES considers the ability of the system to satisfy the load requirements at each individual load point in the BES in addition to meeting the overall system load [14–16]. A WECS could conceivably be connected at any bus in the BES and serve as a generation source. Relatively little work has been done on the incorporation of WECS in quantitative BES reliability evaluation. This chapter extends the concepts presented in [1–16] to include some of the recent work on the reliability of BES incorporating the chronological variability of WECS.

2 Modeling and Simulating Wind Speeds

Each wind site has a unique wind speed profile that depends on its geographic location and the site topology. The variability in the chronological wind speed can be modeled using actual wind speed data for a lengthy period of time or by wind speed time series models developed from the actual data [9]. The wind speed model and data for the Swift Current site located in the Province of Saskatchewan, Canada are used in the initial studies described in this chapter. The mean and standard deviation of the hourly wind speed at the Swift Current site are 19.46 and 9.7 km/h, respectively. The hourly mean and standard deviation of wind speeds from a 20-year database (1 Jan. 1984 to 31 Dec. 2003) for the Swift Current location were obtained from Environment Canada. These data were used to build an auto-regressive and moving average (ARMA) time series model [9]. The ARMA (4, 3) model is the optimal time series model for the Swift Current site and the parameters are shown in (1):

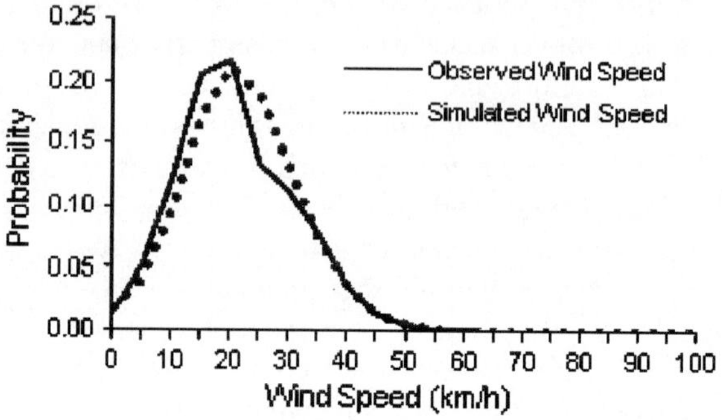

Fig. 1 Observed and simulated wind speed distributions for the Swift Current site

$$y_t = 1.1772y_{t-1} + 0.1001y_{t-2} - 0.3572y_{t-3} + 0.0379y_{t-4} + \alpha_t - 0.5030\alpha_{t-1} - 0.2924\alpha_{t-2} + 0.1317\alpha_{t-3} \quad (1)$$

where $\alpha_t \in \text{NID}(0,0.524760^2)$ is a normal white noise process with zero mean and the variance 0.524760^2.

The wind speed time series model can be used to calculate the simulated time-dependent wind speed SW_t using (2):

$$\text{SW}_t = \mu_t + \sigma_t x y_t \quad (2)$$

where μ_t is the mean observed wind speed at hour t; σ_t is the standard deviation of the observed wind speed at hour t.

The chronological wind speeds generated using (2) can be used directly in a sequential MCS or accumulated and used to create a wind speed probability distribution. Figure 1 shows a comparison of the observed wind speed probability distributions for the original 20 years of data and the simulated wind speed probability distribution obtained using the ARMA (4, 3) model and a large number (8,000) of simulated years. The observed average wind speed is 19.46 km/h, and the simulated value is 19.53 km/h. The observed wind speed probability distribution is not as continuous as the simulated distribution, as it is based on only 20 years of data.

Figure 1 shows that the ARMA (4, 3) model provides a reasonable representation of the actual wind regime. The observation is often made that wind speed can be represented by a Weibull distribution. Simulation results are used to generate the wind speed probability distributions in the system adequacy studies described later in this paper.

3 Modeling WTGs

The power output characteristics of a WTG are quite different from those of a conventional generating unit. The output of a WTG depends strongly on the wind regime as well as on the performance characteristics (power curve) of the generator. Figure 2 shows a typical power curve for a WTG.

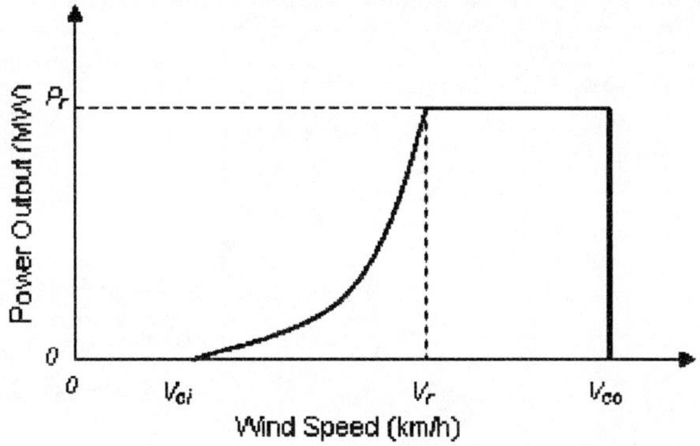

Fig. 2 Wind turbine generating unit power curve

The hourly wind speed data are used to determine the time-dependent power output of the WTG using the operational parameters of the WTG. The parameters commonly used are the cut-in wind speedV_{ci} (at which the WTG starts to generate power), the rated wind speedV_{r} (at which the WTG generates its rated power) and the cut-out wind speedV_{co} (at which the WTG is shut down for safety reasons). Equation (3) can be used to obtain the hourly power output of a WTG from the simulated hourly wind speed:

$$P(\mathrm{SW}_t) = \begin{cases} 0 & 0 \le \mathrm{SW}_t < V_{\mathrm{ci}} \\ (A + B \times \mathrm{SW}_t + C \times \mathrm{SW}_t^2) \times P_{\mathrm{r}} & V_{\mathrm{ci}} \le \mathrm{SW}_t < V_{\mathrm{r}} \\ P_r & V_{\mathrm{r}} \le \mathrm{SW}_t < V_{\mathrm{co}} \\ 0 & \mathrm{SW}_t \ge V_{\mathrm{co}} \end{cases} \tag{3}$$

where P_{r}, V_{ci}, V_{r} and V_{co} are the rated power output, the cut-in wind speed, the rated wind speed and the cut-out wind speed of the WTG, respectively. The constants A, B, and C that depend on V_{ci}, V_{r} and V_{co} are presented in [10]. The WTG units used in the studies in this chapter are considered to have a rated capacity of 2 MW, and cut-in, rated, and cut-out speeds of 14.4, 36 and 80 km/h, respectively.

3.1 The Capacity Outage Probability Table of the WTG

The hourly mean wind speeds and output power for a WTG unit without considering its unavailability or forced outage rate (FOR) are generated using the ARMA time series model and the power curve, respectively. The capacity outage probability table (COPT) of a WTG unit can be created by applying the hourly wind speed to the power curve. The procedure is briefly described by the following steps:

1. Define the output states for a WTG unit as segments of the rated power.
2. Determine the total number of times that the wind speed results in a power output falling within one of the output states.

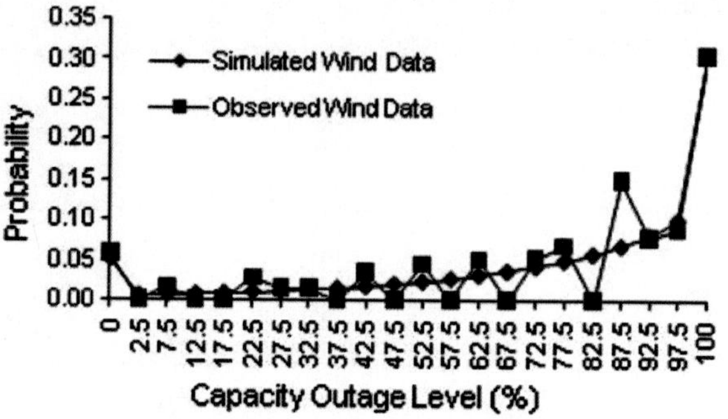

Fig. 3 Capacity outage probability profile for the WTG unit

3. Divide the total number of occurrences for each output state by the total number of data points to estimate the probability of each state.

The WTG COPT can be formed using this approach. Two cases are illustrated in this example. The first case utilizes the actual observed 20 years of Swift Current data. The second case uses the 8,000 simulated years of data. Figure 3 shows the two capacity outage probability distributions. The class interval width is 5% in this figure and the indicated capacity outage level is the midpoint of the class.

Figure 3 shows that the observed data probability profile is discontinuous due to the limited wind data collection and that the simulated wind data provides a reasonable representation for adequacy assessment. The power output characteristics of a WTG are very different from those of conventional generating units. The WTG can be considered as a generating unit with many derated states [17]. Figure 3 shows that the probability of having full WTG output (0% capacity outage) is relatively low for this wind regime. There are many derated states in which the output of a WTG can reside in over the course of its operating history. A basic requirement in practical adequacy assessment is to represent the WTG by an acceptable reduced number of derated states.

3.2 Creating a Multi-State WECS Model Using the Apportioning Method

The apportioning method [7, 18] can be used to create selected multi-state models for a WTG and the WECS. In this approach, the residence times of the actual derated states are apportioned between the completely up, selected derated and completely down states. A detailed analytical procedure that incorporates the WTG FOR is presented and used to build a series of multi-state WECS models in [17]. The probability of a generating unit residing in the full down state in a two-state representation is known as the derated adjusted forced outage rate (DAFOR) [7]. The term DAFOR is used by Canadian electric power utilities. In the United States, the designation for this statistic is the "equivalent forced outage rate" (EFOR). The EFOR or DAFOR is obtained using the apportioning method in

Table 1 Five-state capacity outage probability table for a 20 MW WECS

Capacity outage (MW)	Probability	
	FOR = 0%	FOR = 4%
0	0.07021	0.05908
5	0.05944	0.06335
10	0.11688	0.11475
15	0.24450	0.24408
20	0.50897	0.51875
DAFORW	0.76564	0.77501

Table 2 Five-state WECS models using the Swift Current data

Capacity outage (%)	Probability		
	Annual	Summer	Winter
0	0.07021	0.05385	0.08478
25	0.05944	0.04701	0.06998
50	0.11688	0.09784	0.13187
75	0.24450	0.22871	0.25078
100	0.50897	0.57259	0.46259

which the residence times of the actual derated states are apportioned between the up (normal) and down (outage) states and there are no assigned derated states.

A WECS can contain one or more WTG. A WECS has two basic parts: one is the wind resource and the other is the actual WTG units. If the WECS consists of identical WTG units with zero FOR, the WECS multi-state model is basically the same as that of the single WTG unit. If the FOR of the WTG units is not zero, the WECS derated state COPT is not the same as that of a single WTG unit. An analytical procedure used to create WECS multi-state models including the WTG FOR is described in [17].

As noted earlier, there are many derated states in which the output of a WTG can reside in the course of its operating history. Studies have shown that a five-state COPT can be used to reasonably represent a WTG in a capacity adequacy assessment [17]. This model can also be used to represent a wind farm containing a number of WTG. Table 1 shows the capacity and probability values in a five-state model for a 20 MW WECS containing 10 identical 2 MW WTG. Table 1 presents the five-state model for two cases, one in which the WTG unit FOR is zero and one in which the FOR is 4%. It can be seen from Table 1 that the effect of a 4% WTG FOR is relatively small at this mean wind speed. The WTG FOR can be neglected in many practical situations without creating unreasonable errors in the calculated adequacy indices [17, 19]. This is illustrated further later in this chapter. Table 1 also shows the DAFOR for the two cases. The DAFOR for the wind farm is designated as DAFORW.

Table 2 compares the annual five-state model for a WECS with two seasonal models where winter is from October to March and summer is from April to September. These two models reflect the seasonal variability in the Swift Current wind speed profile.

Table 3 An 11-state WECS model using the Swift Current data

Capacity outage (%)	Probability	
	FOR = 0%	FOR = 4%
0	0.05796	0.03853
10	0.01560	0.02686
20	0.02021	0.02162
30	0.02629	0.02557
40	0.03427	0.03329
50	0.04503	0.04408
60	0.05969	0.05948
70	0.07985	0.07981
80	0.10843	0.10941
90	0.15193	0.15454
100	0.40073	0.40681
DAFORW	0.76564	0.77501

Table 3 shows an 11-state annual WECS model. The five-state model was developed from the 11-state model using the apportioning method. The DAFORW is the same for both models.

4 Adequacy Assessment of Generating Systems Containing Wind Capacity

Considerable work has been done on generating capacity reliability evaluation [1–8]. These studies do not normally include transmission system elements unless dictated by special circumstances or specified constraints such as system interconnection capacities. The most common indices in generating capacity adequacy assessment are the loss of load expectation (LOLE) expressed in h/year or days/year and the loss of energy expectation (LOEE) expressed in MWh/year. References [1–7] describe a wide range of techniques and indices for generating capacity adequacy assessment.

The simplest and most direct method of determining the LOLE and LOEE indices for a generating system is to create a COPT for the system generating units and determine from the table, the probability of having less capacity than load at each hour in the period under study [7]. The hourly loss of load probability values expressed as a LOLE for each 1 h segment are summed to give the LOLE for the period under study. This basic technique is designated as the analytical method and used later in this chapter. Similar results, within the bounds of simulation convergence, etc., can be obtained using sequential or state sampling MCS.

As noted earlier, the most accurate and comprehensive approach to incorporate wind energy in a generating capacity adequacy evaluation is to use sequential MCS and suitable time series wind models. The correlation between the daily load profile and the wind speed at the WECS location is inherently included in the time

Table 4 Generation system adequacy indices for the RBTS

Method	LOLE (h/year)	LOEE (MWh/year)
Analytical	1.0950	9.9032
Sequential MCS	1.0941	9.8531

Table 5 Generation system adequacy indices for the RBTSW

Method	LOLE (h/year)	LOEE (MWh/year)
Analytical	0.7797	6.9561
Sequential MCS	0.6877	5.9278

sequential analysis. Chronology is not retained in most analytical techniques or in a state sampling Monte Carlo approach and therefore load and wind correlation is not inherently incorporated in analyses conducted using these methods. Studies can be done using period analysis, i.e. seasonal periods or monthly periods, and the annual risk obtained by summing the period risks. This approach can be used to include a closer relationship between the hourly wind speed and load profiles. The period approach is used in conventional generating capacity evaluation to include scheduled maintenance considerations [7] and is illustrated in the following by application to a small test system known as the RBTS [20]. The RBTS has 240 MW of generating capacity in 11 generating units represented by two-state models, and a peak load of 185 MW. The annual LOLE and the LOEE obtained using the analytical and sequential MCS techniques are shown in Table 4.

Table 4 shows that the two techniques give virtually identical results when applied to the RBTS. A 20 MW WECS with the Swift Current wind speed regime was added to the RBTS to create the RBTSW. The risk indices for the RBTS obtained using the two techniques are shown in Table 5. The five-state model in Table 1, in which the WTG unit with zero FOR, was used in the analytical analysis.

Table 5 shows that the analytical method using the annual five-state WECS model produces higher risk estimates than the sequential MCS approach in this case. The differences between the two sets of indices will depend on the level of correlation between the chronological wind speed and hourly load profiles. The results and the resulting differences between the two techniques could be quite different for different levels of correlation. The risk indices using different WECS models in the analytical approach are shown in Table 6.

Table 6 shows that the risk indices increase slightly when the WTG units are assigned a FOR of 4%. The other model studies in Table 6 have WTG unit FOR of 0%. Table 6 indicates that the risk indices decrease slightly as the number of states in the WECS model increase. The risk indices also decrease slightly by using two seasonal periods rather than a single annual period. Table 6 also shows that the bulk of the system risk occurs during the winter period. The highest mean hourly wind speeds for the Swift Current site occur in the winter period with the highest values in the month of December. The analysis was extended to consider monthly contributions to the annual risk indices. The ARMA, five-state and 11-state WECS models were created for each month and the resulting monthly system COPT was

Table 6 Generation adequacy indices for the RBTSW using the analytical method with different WECS models

WECS model	LOLE (h/year)	LOEE (MWh/year)
11-state	0.7700	6.8793
11-state (WTG FOR = 4%)	0.7784	6.9554
Five-state	0.7797	6.9561
Five-state (WTG FOR = 4%)	0.7889	7.0398
Five-state winter (4,368 h)	0.5732	5.1364
Five-state summer (4,392 h)	0.1900	1.6511
Summation of winter and summer	0.7632	6.7875
Five-state monthly summation	0.7347	6.4990
11-state monthly summation	0.7240	6.4190

applied to the hourly loads in the related months. The largest monthly risk occurs in the month of December followed by the months of November and June. The RBTS load profile [20] has two peak periods, one in December and one in June. The annual risk indices obtained by summing the monthly risks are shown in Table 6.

The sequential MCS method is the most comprehensive approach to assess generating system adequacy and can incorporate a wide range of operational conditions in the assessment. It also requires more computer solution time than a direct analytical approach or a state sampling MCS method. This is not a serious problem in basic generating capacity assessment but can create difficulties in composite system adequacy studies of large BES. Multi-state WECS models can be readily applied in BES studies using state sampling MCS and this technique presents a practical approach to incorporating WECS in large-scale practical BES adequacy assessment. Both sequential and state sampling methods are applied to composite system adequacy assessment in the following sections of this chapter.

5 Adequacy Assessment of Composite Systems Containing Wind Capacity

A complete adequacy assessment of a composite generation and transmission system includes both load point and system indices [7, 8]. These indices complement each other in an overall appraisal of the system adequacy. The load point indices provide information on the expected adequacy of service at a load point and to the customers served from that load point. The system indices provide valuable information on overall system performance and are used in a wide range of system planning and management decisions. The individual load point indices are highly dependent on the system load curtailment philosophy and each load bus has a different priority in an actual system.

Composite system adequacy assessment is illustrated using the RBTS introduced earlier. The RBTS is a relatively small system with some designed in

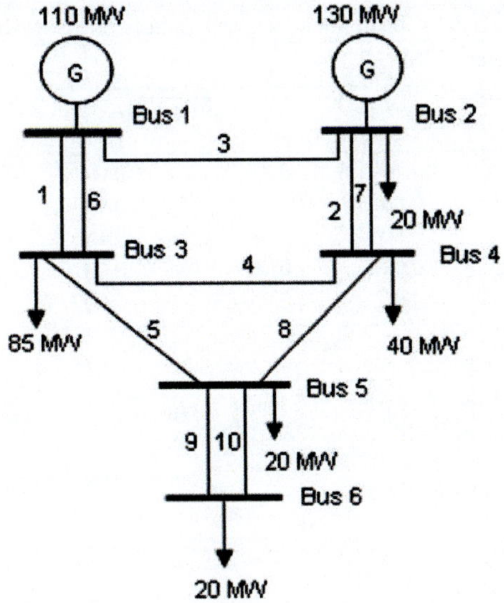

Fig. 4 Single line diagram of the MRBTS

Table 7 System adequacy indices for the MRBTS and MRBTSW obtained using RapHL-II

System	EDLC (h/year)	EENS (MWh/year)	EFLC (occ/year)
MRBTS	1.1267	13.3464	0.2262
MRBTSW	0.7001	7.9974	0.1883

weaknesses, one of which is the radial supply to Bus 6. The RBTS was, therefore, modified by adding a transmission line designated as Line 10 between Bus 5 and Bus 6. Line 10 has the same parameters as Line 9. The modified system is designated as the MRBTS and used as the base case. The single line diagram of the MRBTS is shown in Fig. 4.

The load points and system adequacy at a system peak load of 185 MW were assessed using a sequential MCS program designated as RapHL-II [21]. This program uses the Fast-Decoupled AC load flow technique to calculate load flows and line loadings and to check the system operating constraints. An optimal power flow (OPF) approach is used to conduct corrective actions to alleviate operating constraints. The simulation sample size is 10,000 years in this study. The load shedding philosophy employed in this study is designated as the Pass-I policy in which loads are curtailed at the delivery points that are closest (or one line away from) the element(s) on outage.

The overall system adequacy is shown in Table 7 in terms of the expected duration of load curtailment (EDLC) in h/year, the expected energy not supplied (EENS) in MWh/year and the expected frequency of load curtailment (EFLC) in occ/year. The EDLC and EENS indices used in BES assessment are similar in concept and units to the LOLE and LOEE indices, respectively, used in generating capacity adequacy assessment. Table 7 shows these indices for the MRBTS and

Table 8 Load point adequacy indices for the MRBTS and MRBTSW obtained using RapHL-II

Bus no.	MRBTS			MRBTSW		
	EDLC (h/year)	EENS (MWh/year)	EFLC (occ/year)	EDLC (h/year)	EENS (MWh/year)	EFLC (occ/year)
2	0.1692	0.4762	0.0411	0.0932	0.2568	0.0301
3	0.8615	7.3563	0.1777	0.5199	4.2447	0.1453
4	0.8057	5.3473	0.1531	0.5199	3.3721	0.1317
5	0.0307	0.1098	0.0079	0.0237	0.0786	0.0070
6	0.0119	0.0588	0.0027	0.0100	0.0453	0.0024

Table 9 System and load point EENS for the MRBTS and the MRBTSW obtained using MECORE

Bus no.	Priority order	EENS (MWh/year)	
		MRBTS	MRBTSW
2	1	0.000	0.000
3	5	12.566	8.644
4	2	0.029	0.009
5	3	0.293	0.200
6	4	0.674	0.545
System	–	13.562	9.399

the MRBTSW, in which a 20 MW WECS with the Swift Current wind regime is added at Bus 4.

The load point indices for the two systems are shown in Table 8.

The RapHL-II program has been used in a wide range of studies [21] on composite system reliability evaluation, with particular emphasis on well-being analysis and value-based assessment that includes customer outage cost evaluation. The sequential MCS approach permits the individual sector customer load profiles at each bus to be modeled separately and combined to create a time sequential bus representation. The program includes several load curtailment philosophies and can be used to calculate the adequacy index probability distributions for each load point and for the overall system [21].

The adequacy indices in Tables 7 and 8 can also be calculated using contingency enumeration or state sampling MCS [7, 8]. The most common single index in BES adequacy assessment is the EENS. The EENS aggregates the frequency and duration of outages and the magnitude of outage events into a single physical index. The EENS can be combined with an appropriate interrupted energy assessment rate (IEAR) to estimate the customer outage costs at a load point or for the overall system [7].

Table 9 shows the EENS for the MRBTS and the MRBTSW with a 20 MW WECS with the Swift Current wind regime connected at Bus 4, obtained using the state sampling MCS approach. The 20 MW WECS is represented in the MRBTSW study by the five-state COPT shown in Table 2 with zero WTG FOR. The priority order philosophy in this study is based on ranking the composite generation and transmission system delivery points using a reliability worth index,

Table 10 System EENS in MWh/year for the MRBTSW with variation in WECS capacity and location

WECS location	WECS installed capacity				
	0 MW	10 MW	20 MW	30 MW	40 MW
BUS 1	13.562	10.756	9.449	8.681	8.187
BUS 2	13.562	10.760	9.458	8.692	8.197
BUS 3	13.562	10.720	9.397	8.622	8.123
BUS 4	13.562	10.722	9.399	8.623	8.124
BUS 5	13.562	10.710	9.376	8.590	8.080
BUS 6	13.562	10.567	9.151	8.329	7.806

such as the IEAR expressed in $/kWh [7]. The composite generation and transmission system delivery point with the highest IEAR has the highest priority, and the delivery point with the lowest IEAR has the lowest priority. When the composite generation and transmission system encounters a severe contingency that requires load curtailments, the delivery point that has the lowest priority is initially curtailed. This policy minimizes customer interruption costs due to load curtailments. A commercial software designated as MECORE [22] which utilizes the state sampling MCS method was used in this study. The MECORE software is designed to conduct adequacy evaluation in composite generation and transmission systems and to provide individual load point and system adequacy indices using the economic priority order for load curtailment.

Table 9 indicates that Bus 3 is the least reliable load point in the MRBTS, as Bus 3 has the lowest priority. It can also be seen from Table 9 that the system EENS is dominated by the performance at Bus 3. The effect of the load curtailment philosophy on the load point indices can clearly be seen by comparing Table 9 with Table 8. A comparison of Table 7 with Table 9 shows that the system EENS for the MRBTSW is again higher for the five-state model application in the state sampling MCS study than for the sequential MCS analysis.

The system EENS study shown in Table 9 is extended in Table 10 by assuming that the 20 MW WECS is connected at different locations in the MRBTSW. Figure 4 shows that Buses 1 and 2 are in the northern part of the system, which has relatively little load. The system load center is in the south where there is no generation. The system transmission is adequate at the 185 MW load level and there is relatively little difference between the system EENS values due to the location of the WECS connection point. The system EENS generally decreases as the WECS connection point moves further away from the generation in the northern portion of the system. The system EENS decreases with the addition of WTG capacity. The largest change occurs with the initial injection of 10 MW and saturates as the WECS capacity increases.

The WECS capacity additions shown in Table 10 are assumed to be located in a single wind farm where each WTG is subjected to the same chronological wind speed profile. The incremental reliability benefits of added WTG decrease as the capacity of the wind farm increases.

5.1 Adequacy Assessment Considering Wind Speed Correlation

The studies shown in the previous section clearly illustrate the contribution that wind power can make to the adequacy of an electric power system. The reliability benefits of WECS are highly dependent on many factors including the wind regime at the site location, the wind penetration in the system and the wind speed correlation between multiple wind farms. Reference [23] illustrates the differences in generating capacity credit that can be attributed to added wind capacity when this capacity is added at locations with dependent and independent wind speeds. In actual practice, wind farms are neither completely dependent nor independent but are correlated to some degree if the distances between sites are not very large. The degree of correlation is not only dependent on the distance between the wind sites but also on the geographic dispersion and the uniqueness of the individual wind regimes. The most obvious factor, however, is the distance between the wind sites.

The degree of wind speed correlation between two wind sites can be calculated using cross-correlation. The cross-correlation index (R_{xy}) is a measure of how well two time series follow each other. The value of R_{xy} is near the maximum value of 1.0 if the up and down movements of the two time series occur in the same direction (positively correlated). The value is close to zero if the two time series are basically uncorrelated, i.e. the two time series do not follow each other. The cross-correlation equation is shown in (4):

$$R_{xy} = \frac{(1/n)\sum_{i=1}^{n}(x_i - \mu_x)(y_i - \mu_y)}{\sigma_x \sigma_y} \tag{4}$$

where R_{xy} is a cross-correlation coefficient, x_i and y_i are elements of the 1st and 2nd time series, respectively, μ_x and μ_y are the mean values of the 1st and 2nd time series, σ_x and σ_y are the standard deviations of the 1st and 2nd time series, n is the number of points in the time series.

The ARMA time series model has two parts, one part is the autoregressive (AR) model involving lagged terms in the time series itself and the other part is the moving average (MA) model involving lagged terms in the noise or residuals. It is possible to adjust the wind speed correlation level between two or more different wind locations by selecting the random number seeds (initial numbers) for the random number generator process used in the MA model. Reference [24] introduces the concept and employs a trial and error process to select appropriate random number seeds. This is a relatively straightforward method, but can require considerable time and effort and is not very flexible. Reference [25] extends this application by describing a Generic Algorithm used to select the optimum random number seeds in the ARMA model to adjust the degree of wind speed correlation for two wind sites. Reference [25] illustrates the effects of wind speed correlation on the basic reliability indices used in a generating capacity adequacy assessment. The technique is also applied in composite system adequacy assessment in [26] using sequential MCS. The correlation between wind sites can also be included in generating capacity and composite system adequacy assessment using state sampling MCS. The required approach is quite different from that used in a sequential MCS analysis.

In the state sampling technique, the states of all components are sampled and a non-chronological system state is obtained. The basic state sampling procedure is conducted assuming that the behavior of each component can be categorized by a uniform distribution under {0, 1} and component outages are independent events. Detailed descriptions of a state sampling simulation procedure are provided in [7, 8]. Conventional unit and independent WECS outages are assumed to be independent events in the basic state sampling simulation procedure. This assumption, however, is not applicable to partially dependent WECS. It is, therefore, necessary to generate correlated random numbers, which have a uniform distribution and specified correlations, in the simulation process.

Random numbers distributed uniformly under {0, 1} are divided into two clusters in this approach. Random numbers in the first cluster represent conventional units or independent WECS. Random numbers X1, X2 between 0 and 1 in the second cluster represent correlated WECS. If the second variable vectors X2 are generated from the first independent random number set with probability P and generated from the second independent random number set with probability $(1 - P)$, the cross-correlation coefficient R_{xy} between X1 and X2 in the second cluster is equal to the probability P. This approach was used in the state sampling simulation method to generate correlated random numbers to represent the correlated WECS. A detailed development of this approach is given in [25].

The ARMA model for the Swift Current site is given in (1). A similar model for a wind site in Regina, Saskatchewan is given in (5):

$$y_t = 0.9336y_{t-1} + 0.4506y_{t-2} - 0.5545y_{t-3} + 0.1110y_{t-4} + \alpha_t - 0.2033\alpha_{t-1} - 0.4684\alpha_{t-2} + 0.2301\alpha_{t-3} \quad (5)$$

where $\alpha_t \in \mathrm{NID}(0,0.409423^2)$.

The basic wind speed data for the Swift Current and Regina sites are shown in Table 11.

The independent WECS five-state models are shown together in Table 12.

Table 11 shows that the wind regimes at the two sites have relatively similar hourly mean and standard deviation parameters. The two WECS models in Table 12 are also very similar.

Three cases described as follows are used to illustrate the effect on the adequacy of the MRBTS of adding two wind farms with different wind speed correlation levels. A single transmission line with an unavailability of 0.00114 and an average repair time of 10 h is used to connect each WECS to a MRBTS bus.

Table 11 Wind speed data for the two sites

Sites	Regina	Swift Current
Mean wind speed, μ (km/h)	19.52	19.46
Standard deviation, σ (km/h)	10.99	9.70

Table 12 The independent WECS five-state models

Capacity outage (%)	Probability	
	Regina site	Swift Current site
0	0.07585	0.07021
25	0.06287	0.05944
50	0.11967	0.11688
75	0.23822	0.24450
100	0.50340	0.50897

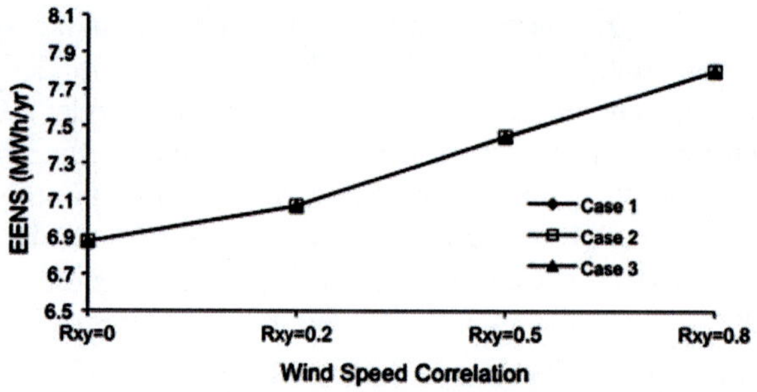

Fig. 5 The system EENS for the three cases

- Case 1: 20 MW WECS is added at Buses 1 and 2.
- Case 2: 20 MW WECS is added at Buses 3 and 5.
- Case 3: 20 MW WECS is added at Buses 3 and 6.

The three cases illustrate the system and load point reliability effects when correlated wind farms are added at various buses. Figure 5 shows the differences in the system EENS when the two wind farms are connected at the selected locations. The system EENS values in Case 2 are basically the same as those in Case 3. The differences between Case 1 and Case 2 when the wind farms are fully independent are smaller than those obtained considering wind speed correlation. Figure 5 also shows that connecting the two wind farms in the southern portion of the MRBTS (Case 2 and Case 3) results in higher reliability improvements than those obtained by adding the wind farms in the northern portion (Case 1). The reason is that load demands are mainly concentrated in the southern portion of the MRBTS.

Case 2 was extended to consider the impact on the MRBTS EENS as a function of wind speed correlation for the two wind farms at different wind penetration levels. The results shown in Fig. 6 are based on the two wind penetration levels of 14.3% (40 MW) and 29.4% (100 MW). The total WECS capacity is divided equally between the two sites. Figure 6 shows that the system reliability is improved by adding 40 and 100 MW of WECS to the MRBTS. The relative difference in the system EENS between the WECS considering high and low wind speed correlation levels increases with increase in the total installed wind capacity.

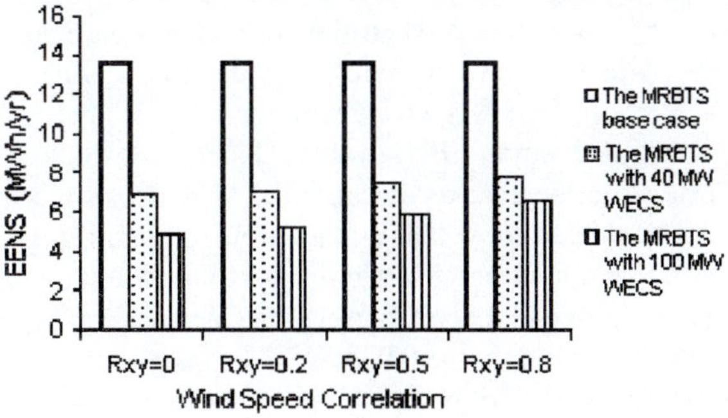

Fig. 6 The system EENS for two wind penetration levels as a function of wind speed correlation level

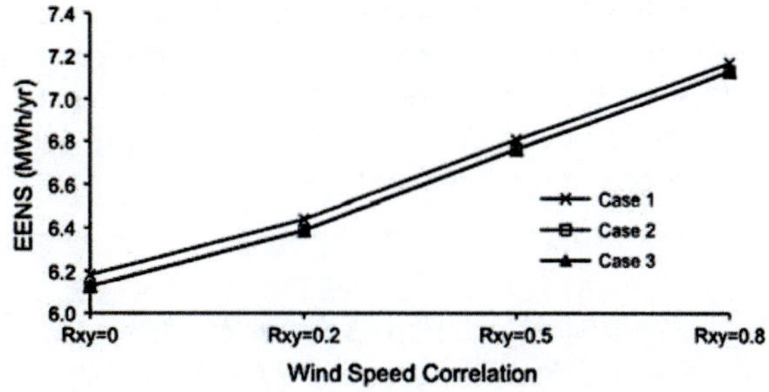

Fig. 7 The EENS at Bus 3 for the three cases

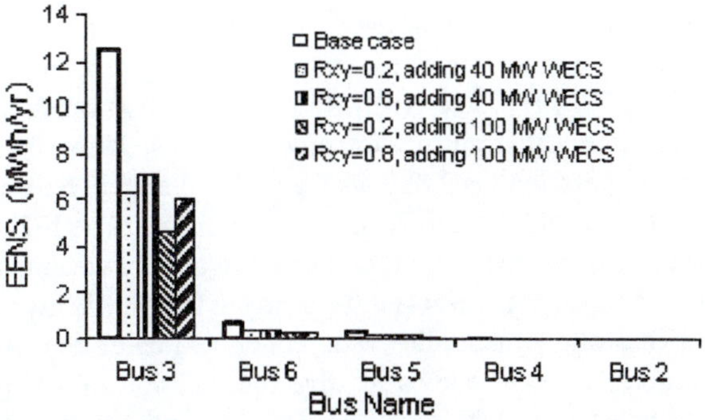

Fig. 8 The load point EENS at different wind penetration levels and degree of wind speed correlation

As noted earlier and shown in Table 9, the system EENS is dominated by the reliability performance at Bus 3. Figure 7 shows the EENS at Bus 3 for the three cases with variation in the wind speed correlation between the two wind sites. It can be seen in Fig. 7 that the EENS at Bus 3 increases as the degree of correlation between the site wind speeds increases. The addition of WECS at the different locations has a relatively small effect on the EENS at Bus 3.

Case 2 was used to illustrate the effect on the load bus EENS of two wind farms with high and low wind speed correlation. The buses are arranged in ascending priority order in Fig. 8. This figure shows how the reliability of each load point changes as wind power is added to the MRBTS. The percentage change in the base case EENS at Bus 3 when the correlation coefficient changes from 0.2 to 0.8 is 5.9% when 40 MW of WECS is added to the MRBTS. The percentage change in the EENS at Bus 3 is 11.1% with the addition of 100 MW of WECS. The changes in the load point EENS values due to considering high and low wind speed correlation increase as the installed WECS capacity increases.

The studies show that it is possible to quantify the general observation that the system and load point reliability benefits of adding wind capacity decrease as the degree of wind speed correlation increases. The resulting benefits are more obvious as the total installed WECS capacity increases. The effects of connecting the correlated WECS at different locations in a composite generation and transmission system are very dependent on the strength of the transmission network. The studies show that there are significant differences in the system reliability indices when the wind generating capacity is located at dependent and independent wind sites. It is, therefore, important to consider the effect of wind speed correlation in the adequacy assessment of composite generation and transmission systems incorporating large-scale WECS.

5.2 Composite System Adequacy Assessment Using the IEEE Reliability Test System (IEEE-RTS)

The wind-integrated BES adequacy studies in the previous section were conducted on a small educational test system designated as the MRBTS. This section illustrates the application of the state sampling approach to adequacy analysis using the IEEE-RTS [27]. The IEEE-RTS is a 24-bus system with 10 generator buses, 17 load buses, 38 transmission lines, 5 transformers and 32 generating units. The system peak load is 2,850 MW and the total generation is 3,405 MW. A single line diagram of the IEEE-RTS is shown in Fig. 9. The IEEE-RTS is considered to have a strong transmission network and a relatively weak generation system. The LOLE and the LOEE are 9.441 h/year and 1,181.7 MWh/year, respectively.

The following studies were conducted using the MECORE software described earlier. The first studies involve the addition of a single 600 MW WECS to the system at two different locations. The wind penetration level with 600 MW of wind power is approximately 15%. The WECS was added at Bus 1 in the southern part of the system and at Bus 19 in the north. In each case, the WECS was connected through a single transmission line with an unavailability and average repair time of 0.00058 and 10 h, respectively. The assumed carrying capacity of the line is the WECS installed capacity.

The system EENS for the IEEE-RTS is 1,593.1 MWh/year. Table 13 shows the system EENS and EDLC with the two WECS additions. There is relatively little difference in the two sets of adequacy indices due to the strong transmission network.

Table 14 presents the system and load point EENS indices for the IEEE-RTS and a wind-assisted system designated as the IEEE-RTSW, in which the 600 MW WECS is connected to Bus 19. The system and load point indices decrease with

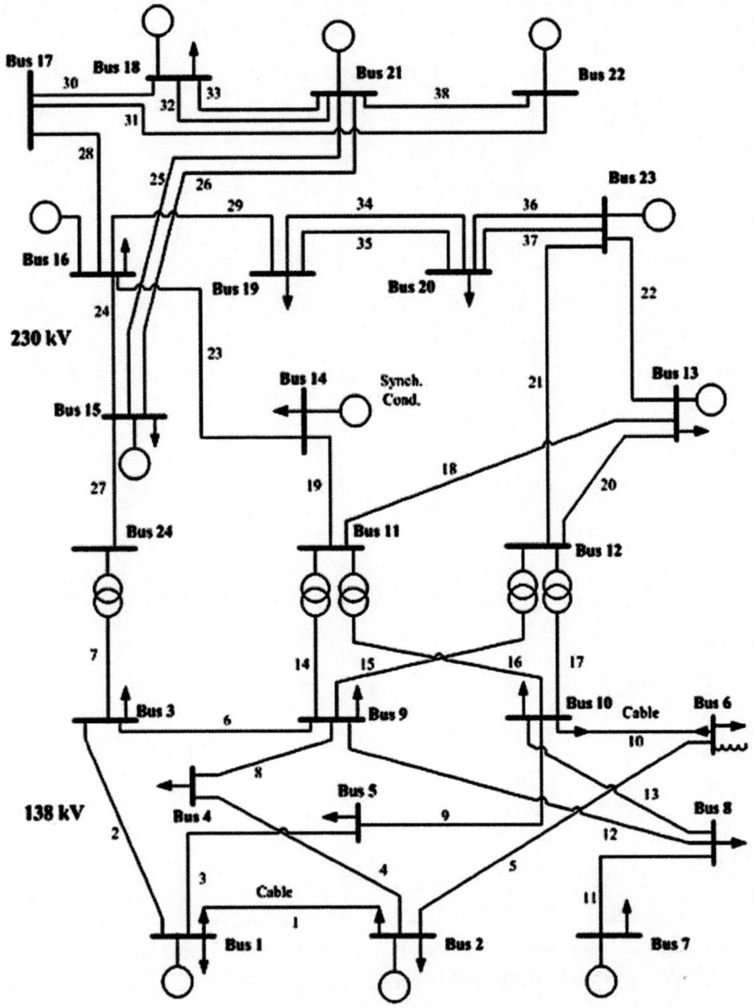

Fig. 9 Single line diagram of the IEEE-RTS

Table 13 The IEEE-RTS EENS and EDLC for a peak load of 2,850 MW with a 600 MW WECS added at two different locations

WECS location	EENS (MWh/year)	EDLC (h/year)
Bus 1	942.303	7.595
Bus 19	939.004	7.535

the WECS addition. The economic priority order is used as the load curtailment policy in this study and the effects can clearly be seen in the load point indices in Table 14. The IEEE-RTS is generation weak and the added wind power has a significant impact on the system EENS, and on the EENS at the low priority buses.

Table 14 The system and load point EENS for the IEEE-RTS and the IEEE-RTSW

Bus name	EENS (MWh/year)	
	IEEE-RTS	IEEE-RTSW-19
Bus 2	0.399	0.204
Bus 3	0.345	0.169
Bus 4	0.000	0.000
Bus 5	0.000	0.000
Bus 6	0.394	0.392
Bus 7	0.008	0.004
Bus 8	0.020	0.008
Bus 9	407.226	240.042
Bus 10	1.857	0.994
Bus 13	0.146	0.060
Bus 14	68.057	38.827
Bus 15	310.220	180.886
Bus 16	18.572	10.485
Bus 18	13.670	7.458
Bus 19	757.805	451.468
Bus 20	14.454	8.051
System	1,593.130	939.004

Table 15 The system EENS for the RTS with the addition of two WECS at different locations

R_{xy}	IEEE-RTS	Case 1	Case 2
0.0	1,593.130	751.189	751.478
0.2		784.810	785.096
0.5		840.307	840.567
0.8		893.648	894.092

The effect of multiple WECS is illustrated by connecting 300 MW WECS with various degrees of wind speed correlation to two buses in the southern portion of the system. The unavailability and average repair time of the WECS connection lines are 0.00058 and 10 h, respectively. The following cases were considered.

- Case 1: 300 MW WECS is added at Buses 1 and 3.
- Case 2: 300 MW WECS is added at Buses 1 and 4.

The system EENS values are shown in Table 15 and Fig. 10, respectively.

As noted earlier, the IEEE-RTS studies in this section were conducted using the state sampling MCS procedure in the MECORE software. Similar studies have been conducted using the sequential MCS approach in RapHL-II. These studies are described in detail in [28].

Table 15 and Fig. 10 clearly show that wind speed correlation is an important factor in wind-integrated power system adequacy assessment. In systems with strong transmission networks, the degree of wind speed correlation between two wind sites can be more important than their connection points in the BES. The effects of wind speed correlation can be masked in systems with weak

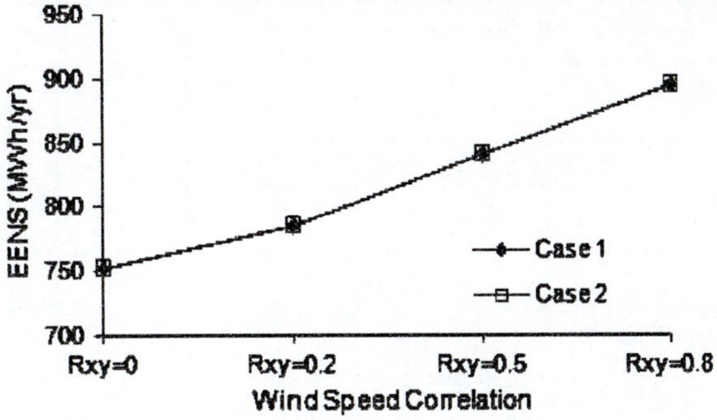

Fig. 10 The system EENS for the two cases

transmission facilities. This is illustrated in [29] where the original IEEE-RTS is modified to create systems with adequate generation and relatively weak transmission networks.

5.3 Well-Being Analysis in Wind-Integrated Composite Systems

References [1–8] clearly show that probability methods have been used in power system reliability assessment for many years. This is particularly true in the area of generation planning. There has been considerably less actual utility application in transmission planning, where most utilities use a deterministic approach. The deterministic criterion commonly applied is known as the $N-1$ approach and requires the system to remain secure under the loss of any one generating unit or transmission line. There is a growing interest in combining deterministic criteria with probabilistic assessment in an integrated approach to composite system reliability evaluation. This approach has the potential to create a bridge between the deterministic and probabilistic methods and provide an effective adequacy and security assessment framework.

A concept to address system security in the form of system operating states is formulated in [30, 31] and quantified in [32]. The operating state framework is simplified in [33] and the resulting process designated as system well-being analysis. The well-being indices are evaluated in [34] in a generating system using MCS. Reference [35] extends the well-being framework for composite generation and transmission systems using a non-sequential Monte Carlo model. The well-being analysis conducted in [36, 37] uses sequential MCS. The simplified well-being operating state framework [33–35] is shown in Fig. 11.

The system states shown in Fig. 11 are categorized as healthy, marginal and at risk. In the healthy state, there is sufficient generation and transmission capacity to serve the total system demand and to meet the N-1 criterion. The system is operating without violation in the marginal state, but there is not enough margin to satisfy the pre-defined deterministic criterion. In the at risk state, system operating constraints are violated and load may be curtailed.

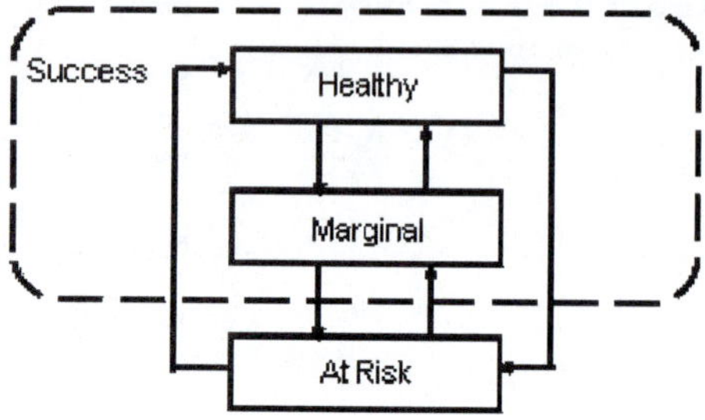

Fig. 11 System well-being analysis model

Table 16 System indices for the three study systems

Study system	EDLC (h/year)	EENS (MWh/year)	EFLC (occ/year)
MRBTS	3.8761	46.9453	0.8415
MRBTSW1	3.8730	46.8670	1.0404
MRBTSW2	3.8872	49.7180	1.3184

The well-being indices include the probability, frequency and average duration of each system operating state.

The sequential MCS technique is used in the following studies to conduct composite system well-being analysis. The N-1 criterion is used as the deterministic criterion in the well-being framework in these studies. The basic procedure for including well-being considerations in the simulation process is described in [36, 37]. Three composite systems designated as the MRBTS, MRBTSW1 and MRBTSW2 are used in this study. The MRBTS is shown in Fig. 4. The load model in the well-being studies, however, is different from that used earlier in this chapter. In the well-being analyses, the individual bus load models are based on customer sector representations [16, 37].

The MRBTSW1 is the MRBTS with a 5 MW generating unit with FOR = 1% removed and replaced by 20 MW of wind power at Bus 4. The Regina wind regime is used in this study. The addition of 20 MW of wind power with this wind regime is required to maintain the EDLC criterion risk when this 5 MW unit is removed. The MRBTSW2 is the MRBTS with a 10 MW unit with FOR = 2% removed and replaced by 65 MW of wind power to meet the EDLC criterion risk.

The basic adequacy indices for the three systems are shown in Table 16. The three systems are equivalent in the sense that they have basically the same level of adequacy at a peak load of 179.3 MW in terms of the EDLC. It is important to note that they are not all equivalent in terms of the EENS and the EFLC.

The well-being indices for the MRBTS with different wind injections are shown in Tables 17 and 18. Table 17 shows the system probability of each operating state expressed in h/year. The risk value in Table 17 with no wind addition is the EDLC value for the MRBTS given in Table 16. The time spent in the healthy state increases and the time spent in the marginal state decreases as the system reliability improves with the added wind power.

Table 17 System probability of each operating state expressed in h/year for the MRBTS with different wind power additions

System state	Wind capacity addition				
	0 MW	10 MW	20 MW	30 MW	40 MW
Healthy	8,558.9100	8,564.3630	8,583.5810	8,599.1310	8,610.1660
Marginal	173.2133	168.6427	149.9660	134.7198	123.8975
Risk	3.8761	2.9942	2.4530	2.1497	1.9360

Table 18 System frequency in occ/year of each operating state in occ/year for the MRBTS with different wind additions

System state	Wind capacity addition				
	0 MW	10 MW	20 MW	30 MW	40 MW
Healthy	33.3762	32.7483	32.8255	31.6508	30.7455
Marginal	34.1422	33.4378	33.4610	32.2463	31.3085
Risk	0.8415	0.7433	0.6820	0.6440	0.6100

Table 19 System probability of each operating state in h/year for the different systems

System state	Systems		
	MRBTS	MRBTSW1	MRBTSW2
Healthy	8,558.9100	8,528.7570	8,551.4450
Marginal	173.2133	203.3697	180.6679
Risk	3.8761	3.8730	3.8872

Table 18 shows the system frequency of each state with the different wind power additions. The risk state frequency value in this table is the same as the EFLC for the MRBTS given in Table 16. The marginal state frequency is slightly larger than that of the healthy state. The frequencies of the three states tend to decrease with increase in wind power addition. The frequencies of the healthy and marginal states remain almost unchanged when the wind injections increase from 10 to 20 MW. This is due to the counteracting effects on system frequency of the added wind power generating capacity and the intermittent performance of wind power.

The basic effects of wind power on the well-being indices tend to be masked in the above analysis, as the system capacity increases with the addition of wind power and the system becomes more reliable. This is examined in the following studies using the equivalent wind capacity systems designated as MRBTSW1 and MRBTSW2.

The well-being indices for the MRBTS, MRBTSW1 and MRBTSW2 are shown in Table 19. The healthy state probability indicates the amount of time that a system spends in the state in which the (N-1) criterion is satisfied. The marginal state indicates the time in which a system resides in the state in which the criterion is not satisfied, but there is no actual load curtailment. The healthy state probabilities for the MRBTSW1 and MRBTSW2 are smaller than that of the MRBTS,

Table 20 System frequency of each operating state in occ/year for the different systems

System state	Systems		
	MRBTS	MRBTSW1	MRBTSW2
Healthy	33.3762	42.8794	47.5189
Marginal	34.1422	43.8585	48.7488
Risk	0.8415	1.0404	1.3184

while the marginal state probabilities of the MRBTSW1 and MRBTSW2 are larger than that of the MRBTS. This table shows that while the risk state probabilities (EDLC) are basically the same, the healthy and marginal state probabilities are different for the three systems.

The system frequencies of the operating states for the three systems are shown in Table 20. The frequencies of the three states are larger for the MRBTSW systems than for the MRBTS. This indicates that the replacement of conventional generating units with WTGs causes more state transitions due to the intermittent performance of wind power. The MRBTSW2 has a larger proportion of wind power than the MRBTSW1 and therefore the intermittent performance of wind power has a larger effect on the operating state frequencies of the three states for the MRBTSW2 as shown in Table 20.

The system reliability is improved by the addition of any suitable form of generating capacity, including wind power. The well-being indices are also affected by the addition of wind power. The healthy state probability increases with increase in wind power in the system, while the marginal and risk state probabilities decrease. In general, the frequency of each operating state decreases slightly with increase in the wind capacity. However, at some wind capacity levels, the healthy and marginal state frequencies actually increase with increase in the wind capacity. The addition of generating capacity in the form of wind power and the intermittent behavior of wind power has opposite effects on the system state frequencies. The added wind capacity tends to reduce the state frequency and can mask the effects of wind power on the state operating frequencies due to its intermittent behavior.

When a conventional unit is replaced with an equivalent amount of wind power to maintain the EDLC or the EENS, the EFLC in the wind power added system will be larger than that in the original system due to the intermittent behavior of wind power. When the system peak load changes, the EDLC and EENS equivalencies no longer apply but the difference may be acceptable for small load changes.

Wind power not only affects the system indices differently from conventional generating units but also affects the well-being indices differently [37]. Even though the EDLC for the equivalent system with wind is the same as the original system, the healthy state probability is smaller and the marginal state probability is larger for the wind-assisted system. This indicates that the equivalent system is more likely to transfer to the risk state than the original system. The state frequencies increase considerably for the equivalent system, which indicates that there are more transitions between states. The operating state frequencies increase as more conventional generating capacity is replaced.

6 Summary and Conclusions

This chapter presents some of the basic concepts, solutions and difficulties associated with incorporating wind power in the adequacy assessment of composite generation and transmission systems. Wind generation behaves quite differently than conventional electric power generation facilities due to its variable, intermittent and diffuse nature, and therefore requires modeling techniques that recognize and incorporate these features. As noted several times in this chapter, time sequential MCS is the most accurate and comprehensive approach to evaluating the adequacy of a generating system or a composite generation and transmission system. It also requires more computer solution time when analyzing large practical systems and there has been some excellent work done on this issue. The application of time sequential MCS to adequacy evaluation in wind-assisted composite systems is illustrated in this chapter using the two test systems.

The chapter also describes a non-sequential (state sampling) MCS approach to adequacy assessment and the development of discrete-step wind power capacity models that can be used in this approach. The WECS model is represented as a large derated state generating unit with a high DAFOR. This model cannot retain the chronological relationship between the wind speed and load profile that is embedded in the time sequential approach. The chapter illustrates this using a basic generating system and a series of sensitivity studies. The numerical results presented are system and data specific and will, and should, vary with changes in these parameters. The studies do, however, illustrate the effect on the predicted system adequacy indices of varying the number of states in the WECS model, incorporating the WTG FOR in the calculation and conducting seasonal and monthly period analyses.

Composite system adequacy evaluation indicates the ability of the overall BES to serve the major load points and therefore involves detailed modeling of the transmission network. The effects of wind power site location on the system and load point adequacy indices are illustrated using the two test systems. The studies clearly indicate that a major factor in the analysis of a wind-integrated power system is the degree of wind speed correlation between the various WECS. The techniques applied in this chapter to introduce wind site correlation in both sequential and state sampling approaches to composite system adequacy evaluation are illustrated by application to the test systems.

The concept of well-being analysis was developed to create a joint deterministic–probabilistic approach to the assessment of system adequacy and steady-state security. The rapid increase in the development and use of wind power dictates that this generation asset should be integrated into the planning and operation of power systems rather than treated simply as an environmental benefit. The studies shown in the chapter illustrate that wind power due to its variable nature can create a higher likelihood of violating the basic planning and operating criteria and that this should be recognized in setting planning and operating standards. The material in this chapter provides a brief look at some of the work that has been done at the

University of Saskatchewan on a challenging, exciting and complex area of electric power system activity.

The reader is encouraged to look at the references attached to this chapter for further details and to read the many publications on this subject in the technical literature.

Acknowledgments The authors would like to acknowledge the research grants provided by the National Science and Engineering Research Council in support of the work reported in this chapter.

References

1. Billinton R (1972) Bibliography on the application of probability methods in power system reliability evaluation. IEEE Trans PAS 9(2):649–660
2. Allan RN, Billinton R, Lee SH (1984) Bibliography on the application of probability methods in power system reliability evaluation. IEEE Trans PAS 103(2):275–282
3. Allan RN, Billinton R, Shahidehpour SM, Singh C (1988) Bibliography on the application of probability methods in power system reliability evaluation. IEEE Trans Power Syst 3(4):1555–1564
4. Allan RN, Billinton R, Breipohl AM, Grigg CH (1994) Bibliography on the application of probability methods in power system reliability evaluation, 1987–1991. IEEE Trans Power Syst 9(1):41–49
5. Allan RN, Billinton R, Breipohl AM, Grigg CH (1999) Bibliography on the application of probability methods in power system reliability evaluation: 1992–1996. IEEE Trans Power Syst 14(1):51–57
6. Billinton R, Fotuhi-Firuzabad M, Bertling L (2001) Bibliography on the application of probability methods in power system reliability evaluation 1996–1999. IEEE Trans Power Syst 16(4):595–602
7. Billinton R, Allan RN (1996) Reliability evaluation of power systems, 2nd edn. Plenum Press, New York
8. Billinton R, Li W (1994) Reliability assessment of electrical power systems using Monte Carlo methods. Plenum Press, New York
9. Billinton R, Chen H, Ghajar R (1996) Time-series models for reliability evaluation of power systems including wind energy. Microelectr Reliab 36(9):1253–1261
10. Giorsetto P, Utsurogi KF (1983) Development of a new procedure for reliability modeling of wind turbine generators. IEEE Trans Power Apparatus Syst 102(1):134–143
11. Billinton R, Chowdhury AA (1992) Incorporation of wind energy conversion systems in conventional generating capacity adequacy assessment. IEE Proc Gener Transm Distrib 139(1):47–56
12. Billinton R, Bai G (2004) Generating capacity adequacy associated with wind energy. IEEE Trans Energy Convers 19(3):641–646
13. Billinton R, Chen H, Ghajar R (1996) A sequential simulation technique for adequacy evaluation of generating systems including wind energy. IEEE Trans Energy Convers 11(4):728–734
14. Ubeda R, Allan RN (1992) Sequential simulation applied to composite system reliability evaluation. IEE Proc C 139(2):81–86
15. Sankarakrishnan A, Billinton R (1995) Sequential Monte Carlo simulation for composite power system reliability analysis with time varying loads. IEEE Trans Power Syst 10(3):1540–1545

16. Billinton R, Wangdee W (2006) Delivery point reliability indices of a bulk electric system using sequential Monte Carlo simulation. IEEE Trans Power Deliv 21(1):345–352
17. Billinton R, Gao Y (2008) Multi-state wind energy conversion system models for adequacy assessment of generating systems incorporating wind energy. IEEE Trans Energy Convers 23(2):163–170
18. Billinton R, Li Y (2004) Incorporating multi-state models in composite system adequacy evaluation. In: Proceedings of the 8th international conference on probabilistic method applied to power systems, Ames, Iowa, September
19. Karki R, Billinton R (2006) Impact of renewable energy unit availability on power system adequacy. Int J Power Energy Syst 26(2):147–152
20. Billinton R, Kumar S et al (1989) A reliability test system for educational purposes—basic data. IEEE Trans Power Syst 4(3):1238–1244
21. Billinton R, Wangdee W (2006) Predicting bulk system reliability performance indices using sequential Monte Carlo simulation. IEEE Trans Power Deliv 21(2):909–917
22. Li W (1998) Installation guide and user's manual for the MECORE Program, July
23. Billinton R, Huang D (2009) Wind power modeling and the determination of planning capacity credit in an electric power system. In: Proceedings of the advances in risk and reliability technology symposium (AR2TS 09), April 21–23
24. Wangdee W, Billinton R (2006) Considering load-carrying capability and wind speed correlation of WECS in generation adequacy assessment. IEEE Trans Energy Convers 21(3):734–741
25. Gao Y, Billinton R (2009) Adequacy assessment of generating systems containing wind power considering wind speed correlation. IET Renew Power Gener 3(2):217–226
26. Billinton R, Wangdee W (2007) Reliability-based transmission reinforcement planning associated with large-scale wind farms. IEEE Trans Power Syst 22(1):34–41
27. Task Force IEEE (1979) IEEE reliability test system. IEEE Trans Power Apparatus Syst 98:2047–2054
28. Wangdee W, Billinton R (2007) Reliability assessment of bulk electric systems containing large wind farms. Int J Electr Power Energy Syst 29(10):759–766
29. Billinton R, Gao Y, Karki R (2009) Composite system adequacy assessment incorporating large-scale wind energy conversion systems considering wind speed correlation. IEEE Trans Power Syst 24(3):1375–1382
30. Fink LH, Carlsen K (1978) Operating under stress and strain. IEEE Spectrum 15(3):48–53
31. EPRI Final Report (1987) Composite system reliability evaluation: Phase I—Scoping study. Tech Report EPRI EL-5290, December
32. Billinton R, Khan E (1992) A security based approach to composite power system reliability evaluation. IEEE Trans Power Syst 7(1):65–72
33. Billinton R, Lian G (1994) Composite power system health analysis using a security constrained adequacy evaluation procedure. IEEE Trans Power Syst 9(2):936–941
34. Billinton R, Karki R (1999) Application of Monte Carlo simulation to generating system well-being analysis. IEEE Trans Power Syst 14(3):1172–1177
35. Leite da Silva AM, Chaves de Resende L, Antonio da Fonseca Manso L, Billinton R (2004) Well-being analysis for composite generation and transmission systems. IEEE Trans Power Syst 19(4):763–1770
36. Wangdee W, Billinton R (2006) Bulk electric system well-being analysis using sequential Monte Carlo simulation. IEEE Trans Power Syst 21(1):188–193
37. Huang D, Billinton R (2009) Effects of wind power on bulk system adequacy evaluation using the well-being analysis framework. IEEE Trans Power Syst 24(3):1232–1240

Strategic Lines and Substations in an Electric Power Network

Alvaro Torres and George J. Anders

1 Introduction

The development of tools for the analysis of critical infrastructure, and particularly of the power system grid, has been an intense activity during the last several years. A number of different techniques have been proposed for vulnerability analysis of power system and for security in cases of catastrophic events. Such methods include controlled islanding or intentional islanding by using decision analysis techniques and also graph algebraic approaches. Also, more sophisticated techniques such as algebraic network theory, polyhedral dynamics and artificial intelligence-based search methods have been proposed to find critical nodes, critical transmission lines and system vulnerability indices.

A number of these methods allow identification of the critical elements in a bulk power system and the quantification of the consequences of their failures. These techniques are intended to go further in security analysis than the traditional contingency analysis or $n - m$ approach. The objective is more to evaluate the structural robustness of a network or to identify the more vulnerable or the more

A. Torres (✉)
Department of Electrical and Electronics Engineering, University of Los Andes, Bogotá, Colombia
e-mail: atorres@uniandes.edu.coa; torres@snclavalin.com

A. Torres
SNC Lavalin, Colombia, Cra. 10 97A-13, Of. 207, Bogotá, Colombia

G. Anders
Department of Electrical and Computer Engineering, Technical University of Lodz, Lodz, Poland
e-mail: george.anders@kinectrics.com; george.anders@attglobal.net

G. Anders
Kinectrics Inc., 800 Kipling Avenue, Toronto, ON M8Z 6C4, Canada

G. Anders and A. Vaccaro (eds.), *Innovations in Power Systems Reliability*, Springer Series in Reliability Engineering, DOI: 10.1007/978-0-85729-088-5_5,

fragile elements in a power system than to determine the response of the power system to a single or several contingencies.

The concepts of reliability, security, survivability and vulnerability of a system to different threats like natural failures, external attacks or operator failures and the means of improving the system response in complex systems like computer and communications networks are included in a broader concept of dependability. Dependability encompasses also concepts of fault-tolerant systems, which accelerated the emergence of the terminology, definitions and a wider view of the problem of system reliability and security.

Dependability in computing and concepts of the fault-tolerant systems has been part of the computer field since its origins but with the formation of the corresponding IEEE-CS TC, this field formally emerged in 1970 [3]. In the field of power systems reliability, security and vulnerability have been systematically studied since the beginning of power grids. However, more recently, a need to consider telecommunications and other infrastructures as interdependent has been recognized and the importance of taking into account cascading and catastrophic events with a broader point of view has been stressed. This led to the application of the concepts of dependability in power systems, which are similar to the ones used in the computer networks [3].

On the other hand, graph theory or network theory and matrix and linear algebra have been applied extensively in power systems since the beginning of the application of computer methods in recognition of the network structure of a power system. In the last decade, many developments in graph theory have been applied to complex networks such as internet, social networks, communication and VLSI circuits. In particular, concepts in spectral graph theory are been used to identify complex network characteristics and to have a deeper insight in the behaviour of such networks.

Spectral Graph theory is a study of the spectra of matrices that identify properties of a graph. Matrices like the node-element matrix, adjacency matrix, Laplacian matrix and in the case of power system, the bus admittance and bus impedance matrices have eigenvalues and eigenvectors that can give important information about the structure of the system and its intrinsic robustness. Also, eigenvalue analysis can lead to important conclusions about possibilities of system partitioning and node criticality.

Spectral graph theory has been applied to chemistry for molecules stability analysis, in quantum mechanics and more recently in communication networks. Spectral analysis have been extensively used to identify the characteristics of the dynamics of linear systems and stochastic processes, in analogous way spectral graph theory could be used to identify the properties of the electrical networks.

Spectral graph theory offers an interesting potential when applied to power systems grids. The study of graph eigenvalues and eigenvectors of a power network provides an insight in the intrinsic structure of the system and allows us to identify structural modes and fragilities. So, graph eigenvalues analysis could be used to design security strategies for the system when affected by massive number of events or cascading as well as catastrophic failures.

The potential of graph theory to identify structural properties of power grids could be advantageously used in the network dependability analysis. In this Chapter, two main aspects will be presented. The identification of principal nodes or a method of ranking the importance of nodes in a power grid and a method of partitioning the grid as a conceptual way to avoid the effect of cascading or catastrophic events by dividing the grid into different self-sustainable subsystems.

This chapter presents the basic concepts of the spectral graph theory as could be applied to power system networks and particularly to the identification of the strategic substations and transmission lines from the point of view of reliability and security of the bulk power system. Some examples are presented using the IEEE test system with different number of nodes to show the simplicity of the applications and the type of results that can be obtained. Further and more complex analysis can be done in reliability and security studies when a power system is affected by catastrophic events. However, the main aim of this chapter is to present a method to identify critical nodes and links within a Bulk Electric Power System (BES), which is an important element of the infrastructure of any country.

Critical or strategic nodes or in this specific case, critical substations are those that, are important to maintain the integrity of the network, or in other words if affected by natural events or terrorist actions, would cause a large disruption by itself or by following cascading events.

On the other hand, strategic transmission lines are those that if open could divide the system in self-sustained islands that could be operated independently in the case of extended events that may cause a system blackout. This separation could be considered as a strategic security action after severe disturbances or cascading events. Intentional-controlled islanding is a strategic action to separate the system into the self-healing islands which generators exhibit slow-coherency behaviour regarding angular stability [16, 20].

Spectral graph theory as applied to power systems also could be used, for example, to identify the number of spanning trees of the graph representing the system and to use the identified trees to define the nodes to minimize the number of phase measurement units (PMUs) in a observability problem.

2 Graphs Linear Algebra

2.1 Definitions of Matrices Associated with a Graph G

In creating the ranking measures for electric power system substations, we will use the following definitions concerning various matrices associated with a graph G [6, 9, 11].

- A graph G: a simple undirected graph whose vertices are positive integers. The order of G is the number of vertices.
- Node degree: the degree of vertex k, $\deg_G k$ is the number of edges incident with k. The graph G is regular of degree r if every vertex has degree r.

- External node degree due to node loads external factors for nodes (EFN): graph node degree could be affected by other ranking criteria that consider aspects of node importance beyond the network connectivity.
- External node degree due to node connecting transmission lines. External factor for lines (EFL): graph node degree could be affected by other criteria applied upon the importance of connecting transmission lines that consider aspects of node importance beyond the network connectivity.
- Adjacency matrix **A**: $\mathbf{A} = [A_{ij}]$, where $A_{ij} = 1$ if $\{i, j\}$ is the edge of G and $A_{ij} = 0$ otherwise.
- Degree matrix **D**: is a diagonal matrix $\mathbf{D} = \mathrm{diag}(\deg_G 1,\ldots,\deg_G n)$
- Node-element (vertex-edge)incidence matrix **N**: is the n (number of vertices) $\times$ m (number of edges) 0, 1-matrix with rows indexed by the vertices of G and columns indexed by the edges of G, such that the i, j entry of **N** is 1 if edge j is incident with vertex i, and 0 if not.
- Orientation of G_p: is the assignment of a direction to each edge, converting edge $\{i, j\}$ to either arc $\{i, j\}$ or arc $\{j, i\}$.
- Oriented incidence matrix: the oriented incidence matrix $\mathbf{N}_p$ of an oriented Graph G_p with n vertices and m arcs is the $n \times m$. 0, 1, -1 matrix with rows indexed by the vertices of G and columns indexed by the arcs of G such that the i, k entry of $\mathbf{N}_p$ is 1 if arc is directed leaving node i and -1 if arc is directed toward node i and all other entries are 0.
- Transformation matrix **T**: $\mathbf{T} = \mathrm{diag}((\deg_G 1)^{1/2},\ldots,(\deg_G n)^{1/2})$
- The Laplacian matrix **L**: $\mathbf{L} = \mathbf{D} - \mathbf{A}$
- Normalized adjacency matrix An: $An = T^{-1}AT^{-1}$
- The normalized Laplacian matrix Ln: $Ln = T^{-1}L\,T^{-1} = I - An$
- The signless Laplacian matrix Ls: $Ls = D + A$
- The normalized signless matrix Lns: $Lns = T^{-1}(D + A)\,T^{-1} = I + An$
- Transition matrix of the random walk M: $M = AD^{-1}$
- Weighted graphs and matrices: all definitions given for a simple graph can be applied to weighted graphs. A weighted undirected graph G has an associated weight function satisfying $w(i, j) = w(j, i)$ with $w(i, j) \geq 0$. Unweighted graphs are a special case where all weights are 0 or 1. In the case of an electrical power grid where we will consider external criteria for node ranking, weights are to be considered not only on the links but also for the nodes. So, we have a special case of weighted graph. A graph with node weight. Node weight will be converted to edge weights adding links between the node and the reference node. In this way, we can have the corresponding matrices A_{ext}, D_{ext}, L_{ext}, L_{next} and Lns_{ext}.

2.2 Eigenvalues of a Graph

One of the main issues of the spectral theory is the analysis of the impact of eigenvalue bounds of a graph matrix and the interpretation of the eigenvalues regarding the structure and the properties of the graph.

Considering that all graph matrices are real and symmetric, we can define the ordered spectrum $\sigma(A)$ of an $n \times n$ matrix **A** as the list of eigenvalues, repeated according to multiplicity in a non-decreasing order. Thus, $\sigma(\mathbf{A}) = (\alpha_1, \alpha_2,\ldots,\alpha_n)$ with $\alpha_1 \leq \alpha_2 \leq \cdots \leq \alpha_n$.

The eigenvalues of the Laplacian matrix are represented by the spectrum

$$\sigma(\mathbf{L}) = (\lambda_1, \lambda_2, \ldots \lambda_n) \tag{1}$$

and the eigenvalues of the sign-less Laplacian matrix:

$$\sigma(\mathbf{Lns}) = (\mu_1, \mu_2, \ldots \mu_n) \tag{2}$$

The matrices **A**, **An**, **L** and **Ln** are all non-negative, and if G is connected, they are all irreducible. Also, the matrices **A**, **D**, **An**, **L**, **Ln**, **Ls** and **Lns** are all connected through the incidence matrix **N**.

By using the Sylvester's Law of Inertia [14], one can prove that the spectral radius (the supremum among the absolute values of the elements in its spectrum), is 2, it is to say that $\sigma(\mathbf{L}) \subset [0, 2]$, $\sigma(\mathbf{Lns}) \subset [0, 2]$ and $\mu_{\mathrm{ns\ n}} = 2$, and $\lambda_1 = 0$.

3 Level of Importance of Substations in a Bulk Power System

3.1 General Definitions for Node Importance

One step toward security strategies is to identify critical grid elements, it is to say, a node or link that is of a paramount importance due to its role in the interconnection, its role in services to customers in the grid or other intrinsic characteristic like its cost or repair difficulty.

A substation will be called "strategic substation" when it plays a significant role in the connectivity of the grid for the normal operation of the Bulk Power System. It means that the importance of a substation depends of its location within the grid and how it is related to the rest of the power system, so the criteria to determine this importance include parameters that characterize the substations and the system in which the substations of interest are located.

The criteria for importance here is how a substation is necessary or essential for the integrity of the connectivity or interconnection in the BES. This will be considered as a network based criterion. Other criteria could be formulated to define the importance of a node in a BES due to the relative importance of the loads or the relative importance of the connecting transmission lines. So, we could include in the ranking aspects regarding substation loads and connecting transmission lines.

Examples regarding substation loads are (EFN):

- public health and safety,
- national security,

- regional economy,
- location in respect to population centres, transportation corridors and other lifelines,
- and so on.

Examples regarding connecting transmission lines are (EFL):

- underground cables,
- double circuit transmission lines,
- special structure transmission line,
- transmission lines with limited or congested right-of-way.

In general, the performance measures should cover a variety of station-related performance aspects such as reliability, utilization, security and safety to personnel. In this section, we also consider network external factors for node ranking besides purely connectivity properties of the electric power networks [2].

The measures described in this section use the well-known centrality aspects of the spectral graph theory and a concept of weighted centrality due to network external factors and do not require development of any new analytical tools nor do they use complex computer programs. The new measures are based on the analysis of the structure of the electric power network and weights or factors given by experts. Required data can be readily found in the raw data file for the power flow programs used by the electric utilities and weights for externalities could be established independently.

3.2 Network Centrality

Spectral graph theory is a study of the spectra of matrices that identify properties of a graph. Matrices like the node-element matrix, adjacency matrix, Laplacian matrix and in the case of power system, the bus admittance and bus impedance matrices have eigenvalues and eigenvectors that can give important information about the structure of the system and its intrinsic robustness. Also, eigenvalue analysis can lead to important conclusions about node criticality.

Network centrality defined as the relative importance of a vertex within a graph could be measured with several different indices, such as:

- Eigenvector centrality: is a measure of the importance of a node in a network. It assigns relative scores to all nodes in the network based on the principle that connections to high-scoring nodes contribute more to the score of the node in question than equal connections to low-scoring nodes. This means that the importance of a given node depends not only on the number of incident edges but also on the relative importance of the nodes to which it is connected.
- Electrical centrality: recently, Hines and Blumsack [8] proposed an "Electrical Centrality Measure" calculated from the Zbus Matrix that can be considered a betweenness centrality measure. The importance of a substation is inversely

proportional to the Thévenin equivalent circuit seen in the substation. In other words, the electrical equivalent distance (Z Thévenin) between the generation nodes and the substation determines its importance. So, the importance of a substation is represented by its maximum level of the short circuit power.

3.3 Criteria for Node Importance or Node Centrality

For a power system grid, we define the matrix **A** and all other system matrices including the reference node, so that generators represent a link in the network. Also, for a power system network, each substation is represented by a node and each transmission line by a link. We can have multiple links between nodes. Also, further approximations or simplifications could be applied. We could consider only a system with one voltage level, so for example, if the power system to be represented includes voltage levels of 230 and 500 kV, every link at higher voltage than 230 kV should be represented by the 230 kV equivalent links. Transformers will not be represented as links since the total substation is represented by one node. However, we may like to include factors exogenous to the grid structure itself to have a node ranking that not only include the node centrality concept but the node service importance and the relative physical importance of links.

To determine the external factor or weight for a node the following considerations could be applied:

- Substations: for substations with two levels of high voltage (500/230/220 kV), the two nodes could be combined as a "super node" considering all the links for both of them. Substations with special loads could have a weight proportional to the load importance. For example, a substation with an important load due to National Security could have a weight of 2 or 3 instead of the standard weight of 1 for all substations.
- Equivalent links: a 500 kV transmission line is equivalent to two 230 kV lines. So the total number of links for each substation is referred to 230 kV. In this case, a weight of 2 could be associated with the line.
- Generators: all machines connected to a low voltage node are represented as an equivalent generator connected to a node
- Transformers: they are ignored as links for the nodes. However, we could include an external factor in the node to consider special loads or external criteria for node ranking.
- Compensation elements: they are ignored as links for the nodes. However, we could include an external factor in the node to consider special cases or external criteria for node ranking.

Association of a node weight as explained above allow us to identify strategic or important substations regardless of the voltage level or to classify substations that could have two or more voltage levels and substations with important loads and transmission lines. Eigenvector centrality intrinsically recognizes the relative importance of the linked nodes due to the location of a node within the network.

An eigenvector $\mathbf{x} = (x_1, x_2,\ldots,x_n)$ and the corresponding eigenvalue λ, of a matrix **A** satisfy the following equation:

$$\lambda \mathbf{x} = \mathbf{A}\mathbf{x} \tag{3}$$

For node i, we can write

$$x_i = \frac{1}{\lambda}\sum_{j=1}^{n} A_{i,j}x_j \tag{4}$$

If we define the centrality index for a given node i as proportional to the average of the centrality indices of neighbouring nodes, then the element x_i of the eigenvector **x** of the adjacency matrix is the centrality index for node i [8]. A centrality index should be a non-negative number. So, all elements of the eigenvector **x** should be non-negative numbers.

According to the Perron–Frobenius Theorem for square non-negative matrices [14], an $n \times n$ non-negative matrix **A** has a real eigenvalue r such that any other eigenvalue satisfies $\lambda \leq r$ and its corresponding eigenvector has non-negative entries. Every other eigenvector for smaller eigenvalues has negative components. Therefore, λ should be the greatest eigenvalue r of the adjacency matrix **A** and $\boldsymbol{x}$ is its eigenvector, it is to say, the eigenvector corresponding to the first eigenvalue when they are ordered.

With this criterion the importance of a node is not only determined by the node degree but by the number and quality of its connected nodes as well [18].

3.4 Examples

Three examples are presented below to show the application of the concepts presented above. First example uses a five-node system and matrices are presented together with the results. A second example uses the IEEE-39 node test system. For these two examples we show the effect of considering weighted nodes and edges. The third example considers a larger real-life system.

3.4.1 Example 1: Five-Node System

Figure 1 shows a five-node system. It is easy to see that considering network centrality, node one is the most important and that all other nodes are equally important since their role in the interconnection is similar and there is no other criterion for discrimination.

Matrices for system shown in Fig. 1 are shown in Fig. 2. Now, let us assume that link 1–2 is of higher relative importance than any other for some reason (voltage, underground cable, etc.) and due to this fact we assign a factor 3 for this link, as shown in Fig. 3.

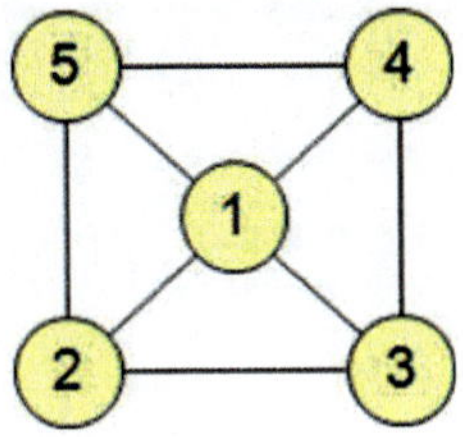

Fig. 1 Five-node graph

$$\mathbf{A} = \begin{pmatrix} 0 & 1 & 1 & 1 & 1 \\ 1 & 0 & 1 & 0 & 1 \\ 1 & 1 & 0 & 1 & 0 \\ 1 & 0 & 1 & 0 & 1 \\ 1 & 1 & 0 & 1 & 0 \end{pmatrix} \qquad \mathbf{An} = \begin{pmatrix} 0 & 0.289 & 0.289 & 0.289 & 0.289 \\ 0.289 & 0 & 0.333 & 0 & 0.333 \\ 0.289 & 0.333 & 0 & 0.333 & 0 \\ 0.289 & 0 & 0.333 & 0 & 0.333 \\ 0.289 & 0.333 & 0 & 0.333 & 0 \end{pmatrix}$$

$$\sigma(\mathbf{A}) = \begin{pmatrix} -2 & -1.236 & 0 & 0 & 3.236 \end{pmatrix} \qquad \sigma(\mathbf{An}) = \begin{pmatrix} -0.667 & -0.333 & 0 & 0 & 1 \end{pmatrix}$$

$$\mathbf{L} = \begin{pmatrix} 4 & -1 & -1 & -1 & -1 \\ -1 & 3 & -1 & 0 & -1 \\ -1 & -1 & 3 & -1 & 0 \\ -1 & 0 & -1 & 3 & -1 \\ -1 & -1 & 0 & -1 & 3 \end{pmatrix} \qquad \mathbf{Ln} = \begin{pmatrix} 1 & -0.289 & -0.289 & -0.289 & -0.289 \\ -0.289 & 1 & -0.333 & 0 & -0.333 \\ -0.289 & -0.333 & 1 & -0.333 & 0 \\ -0.289 & 0 & -0.333 & 1 & -0.333 \\ -0.289 & -0.333 & 0 & -0.333 & 1 \end{pmatrix}$$

$$\sigma(\mathbf{L}) = \begin{pmatrix} 0 & 3 & 3 & 5 & 5 \end{pmatrix} \qquad \sigma(\mathbf{Ln}) = \begin{pmatrix} 0 & 1 & 1 & 1.333 & 1.667 \end{pmatrix}$$

$$\mathbf{Ls} = \begin{pmatrix} 4 & 1 & 1 & 1 & 1 \\ 1 & 3 & 1 & 0 & 1 \\ 1 & 1 & 3 & 1 & 0 \\ 1 & 0 & 1 & 3 & 1 \\ 1 & 1 & 0 & 1 & 3 \end{pmatrix} \qquad \mathbf{Lns} = \begin{pmatrix} 1 & 0.289 & 0.289 & 0.289 & 0.289 \\ 0.289 & 1 & 0.333 & 0 & 0.333 \\ 0.289 & 0.333 & 1 & 0.333 & 0 \\ 0.289 & 0 & 0.333 & 1 & 0.333 \\ 0.289 & 0.333 & 0 & 0.333 & 1 \end{pmatrix}$$

$$\sigma(\mathbf{Ls}) = \begin{pmatrix} 1 & 2.438 & 3 & 3 & 6.562 \end{pmatrix} \qquad \sigma(\mathbf{Lns}) = \begin{pmatrix} 0.333 & 0.667 & 1 & 1 & 2 \end{pmatrix}$$

Fig. 2 Graph matrices for system in Fig. 1

Figure 4 shows node ranking for the original graph and with the weight due to external factor considering an external importance of line 1–2. We can see that node 2 is now more important than nodes 3, 4 and 5. External factors signify importance not due to the network structure itself.

3.4.2 Example 2: IEEE 39-Node Test System

Figure 5 shows an example with a larger network, the IEEE 39-node test system. The first six nodes in the ranking are shown for two cases. One with all lines having the same importance and the second when the lines between nodes 3–4, 3–2 and 2–25 have special weights assigned. The analysis is conducted with a weighted graph. Matrices are not shown due to their sizes.

$$EFL=\begin{pmatrix}1&2&3\\1&3&1\\1&4&1\\1&5&1\\2&5&1\\2&3&1\\3&4&1\\4&5&1\end{pmatrix}\quad EFN=\begin{pmatrix}1\\1\\1\\1\\1\end{pmatrix}\quad A=\begin{pmatrix}0&1&1&1&1\\1&0&1&0&1\\1&1&0&1&0\\1&0&1&0&1\\1&1&0&1&0\end{pmatrix}\quad Aext=\begin{pmatrix}0&3&1&1&1\\3&0&1&0&1\\1&1&0&1&0\\1&0&1&0&1\\1&1&0&1&0\end{pmatrix}$$

$$D=\begin{pmatrix}4&0&0&0&0\\0&3&0&0&0\\0&0&3&0&0\\0&0&0&3&0\\0&0&0&0&3\end{pmatrix}\quad Dext=\begin{pmatrix}6&0&0&0&0\\0&5&0&0&0\\0&0&3&0&0\\0&0&0&3&0\\0&0&0&0&3\end{pmatrix}$$

$$L=\begin{pmatrix}4&-1&-1&-1&-1\\-1&3&-1&0&-1\\-1&-1&3&-1&0\\-1&0&-1&3&-1\\-1&-1&0&-1&3\end{pmatrix}\quad Lext=\begin{pmatrix}6&-3&-1&-1&-1\\-3&5&-1&0&-1\\-1&-1&3&-1&0\\-1&0&-1&3&-1\\-1&-1&0&-1&3\end{pmatrix}$$

$$Ln=\begin{pmatrix}1&-0.289&-0.289&-0.289&-0.289\\-0.289&1&-0.333&0&-0.333\\-0.289&-0.333&1&-0.333&0\\-0.289&0&-0.333&1&-0.333\\-0.289&-0.333&0&-0.333&1\end{pmatrix}\quad Lnext=\begin{pmatrix}1&-0.548&-0.236&-0.236&-0.236\\-0.548&1&-0.258&0&-0.258\\-0.236&-0.258&1&-0.333&0\\-0.236&0&-0.333&1&-0.333\\-0.236&-0.258&0&-0.333&1\end{pmatrix}$$

$$x_{Lns}(n)=\begin{pmatrix}0.5&0.433&0.433&0.433&0.433\end{pmatrix}\quad x_{Lnsex}(n)=\begin{pmatrix}0.548&0.5&0.387&0.387&0.387\end{pmatrix}$$

Fig. 3 Graph matrices for system in Fig. 1 with external factors for link 1–2

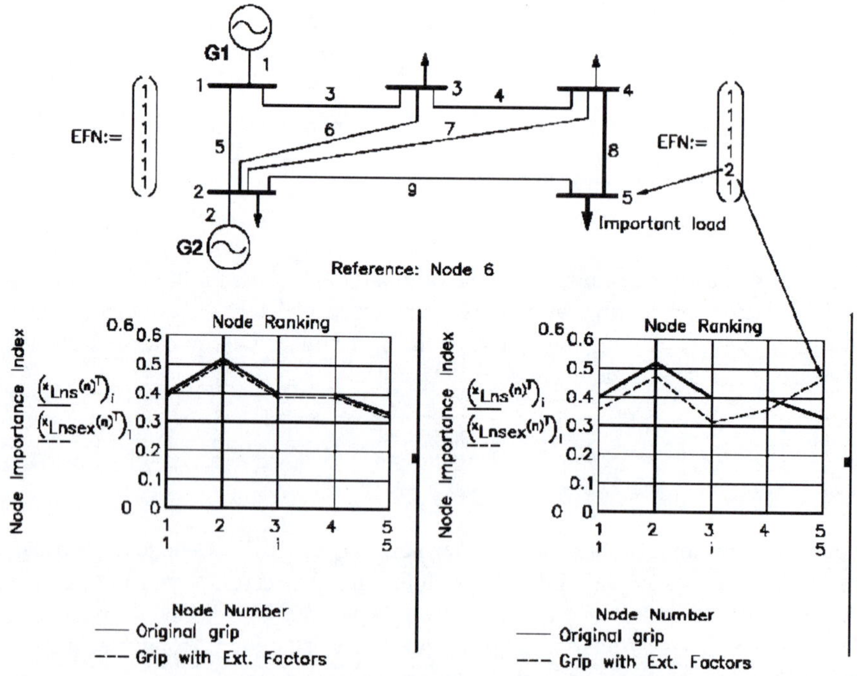

Fig. 4 Node order for the five-node system considering external factors

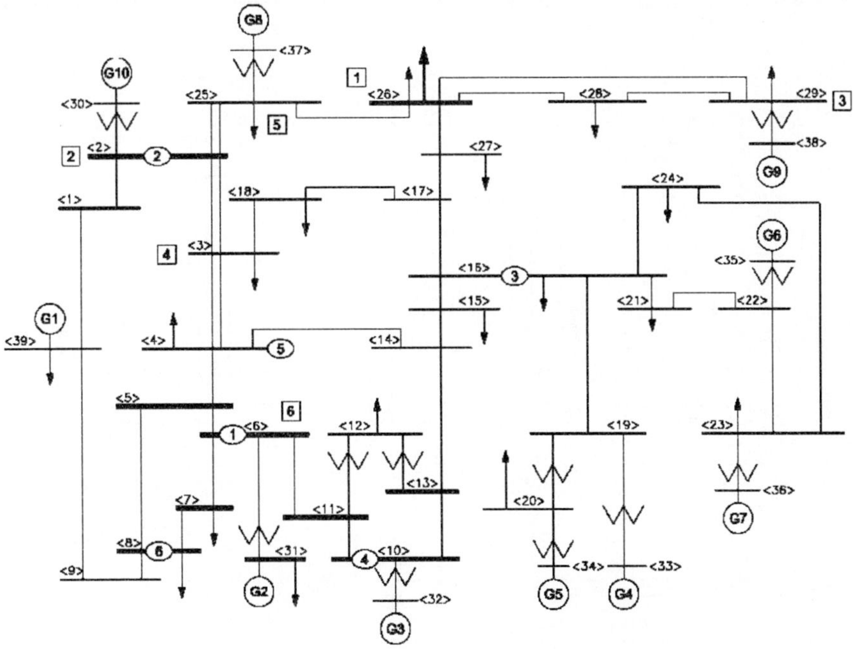

Fig. 5 IEEE 39-node test system. Importance node ranking without external factors (*squares with numbers*) and considering the importance of the transmission lines between nodes 3–4,3–2,2–25 (*ovals with numbers*)

3.4.3 Example 3: A Real Bulk Power System

The criteria presented in this chapter have been used to rank the importance of the substations in a real 87-node 230–500 kV Bulk Power System shown in Fig. 6. The studies were conducted to determine its role in the interconnection or integrity of the interconnected system.

Several ranking methods using the centrality concept will be investigated. Ranking method may be simple or sophisticated; their use depends on the information available and the desired precision of the required results. The proposed methodology looks for a compromise between application, in terms of tools and needed information, and the precision of the results.

Since the eigenvector centrality considers not only the number of edges incident on a node but also the relative importance of the other connected nodes, and it is a method easy to implement, it is a good method to recommend. Also, with this method, a weighted adjacency matrix could be considered by using the imaginary part of the electrical admittances. However, using the un-weighted adjacency matrix we can obtain satisfactory results as well. Figure 7 shows the ranking of the substations for this network and a comparison of the ranking given by different criteria.

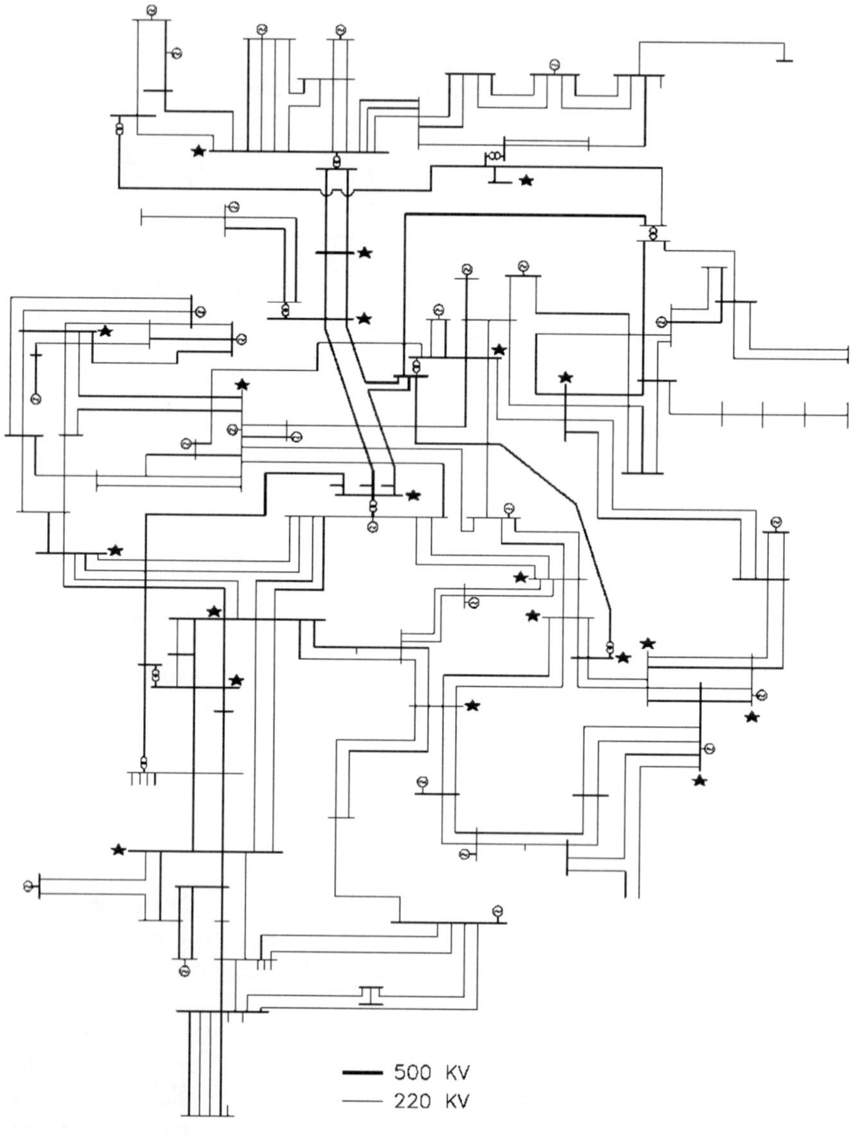

Fig. 6 87-node system

4 Graph Partitioning

Graph partitioning is a problem in graph theory related to a division of the vertices of a graph into two or more sets, while cutting as few edges as possible. There are many restrictions that can be introduced to the problem to consider the number of edges or weights or sizes of the sub-graph to be produced by a partitioning.

NODE IMPORTANCE RANKING BY DIFFERENT CRITERIA					
			SELECTED		
SUB. NAME	Short Circuit (MVA)	Number Equivalent Edges	Ranking By Eigenvector Centrality	Ranking By Degree Centrality	Ranking By Connectivity
SAB	8612	16	1	1	2
PRI	7888	16	2	1	6
SANCA	12748	14	3	3	4
ESME	7211	11	4	4	1
CERRO	7275	10	5	5	5
GUAT	10995	10	6	5	3
VIR	6175	9	7	7	8
GUA	10716	9	8	7	11
PUR	7211	8	9	9	7
MES	8007	8	10	9	13
CHIN	6322	8	11	9	16
COP	4070	7	12	12	9
CHIV	9720	7	13	12	10
GUAT	4103	7	14	12	12
NORO	8645	6	15	15	17
SANMA	6972	6	16	15	43
BAC	8605	6	17	15	48
YUM	7171	6	18	15	19
ANCO	7011	6	19	15	30
TOR	7967	6	20	15	18
BAR	6821	6	21	15	23
SANBE	3227	6	22	15	25
FUND	6097	6	23	15	29
JAM	2669	6	24	15	26
BET	3546	6	25	15	24

Fig. 7 Ranking of the first 25 most important substations for the 87-node system

A graph partitioning could be done by a progressive bi-sectioning of a graph or by a direct *k*-way partitioning. Both procedures could be obtained by matrix spectral analysis, so called spectral partitioning. These techniques have been used intensively in image partitioning when an image is considered as a large connected network.

A high interest in power network partitioning dates to the early days of diakoptics methods introduced by Gabriel Kron to break a matrix problem into small sub-problems, which can be solved independently and then to associate the solutions to obtain the global solution. Also, the problem of partitioning appears in the decomposition methods for state estimation or computational parallel processing for very large systems. The spectral graph theory has been applied to image segmentation and digital image processing [15].

More recently, the problem of power system network partitioning has appeared in security problems. On the one hand, we can see problems where early detection of island formation in power system networks under multiple line outages is important for the study of security and control strategies. On the other hand, intentional islanding and adaptive load shedding schemes have been proposed to avoid cascading outages due to large catastrophic events or simultaneous events that usually precede the occurrence of blackouts [1, 16, 20].

The study of predetermined islanding scenarios with load-shedding schemes and distributed generation strategies may constitute the basis for the predetermined action plans against system collapse when system security is compromised due to a large number of line outages. To detect the formation of islands inside a power system subjected to multiple line outages and for controlled islanding to prevent blackouts, several methods based on minimum cut sets, decision trees and generator coherency have been proposed [4, 7, 10, 16].

Graph spectral analysis allows identification of the natural structure of the system and how it can be partitioned by minimizing a specially defined performance index. The aim of this section is to show the application of the graph spectral theory for power system partitioning to be used in controlled islanding strategies for security enhancement.

4.1 Graphs Cut Problem

In this section, we analyze a graph partitioning problem from the point of view of its spectral characteristics to determine how to divide a graph cutting as few edges as possible. In the case of a power system, the structure of the resulting subsystems includes slow coherent generators under transient system stability conditions given that generator connections are properly defined.

The graph bisection problem is to find a set S of a graph G, $S \subset G$ such that approximately:

$$|S| = \left\lceil \frac{|S|}{2} \right\rceil \tag{5}$$

This problem can be formulated as follows:

$$\min|\{(i,j) \in G : i \in S, j \notin S\}| \tag{6}$$

Set $S \subseteq G$ is referred to as a cut, which represents the partition of G into S and its complement.

Since a perfect bisection may not be possible, it could be good enough to obtain an acceptable value of a performance index or cut quality index like the number of edges cut to the number of vertices or edges removed, the so called ratio of a cut defined, or the conductance of a cut.

The conductance of a set S is defined as:

$$\phi(S) = \frac{|\partial(S)|}{d(S)d(G-S)} \tag{7}$$

where $|\partial(S)|$ is called the boundary of S, which means all edges with one end point in S and the other outside of S and $d(S)$ is the number of edges attached to each side of the graph. It is to say that $d(S)$ is the sum of the degrees of vertices in S. In a weighted graph, the weighted degree of a vertex is the sum of the weights of the edges attached to it.

So, the problem is to find the set S of minimum conductance:

$$\phi(G) = \min_{S \subset G} \phi(G) \tag{8}$$

We have to find a relationship between the graph conductance and the graph eigenvalues and eigenvectors. This is given by the Cheeger inequality, which could be formulated as [5]:

$$\varphi(G) \leq 2\sqrt{2\lambda_2(\mathbf{L}n)} \tag{9}$$

where $\lambda_2(\mathbf{L}n)$ is the second smallest eigenvalue of the normalized Laplacian matrix $\mathbf{L}ns$, which is considered as the most important information in the spectrum of a graph [6]. It is to say that for some j

$$\varphi(S_j) \leq \sqrt{2\lambda_2(\mathbf{L}n)} \tag{10}$$

This means that for a small $\lambda_2(\mathbf{L}n)$ we will have a cut of small conductance, but is difficult to prove that it is the minimum conductance. However, we could consider it as a good solution for a bi-section of a graph.

The procedure is to compute the eigenvectors and to partition the graph into two subgraphs using the second smallest eigenvector. In some cases, the eigenvector takes negative and positive values and the signs will give the exact partition of the graph. However, we can choose the splitting point for better convenience.

To divide a graph in more than two parts, we can use recursive bi-partitioning with the second smallest eigenvalue and its corresponding eigenvector for each sub-graph successively. One disadvantage of the recursive bi-partitioning is the treatment of oscillatory eigenvectors. However, one can decide the splitting point to reduce their effect.

Although there is no complete knowledge about the meaning of many spectral characteristics of a graph, the other eigenvectors besides the corresponding to the second smallest eigenvalue also contain useful information. We can use the higher eigenvectors for a direct k-way partitioning of a graph. The higher eigenvectors are indicators of how to partition a graph in more than two cuts. However, avoiding the effect of oscillatory eigenvectors is an important task to [15].

4.2 Examples

We will discuss three examples to show the application of the presented concepts. First example uses a nine-node graph as shown in Fig. 8 with 12 links.

Second example uses the IEEE 39-node Reliability Test System. We can consider weighted edges with line and generator admittances by using the $\mathbf{Y}$ admittance matrix with the modification of considering the positive values instead of the normal $\mathbf{Y}_{\mathrm{BUS}}$ admittance matrix. In this example, we go further to test the slow coherency of the subsystems once they are intentionally separated after a severe set of contingencies.

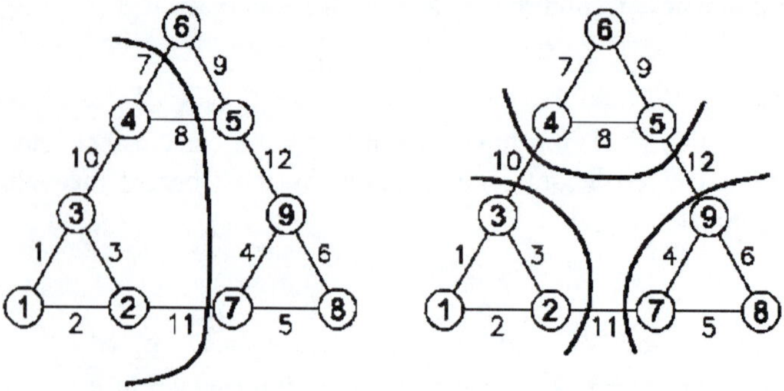

$$\sigma(Lns) = (0.354 \quad 0.401 \quad 0.5 \quad 0.5 \quad 0.734 \quad 0.905 \quad 1.698 \quad 1.908 \quad 2)$$

$$\sigma(Ln) = (0 \quad 0.092 \quad 0.302 \quad 1.095 \quad 1.266 \quad 1.5 \quad 1.5 \quad 1.599 \quad 1.646)$$

$$x_{Ln2} = (-0.404 \quad -0.404 \quad -0.403 \quad -0.108 \quad 0.108 \quad 0 \quad 0.404 \quad 0.404 \quad 0.403),$$

$$x_{Ln3} = (-0.296 \quad -0.296 \quad -0.144 \quad 0.425 \quad 0.425 \quad 0.497 \quad -0.296 \quad -0.296 \quad -0.144),$$

Fig. 8 Graph partitioning for a nine-node, 12-link graph

The third example shows the IEEE 69-node system. In this example, we will test the effect of network partitioning to be used for splitting power system when a disturbance occurs. Power system analyses were performed with the Power System Analysis Toolbox PSAT [10].

4.2.1 Example 1: Nine-Node System

Figure 8 shows the nine-node system. Eigenvalues for the Laplacian matrix **L***n* are shown with the eigenvector for the second smallest and third smallest eigenvalues. Eigenvector for the second smallest eigenvalue allow us to define the most suitable bisection of the graph and the eigenvector entries for the third smallest eigenvalue allow us to define a suitable partition in three sub graphs. Graph partition defines link importance.

To find a cut to separate the graph into two parts, we use the eigenvector corresponding to the second smallest eigenvalue. This approach gives the result shown in Fig. 8.

When using the weighted matrices, we may have slightly different results. However, the results obtained with the un-weighted matrices are good enough considering that simpler calculations and less information are required.

4.2.2 Example 2: IEEE 39-Node Reliability Test System

The same approach was applied to a larger system, the IEEE Reliability Test System. Results for a two-way partitioning and important nodes are presented in

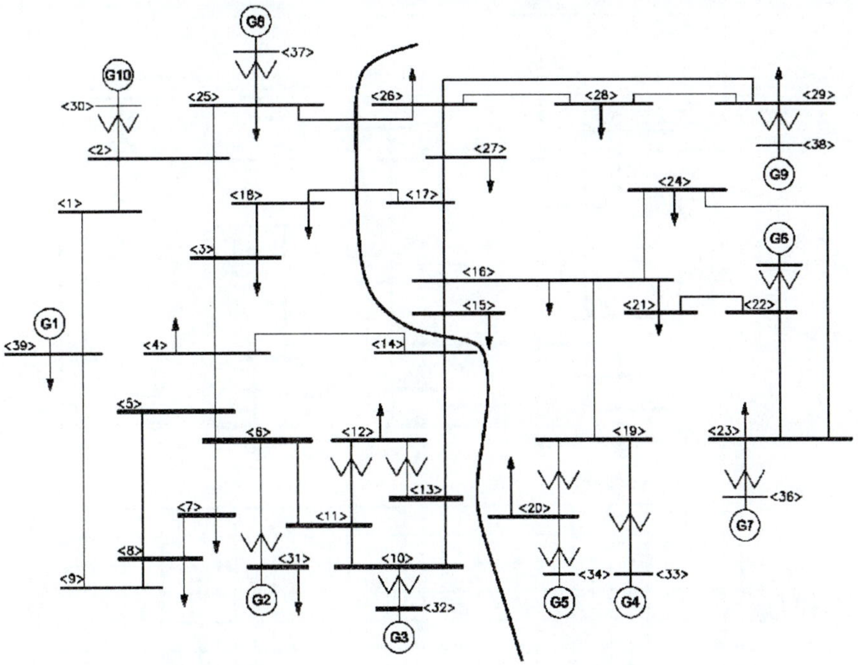

Fig. 9 Two-way partitioning for the IEEE 39-node reliability test system. System diagram taken from [19]

Fig. 9. Figure 10 shows the application of a higher eigenvector for a three-way partition problem.

The three-way partitioning obtained by using the eigenvectors of the third highest eigenvalue is similar to the partition obtained by the slow coherency methods [21]. Figure 10 shows the three-way partition and the entries of the eigenvector for the third highest eigenvalue. The main difficulty may reside in defining the dividing value for the eigenvector entries to obtain the desired partitioning.

The slow coherency in a power system is defined by the connection matrix of the system and the generator inertias. If we consider the linear electromechanical model for an n-machine power system and neglecting damping and the off-diagonal conductance terms, we have:

$$\begin{aligned} \frac{d^2\Delta\delta}{dt^2} &= \Delta\ddot{\delta} = \mathbf{M}^{-1}\mathbf{K}\Delta\delta \\ K_{ij} &= E_iE_jY_{ij}\cos(\delta_i - \delta_j) \end{aligned} \tag{11}$$

where δ is the vector of machine rotor angles, $\mathbf{M}$ is the $n \times n$ diagonal matrix of machine inertias and $\mathbf{K}$ is the $n \times n$ connection matrix whose (i, j) entry is given in

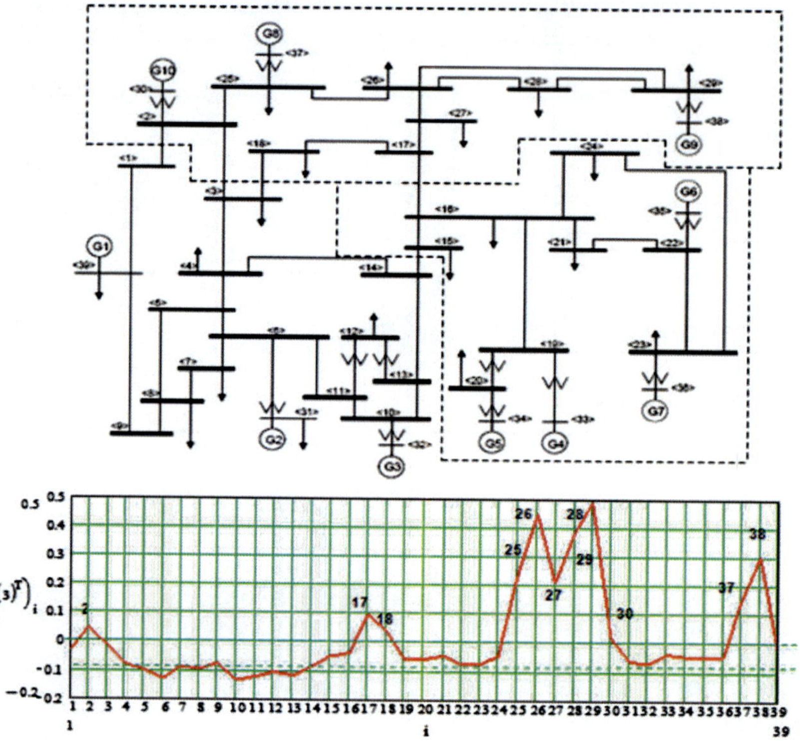

Fig. 10 Three-way partitioning for the IEEE 39-node reliability test system. System diagram taken from [19]

terms of the bus voltages and the admittance matrix. Matrix **K** is dominated by the element-node connection or the node-element incident matrix of the power network.

The slow-coherency method for coherency determination requires the calculation of the slow eigenbasis matrix of the electromechanical model of the power system where the Jacobian matrices allow for the linearization of the nonlinear system equations around an operating point [4].

Once the system state matrix is calculated and the number r of desired coherent areas is chosen, the eigenvector of the r smallest eigenvalue is calculated. It means that the eigenvector corresponding to the r smallest eigenvalue of the system state equation will allow us to obtain r slow coherent areas in the power system [17]. Although to obtain coherent areas the matrix of machine inertias is required, the main component is the connectivity matrix which represents the structural characteristic of the network. It is to say that obtaining coherent groups by eigenbasis analysis of the electromechanical model is similar to the k-way network partition using the spectral graph theory approach, as shown in Fig. 10 in which the obtained partitioning is approximately equal to the one shown in [21].

4.3 Graph Partitioning and Controlled Islanding for System Security

The determination of the strategic transmission lines in a power system as proposed in this Chapter has the main aim of identifying those circuits which will allow us to separate the system into islands in a controlled way to reduce the possibility of system collapse when subjected to severe disturbances.

The strategy of controlled island separation of a power system is becoming of paramount importance considering the need to prevent blackouts. System blackouts could occur as a consequence of severe disturbances or cascading events that could be produced by catastrophic atmospheric conditions or man-produced events such as terrorism acts.

The island-separated system provides temporal conditions that will allow for a controlled operation and response to the extreme conditions. Once the effects of the disturbances no longer present a threat to the system integrity and security, the system will go back to the interconnected operation.

To show this strategy, the 68-node system shown in Fig. 11 has been used. This system is a reduced equivalent of the interconnected New England test system (NETS) and New York power system (NYPS). There are five geographical regions; areas 3, 4 and 5 are approximated by equivalent generators models [13]. Network partitioning is obtained by using the second smallest eigenvalue and the results show the strategic lines for security islanding. The strategic circuits are connecting busses 60–61(2), 27–53 and 53–54(2) [17].

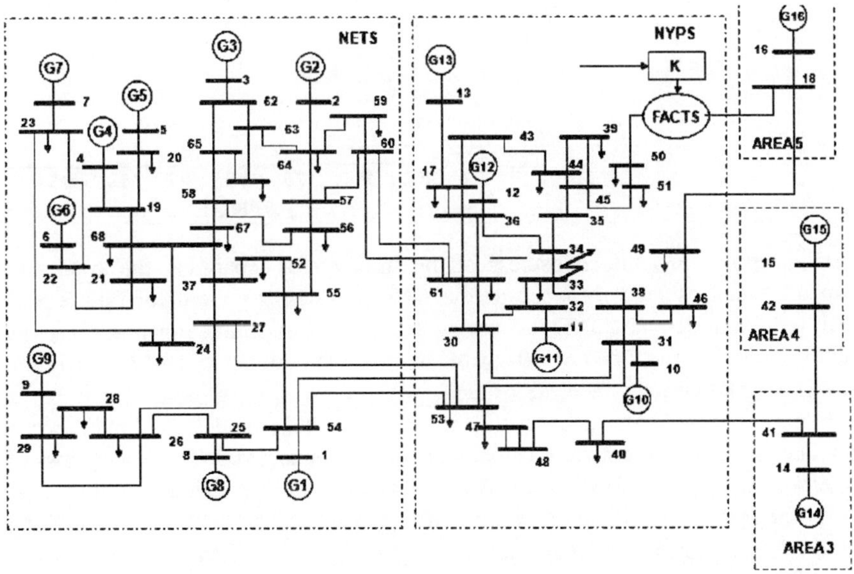

Fig. 11 68-Node system

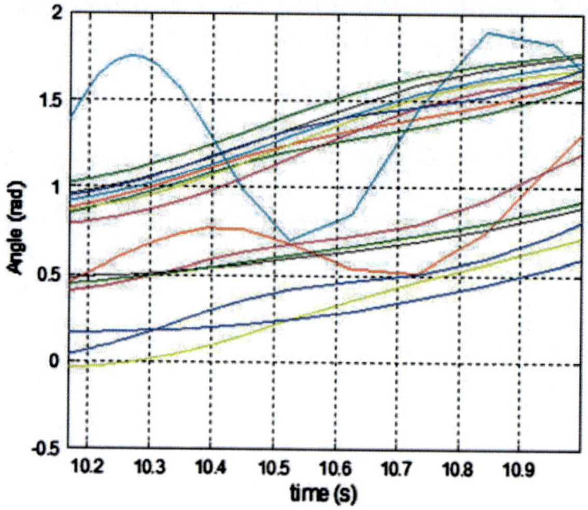

Fig. 12 Angular swings for generators in the 68-node system during a fault. Taken from [17]

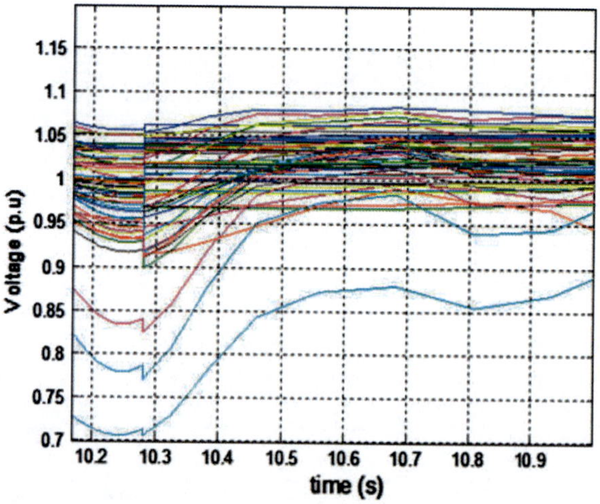

Fig. 13 Bus voltages in the 68-node system during a fault. Taken from [17]

In order to show the robustness of the strategy for separating the system to respond to a fault, the response of the 68 bus system to a critical failure was analyzed. A three-phase fault occurs near bus 33. For simulation purposes, the fault happened at the time $t = 10$ s, time to clear the fault is $t = 10.17$ s, and time for system restoration is $t = 50$ s.

Figures 12 and 13 show the angular rotor swings and voltages of all buses when the fault occurs and the system is maintained interconnected. We can see a loss of synchronism of generators and important deviations on the voltage profiles.

Power system is split into two islands by opening the strategic lines. Time required to perform system islanding by opening all selected strategic lines is $t = 10.28$ s. Machines oscillate coherently and bus voltages behave much better. Figure 14 shows bus voltages after system interconnection is reestablished.

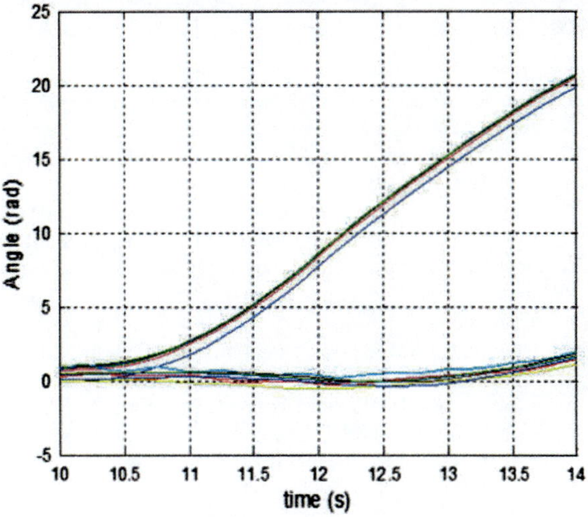

Fig. 14 Bus voltages after interconnection is re-established. Taken from [17]

5 Conclusions

The bulk transmission system is made up of different components such as lines, transformers, and so on, and these components are connected together to perform the function of transmitting electric energy, economically, reliably and safely from generating stations to distribution systems. In a competitive electricity market, owners of the transmission systems are required by regulatory rules to manage effectively their assets and resources and to maintain their systems to specific performance standards. Meeting these standards could be a challenging task for transmission owners or providers particularly if they want to maintain a good investment rate of return for their shareholders. In order for utilities to meet such challenges, new power assessment methods are required.

The identification of strategic substations will allow the transmission system owners and regulators to provide security standards and requirements that will focus on those substations that are more important for the system security and interconnection integrity. The concept of network centrality based on spectral graph theory provides a method to rank substation based on the intrinsic structural characteristics of a network.

The identification of strategic transmission lines from the point of view of the network properties and not from the considerations of transmission capacity or line cost or its properties, provides a method for security enhancement by splitting the power system as a defensive strategic response for severe or catastrophic events.

Graph theory has been used traditionally for matrix analysis in power systems. Spectral graph theory is a promising field of research in reliability, security, observability and state estimation.

Acknowledgments We would like to thank our student Ricardo Moreno for the fault simulations of a power system results of which are included at the end of this Chapter.

References

1. Amin M, Wollenberg BF (2005) Toward a smart grid. IEEE Power and Energy Magazine, Sept–Oct 2005, pp 34–41
2. Anders GJ, Torres A, Da Silva A (2010) Guide to identify and rank critical facilities. CEATI report T083700-3357
3. Avizienis A, Laprie J, Randell B, Landwehr C (2004) Basic concepts and taxonomy of dependable and secure computing. IEEE Trans Dependable Secur Comput 1(1):11–33
4. Chow JH, Cullum J, Willoughby R (1984) A sparsity-based technique for identifying slow-coherent areas in large power systems. IEEE Trans Power Apparatus Syst PAS-103(3):463–473
5. Chung Fan (2007) Four proofs for the Cheeger inequality, graph partition algorithms. ICCM II:1–4
6. Fiedler M (1973) Algebraic connectivity of graphs. Czech Math J 23(98):298–305
7. Güler T, Gross G (2007) Detection of island formation and identification of causal factors under multiple line outages. IEEE Trans Power Syst 22(2):505–513
8. Hines P, Blumsack S (2008) A centrality measure of electrical networks. In: Proceedings of the 41st Hawaii international conference on system sciences
9. Hogben L (2005) Spectral graph theory and the inverse eigenvalue problem of a graph. Electron J Linear Algebra 14:12–31
10. Liu Y, Liu Y (2006) Aspects on power system islanding for preventing widespread blackout. In: Networking, sensing and control. ICNSC '06. Proceedings of the 2006 IEEE international conference, pp 1090–1095
11. Milano F (2010) Power systems analysis toolbox, PSAT. University of Castilla. http://www.power.uwaterloo.ca/~fmilano/psat.htm
12. Ostrowski AM (1960) A quantitative formulation of Sylvester's law of inertia. Proc Natl Acad Sci 46(6):859–862
13. Pal B, Chaudhuri B (2005) Robust control in power systems. Springer, New York, pp 40–45
14. Perron-Frobenius Theorem (2010) http://www.prenhall.com/divisions/esm/app/ph-linear/leon/html/perron.html
15. Shi J, Malik J (2000) Normalized cuts and image segmentation. IEEE Trans Pattern Anal Mach Intell 22(8):888–905
16. Senroy N, Heydt G, Vittal V (2006) Decision tree assisted controlled islanding. IEEE Trans Power Syst 21(4):1790–1797
17. Torres A, Moreno R (2010) A strategy of security in power system: controlled islanding of PowerSystems. University of Los Andes (in press)
18. Torres A, Anders GJ (2009) Spectral graph theory and critical nodes. Fourth international conference on dependability of computer systems DepCoS-RELCOMEX 2009, Wrocław University of Technology, Institute of Computer Engineering, Control and Robotics, June 2009
19. University of Winsconsin, Electrical and Computer Engineering, IEEE benchmarks (2010) http://psdyn.ece.wisc.edu/IEEE_benchmarks/index.htm
20. You H, Vittal V, Wang X (2004) Slow coherency-based islanding. IEEE Trans Power Syst 19(1):483–491
21. Yusof SB, Rogers GJ, Alden RTH (1993) Slow coherency based network partitioning including load buses. IEEE Trans Power Syst 8(3):1375–1382

Transmission Expansion Planning: A Methodology to Include Security Criteria and Uncertainties Using Optimization Techniques

Armando M. Leite da Silva, Leandro S. Rezende and Luiz Antônio F. Manso

1 Introduction

The main objective of the multi-stage transmission expansion planning (TEP) is to define where, when, and what reinforcements should be placed in the electrical network, to ensure an adequate quality level of energy supply to customers. In a competitive energy market, TEP is a complex optimization task to ensure that the power system will meet the predicted demand and the security criteria, along the planning horizon, while minimizing investment, operational, and interruption costs. This practice is the only rational response to conflicting customer and regulatory demands [1, 2].

The multi-stage nature of the TEP problem requires consideration of multi-time periods and determining possible sequences of transmission reinforcements. To circumvent the multi-stage nature, simplified studies (also known as static analyses) determine, for just one stage, where new transmission facilities should be installed. Several works to solve the TEP problems can be found in the literature [2–16]. However, only a few have considered the multi-stage nature of the TEP problem [7, 9, 12, 15, 16].

A. M. Leite da Silva (✉) and L. S. Rezende
Institute of Electric Systems and Energy, Federal University of Itajubá – UNIFEI, Itajubá MG, Brazil
e-mail: armando@unifei.edu.br

L. S. Rezende
e-mail: leandrorezende@unifei.edu.br

L. A. F. Manso
Department of Electrical Engineering, Federal University of São João Del Rei – UFSJ, São João Del Rei MG, Brazil
e-mail: lmanso@ufsj.edu.br

G. Anders and A. Vaccaro (eds.), *Innovations in Power Systems Reliability*, Springer Series in Reliability Engineering, DOI: 10.1007/978-0-85729-088-5_6,

Optimization approaches based on metaheuristics [5–18] have demonstrated the potential of finding high-quality solutions. Numerous advantages can be linked to these tools: the algorithms are relatively simple, they are able to mix integer and non-integer variables, and also present relatively faster time-responses. Their success is related to the ability to avoid local optima by exploring the basic structure of each problem. Several metaheuristics have been proposed in the last decade to solve the TEP problems, e.g., simulated annealing (SA) [5], Tabu search (TS) [6, 7, 15], genetic algorithms (GA) [8, 9, 18], greedy randomized adaptive search procedure (GRASP) [10, 11], evolution strategies (ES) [12, 15], differential evolution (DE) [13], particle swarm optimization (PSO) [14], ant colony optimization (ACO) [15], and artificial immune systems (AIS) [16].

Owing to today's power network dimensions, random behavior of transmission and generation equipments, load growth uncertainties, new generator source types and locations, market aspects, etc., the TEP problem has become combinatorial, stochastic, and highly complex. When uncertainties and chronological aspects are added to these problems, the optimal solution becomes almost inaccessible, even when using the previously mentioned metaheuristics.

Traditionally, in order to ensure security criteria, TEP problems have been solved by using deterministic approaches, such as the so-called "N-1" and "N-2" criteria, which are based on the worst-case analyses (drawn from single or double contingencies) [18, 19]. In many cases, however, the obtained plan may lead to over investments. On the other hand, although probabilistic approaches [20, 21] are able to measure the quality of a power network, there is no definition of how reliable a power system must be based on the reliability indices, such as loss of load expectation (LOLE), expected power not-supplied (EPNS), and loss of load cost (LOLC), mainly because the indices are dependent on, for e.g., the load model. Since electric companies have not reached a consensus about this subject, they have preferred deterministic rather than probabilistic approaches.

The treatment of external uncertainties corresponds to an extremely complex task, since it substantially contributes to the increase of the expansion plans combinations. Another complicated aspect is how to model these uncertainties, mainly when one cannot access past information to define their future behavior. In this case, these uncertainties can be represented by scenario techniques, decision trees, and/or fuzzy sets [22–24].

This chapter proposes a new methodology to solve the multi-stage TEP problem considering security criteria and the treatment of external uncertainties, as the load/generation growth. Optimization approaches based on metaheuristics, Monte Carlo simulation, and scenario techniques are among the tools selected in the developed procedure. In addition, a discussion about how to ensure security criteria using deterministic and probabilistic approaches is presented through a case study on the Garver system [25]. A real transmission network is used to illustrate the developed methodology.

2 Transmission Expansion Planning Problem

Different from most works that perform a static planning, the present chapter solves the TEP problem considering the chronology of reinforcements, the interest of which is not only to define what reinforcements should be placed in the electrical network and their corresponding locations, but also when they will be added along with the planning horizon to ensure an adequate level of energy supply to the customers. At the end, the best expansion plans must be selected in order to minimize the present value costs involved in the objective function.

The next sections are divided as follows. First, the mathematical formulation of the multi-stage TEP problem is presented, including the solution matrix definition and the linear programming algorithm, based on a DC flow model, used to evaluate the candidate plans built by the optimization tool. Then, a discussion on how to insert the operational costs associated with the ohmic losses in the objective function using the linear DC flow model is presented. Next, security criteria are discussed, focusing on the deterministic and probabilistic approaches. Finally, the challenges of considering external uncertainties and the tools available to treat them are presented.

2.1 TEP Problem Representation

The first step, considering the mathematical formulation of the TEP problem, is to define the representation of the solution matrix S^k (also named sequence), which corresponds to the candidate plan k, as follows:

$$S^k = \begin{bmatrix} s_{11}^k & s_{12}^k & \cdots & s_{1l}^k & \cdots & s_{1n}^k \\ s_{21}^k & s_{22}^k & \cdots & s_{2l}^k & \cdots & s_{2n}^k \\ \vdots & \vdots & \vdots & \vdots & \vdots & \vdots \\ s_{t1}^k & s_{t2}^k & \cdots & s_{tl}^k & \cdots & s_{tn}^k \\ \vdots & \vdots & \vdots & \vdots & \vdots & \vdots \\ s_{y1}^k & s_{y2}^k & \cdots & s_{yl}^k & \cdots & s_{yn}^k \end{bmatrix} \tag{1}$$

where n indicates the number of possible network branches allowed to receive reinforcements, y corresponds to the number of stages along the planning horizon, and s_{tl}^k refers to the total number of reinforcements at stage t and branch l in relation to the base system network configuration. If only one stage is considered (static problem), one has the following solution vector:

$$S_t^k = \begin{bmatrix} s_{t1}^k & s_{t2}^k & \cdots & s_{tl}^k & \cdots & s_{tn}^k \end{bmatrix} \tag{2}$$

During the optimization process, the selection of candidate plans is carried out taking into account the entire horizon, aiming at minimizing the present value of the objective function to be defined later in this chapter.

To build a sequence, the maximum number of reinforcements in each branch must not be exceeded. In addition, it is important to ensure a coordination scheme of the added reinforcements. For example, reinforcements inserted in stage t must be mandatorily included in the following stages: $t+1$, $t+2$, ..., y. These constraints are defined as follows:

$$\begin{aligned} s_{tl}^{k} &\leq N_{l\max} \quad \forall l \in \{1,\ldots,n\}; \quad \forall t \in \{1,\ldots,y\}. \\ s_{tl}^{k} &\leq s_{(t+1)l}^{k} \quad \forall l \in \{1,\ldots,n\}; \quad \forall t \in \{1,\ldots,y-1\}. \end{aligned} \tag{3}$$

where $N_{l\max}$ refers to the maximum number of reinforcements allowed in branch l.

Every time a candidate plan k is obtained, an evaluation is performed through a linear programming (LP) based on a DC power flow model, as described below:

Minimize:

$$z = \alpha^T r \tag{4}$$

subject to:

$$g + r + B\theta = d \tag{4a}$$

$$0 \leq g \leq g_{\max} \tag{4b}$$

$$0 \leq r \leq d \tag{4c}$$

$$|f| \leq f_{\max} \tag{4d}$$

where α refers to the load shedding penalty vector; r is the load not-supplied vector; g is the generation bus vector; B represents the susceptance matrix; θ is the voltage angle vector; d is the load vector; $g_{\max}$ refers to the generation limit bus vector; f is the power flow vector; and $f_{\max}$ represents the power flow limit vector.

This LP algorithm is applied to each stage t of the planning horizon and can be efficiently solved by the interior point method. The Lagrange multipliers associated with the constraints of (4a) are of high interest, since they can assist in the construction process of better quality initial sequences for the metaheuristic tools, as it will be discussed in Sect. 3.4. These multipliers measure the benefits in terms of load not-supplied index concerning changes in the circuit by the addition of new reinforcements. Denoting by π^d this Lagrange multiplier vector, the benefits can be estimated by [10]:

$$\pi_{ij}^{d} = (\theta_i - \theta_j)\left(\pi_i^d - \pi_j^d\right) \tag{5}$$

where π_{ij}^{d} is the Lagrange multiplier associated with the branch susceptance connecting buses i and j.

The proposed heuristic function also considers the investment costs associated with the new reinforcements, and it will assist the metaheuristics in finding better quality sequences. It is defined as follows [15]:

$$\eta(i,j) = \frac{\pi_{ij}^{d}}{\mathrm{Cinv}_l} \tag{6}$$

where $Cinv_l$ is the unit investment cost for the reinforcements added in branch l that connects buses i and j.

Once sequence k is found, for a specific planning horizon composed of y stages, the objective function of the multi-stage TEP problem can be described by:

$$\mathrm{Min}\, f(S^k) = \sum_{t=1}^{y} \frac{\left(\sum_{l=1}^{n} \mathrm{Cinv}_l m_{tl}^{k} + \mathrm{Closs}_t^{k} + \alpha^T r_t\right)}{(1+e)^{h(t)}} \tag{7}$$

where $f(S^k)$ represents the total cost function in present value terms, associated with sequence k; e indicates the discount rate; $h(t)$ corresponds to a function that informs the numerical difference between the year of stage t and the base year; $Cinv_l$ was already defined in (6); m_{tl}^{k} refers to the number of reinforcements located at stage t and branch l of sequence k, i.e., $m_{tl}^{k} = s_{tl}^{k} - s_{(t-1)l}^{k}$ (if t represents the initial stage, then $m_{tl}^{k} = s_{tl}^{k}$); α and r_t are the same as defined in (4); and $Closs_t^k$ represents the operational costs associated with ohmic losses at stage t of sequence k, which will be described in the next section.

2.2 Ohmic Losses

To include the operational costs associated with the ohmic losses in the optimization process, a special DC flow model is used. Basically, the losses are calculated using the voltage angle vector obtained by the LP solution of a given configuration. Then, these losses are distributed as loads, where terminal buses i and j receive half of the ohmic losses found in the circuit that connects these buses. Again, a new LP is solved considering this new increased load and a new voltage angle vector is found. This corresponds to the solution for a given configuration.

The amount of losses associated with the circuit between buses i and j can be approximated as:

$$P_{ij} = \left(r_{ij} \times f_{ij}^{2}\right) \tag{8}$$

where r_{ij} is the resistance of the circuit and f_{ij} is the active power flow, and all quantities are in pu. The total operational ohmic losses cost (*C*loss) is given by [15]:

$$Closs = 8,736 \times C_{kWh} \times LF \times \sum_{\forall ij} P_{ij} \tag{9}$$

where C_{kWh} represents the loss unit cost in US$/kWh, LF is the loss factor, which modulates the load curve, and the value 8,736 (i.e., 52 weeks) aims at converting the incremental loss costs into annualized costs.

2.3 Security Criteria

The objective of the TEP problem is not only to obtain a plan that minimizes the investment and operational costs, as shown in (7), but also to ensure an adequate quality level of energy supply to customers. Although the consideration of security criteria is essential in real networks, most research works available in the literature solve the multi-stage TEP problem only for the intact network [5–16]. Deterministic approaches, as the "N-1", are taken into account only in a few works [18, 19].

Another possibility to consider security criteria is through the evaluation of the reliability indices. However, the difficulty that appears in probabilistic approaches is the high computational effort needed to evaluate the candidate plans. In the following sections, a detailed explanation about how to include security criteria in multi-stage TEP problems is described considering the deterministic and probabilistic approaches.

2.3.1 Deterministic Approaches

Traditionally, security criteria are accomplished by the well-known "N-1" and "N-2" deterministic approaches. One of the strategies consists in applying, for example, the "N-1" criterion only to the best plans found by the optimization tool, i.e., those that satisfy the *intact network criterion*: i.e., all equipment operating within their limits with loads preserved. Nevertheless, one cannot guarantee that the final plans, after considering the "N-1", refer to the best ones.

Another strategy is to take into account the "N-1" criterion, as a constraint, during the optimization process. In this case, the best plans found by the optimization process have not only to satisfy the intact network criterion, but also the removal or contingency of any single transmission element. However, for large networks, the computational effort can become prohibitive and the obtained plans may lead to over investments. An alternative would consist in applying a contingency list formed by only the most important circuits located at interest areas.

The advantage of using contingency lists is that the circuits can be included based not only on static analyses information, but also on their dynamic behavior analyses. However, bearing in mind the chronology of the TEP problem, it may be difficult to create a general contingency list, since the system conditions, generation, and load points may be completely different along the planning horizon and, consequently, the importance of the circuits may change.

2.3.2 Probabilistic Approaches

An alternative way to consider security criteria is to measure the quality of the network through the probabilistic approaches. In using such approaches, it is possible to include, for example, costs related to system reliability in the objective function. A new objective function is presented considering the reliability index LOLC, as follows:

$$\text{Min}\, f(S^k) = \sum_{t=1}^{y} \frac{\left(\sum\limits_{l=1}^{n} \text{Cinv}_l m_{tl}^k + \text{Closs}_t^k + \alpha^T r_t\right) + \text{LOLC}_t^k}{(1+e)^{h(t)}} \tag{10}$$

where LOLC_t^k refers to the energy interruption cost considering the stage t of sequence k and all others terms were already defined in (7).

However, the use of the probabilistic approaches may also become prohibitive during the optimization process, due to the high computational effort needed to evaluate the candidate plans, mainly if large networks are considered. In these cases, an interesting option may be the use of approximated models.

In [20], a state enumeration technique is proposed, where only the "N-1" probabilistic criterion is considered together with a peak load model to select the candidate plans. Although the estimated reliability indices may not be exact, this is a good strategy to handle security criteria. After the optimization process, in an external loop, the best plans selected can be reevaluated using an hourly load model and a more complete probabilistic approach, such as the pseudo-chronological Monte Carlo simulation [21].

Although probabilistic approaches are able to measure the quality of a power network, there is no clear definition of how reliable a power system must be, since the reliability indices are dependent on the load model. Some works [7, 12, 15] have included the evaluation of reliability indices just after the optimization process, in an external loop. In this case, although the best selected plans have lower investments costs, they may not ensure any security criteria.

In Sect. 4, a case study on the Garver system is presented showing the results achieved with and without consideration of the "N-1" deterministic approach during the optimization process. The purpose is to make a comparison among the best sequences found in the two situations based on the objective function described by (10), where the LOLC indices will be calculated in an external loop using the pseudo-chronological Monte Carlo simulation.

As a conclusion, a discussion on how to consider security criteria using deterministic and probabilistic approaches is presented.

2.4 External Uncertainties

During recent years, electric power companies have been submitted to structural changes. Among these changes, one can mention the deregulation of electrical energy generation, a higher access to transmission systems, changes in the economical regulations, increasing concerns regarding environmental impacts, etc. [22]. Furthermore, substantial uncertainties have been observed related to: future load growth, fuel prices, generator source types and locations, costs of different reinforcements in the systems, etc. Since such external uncertainties do not follow a behavior that can be described based on past information or statistical rules, there is a higher difficulty in modeling them. Therefore, one of the main challenges imposed on power system planers is to identify the most relevant uncertainties and provide instruction on how to treat them.

One solution to treat uncertainties consists of adopting flexible plans, which must be able to quickly adapt the original plan with reasonable costs, independently of any system change. For this purpose, the uncertainties can be represented by scenario techniques, decision trees, and/or fuzzy sets [23, 24].

In Sect. 5, a procedure to solve a real multi-stage TEP problem is presented, where the scenario technique to treat the load/generation growth uncertainty is included. In this method, the possible values that some parameters can assume in the future (e.g., load growth rate) define the scenarios to be evaluated. The strategies (expansion plans) are, therefore, evaluated under the conditions of each scenario and the decision is taken based on the set of strategies and scenarios using the Minimum Expected Cost or the MiniMax Regret approaches [23].

3 Optimization Approaches Based on Metaheuristics

Metaheuristics [5–18] have demonstrated the potential to find high-quality solutions to solve a range of problems, including the TEP. Numerous advantages can be linked to these approaches: the software complexity is acceptable, they are able to mix integer and non-integer variables, and also present a faster time-response. The success of such approaches is related to their ability to avoid local optima by exploring the basic structure of each problem. Traditionally, the metaheuristics can be classified in two major groups: evolutionary algorithms (EA) and swarm intelligence (SI). However, some approaches do not share the same characteristics of the previously mentioned groups and, thus, they must be classified separately, as the Tabu Search.

3.1 Evolutionary Algorithms

Evolutionary algorithms refer to a subset of the evolutionary computation (EC) that use mechanisms inspired by biological evolution, such as reproduction, mutation, recombination (crossover), and selection. In these algorithms, the solutions, which represent a population of individuals, are modified along the evolution of generations by application of the operators previously mentioned. Nowadays, there are a wide variety of algorithms sharing these characteristics, such as the evolutionary programming (EP), GA, ES, and DE. Since the AIS comprises the three major principles of Charles Darwin's theory of evolution, i.e., repertoire diversity, genetic variation, and natural selection, this approach can also be included in the evolutionary algorithms class. These approaches differ among themselves regarding the evolution mechanisms. Nevertheless, each one has copied from and exchanged ideas with each other becoming quite similar.

Considering the TEP problem, several works can be found in the literature using GA [8, 9, 18], ES [12, 15], DE [13], and AIS [16] metaheuristics. Although there are important applications using GA, in the following sections, a brief model description is provided taking into account only the ES, DE, and AIS that refer to the most recent approaches used to solve the TEP problems.

3.1.1 Evolution Strategies

Unlike most GA, this metaheuristic does not need a codification process since it works with floating point representations. Also, contrary to GA that incorporate both recombination and mutation, ES just use the mutation operator that consists of adding to the sequence k, defined by (1), a normally distributed perturbation Z_t^k at all t stages of the planning horizon, as follows:

$$\tilde{S}_t^k = S_t^k + Z_t^k \tag{11}$$

$$Z_t^k = \sigma \times [N_{t1}(0,1) \quad N_{t2}(0,1) \quad \cdots \quad N_{tl}(0,1) \quad \cdots \quad N_{tn}(0,1)] \tag{12}$$

where $\tilde{S}_t^k$ represents a new individual obtained through the S_t^k mutation; σ is the mutation magnitude; $N_{tl}(0, 1)$ corresponds to the normal distribution. As the variables of the TEP problem are discrete, a "Round" function needs to be applied to each element $\tilde{s}_{tl}^k$ of the new individual.

Bearing in mind the TEP problem, the $(\mu + \lambda)$ ES model has been used with a high success [12, 15], where the best sequences between the parent (μ) and offspring (λ) populations are chosen to the next generation using the selection operator. As the best individuals are always selected, an evolution of population quality along the generations can be observed. This procedure is repeated until a termination criterion is reached, as the maximum number of generations.

3.1.2 Differential Evolution

Unlike the GA, ES, and AIS metaheuristics that follow a probability distribution function to perform a perturbation in the individuals, the DE is based on a weighted difference among individuals, i.e., arithmetical combinations. Basically, at each generation, the offspring population is obtained by application of mutation and recombination operators. Among a range of mutation rules [13] and considering the TEP problem, the best results have been achieved when the following mutation rule is applied to all t stages of the planning horizon:

$$X_t^k = \text{Round}[S_t^{r_1} + F(S_t^{r_2} - S_t^{r_3})] \quad (13)$$

where X_t^k corresponds to the k mutated individual; r_1, r_2, r_3, and k are indices randomly chosen, where $r_1 \neq r_2 \neq r_3 \neq k$; $S_t^{r_1}$, $S_t^{r_2}$, and $S_t^{r_3}$ are the respective selected individuals; F refers to a mutation factor; and "Round" is a function that returns a rounded integer number.

The recombination operator consists of exchanging the information between mutated and original individuals. The following equation shows how this operator works considering all l branches and t stages of the planning horizon:

$$w_{tl}^k = \begin{cases} x_{tl}^k & \text{if } (\text{rand}_l \leq \text{CR}) \text{ or } (l = \gamma) \\ s_{tl}^k & \text{otherwise} \end{cases} \quad (14)$$

where rand_l is a random variable that follows a uniform distribution (0,1); γ is randomly chosen among the eligible branches to receive reinforcements; CR refers to the recombination rate; and w_{tl}^k corresponds to each position of the new individual after the application of the recombination operator.

Finally, the selection operator compares each new individual W^k obtained by the mutation and recombination operators with its respective original individuals from the parent population S^k, shown as follows:

$$S^{k(\text{new})} = \begin{cases} W^k & \text{if } f(W^k) \leq f(S^k) \\ S^k & \text{otherwise} \end{cases} \quad (15)$$

where $S^{k\,(\text{new})}$ refers to the k individual selected to the next generation; and $f(.)$ corresponds to the evaluation of the objective function.

Therefore, this selection operator performs a pair comparison, which is different from the selection operator of all others EA algorithms, where the next generation is formed by the best individuals among the parent and offspring populations.

3.1.3 Artificial Immune Systems

The AIS metaheuristic intends to capture some principles of the natural immune system (NIS). One of the main algorithms presented in the literature is the

CLONALG [26], which is based on the following concepts: reproduction, hypermutation, selection, and receptor editing. In Ref. [16], an adapted CLONALG algorithm is presented to solve the TEP problem, where the sequence S^k, described by (1), corresponds to each antibody of the immune system. In addition, each element s_{tl}^k, which refers to the reinforcement options, corresponds to a position of this antibody.

The role of the reproduction operator is to clone antibodies, i.e., to perform identical copies. As commented in [16], it is more interesting to select all antibodies of the parent population to generate clones if the optimization process aims at locating multiple optima; the objective of the TEP problem is to identify the set of best sequences and not only the best one. In addition, when the number of copies provided by each antibody is the same, a better exploration of the search space may be reached when the hypermutation operator is used [26].

Regarding hypermutation, this operator aims to achieve higher affinity antibodies, i.e., to provide better quality antibodies, by adding a perturbation Z_t^k to each clone, at all stages t of the planning horizon. This operator is the same mutation operator described in the ES metaheuristic.

Later, the selection operator is applied with the purpose of keeping the n_b best antibodies from each group, composed by the parent antibody and its respective mutated clones. It is important to mention that if just the best clone is selected from each group, all others will be disregarded and valuable information may be lost.

Finally, the receptor editing operator must choose the best antibodies obtained from the application of the selection operator. In addition, the selection of identical antibodies is not allowed. This process ensures a better quality and diversified population for the next generation, keeping the same size of the parent population. In the original CLONALG [26], the receptor editing operator aims at substituting the lower affinity antibodies by new ones. The objective is to escape from local optima and provide a better exploration of the search space. The CLONALG proposed in [16], and presented in this chapter, achieves the same objective at assuring a selection of distinct antibodies for the next generation. Better performance is reached by this modified CLONALG algorithm when applied to TEP problems, as shown in [16].

3.2 Swarm Intelligence

Swarm Intelligence is a type of artificial intelligence based on the collective behavior of organized agents. The metaheuristics included in this group, PSO [14] and ACO [15], are typically composed by a population of simple agents that interact locally with each other and with the environment. Although the agents follow quite single rules and there is no control structure responsible for guiding the behavior of the agents, the emergence of a global intelligent behavior is observed, unknown to individual agents.

3.2.1 Particle Swarm Optimization

The particle swarm optimization metaheuristic was inspired by the synchronized movement of flocks of birds. Although birds fly independently, some information may be shared among their members, emerging a social behavior. In the optimization algorithm, a velocity is attributed to each particle and then they fly through the state space problem following the best ones. Each particle is affected by three factors: their own velocity, their best position reached until that moment, and the best overall position achieved by all particles.

Considering the TEP problem [14], k sequences, (1), refer to the position of the particles. The velocity of each particle is updated for all t stages of the planning horizon, as follows:

$$V_t^{k(i_{\text{ter}}+1)} = wV_t^{k(i_{\text{ter}})} + c_1 u_1^k \left[P_t^k - S_t^{k(i_{\text{ter}})}\right] + c_2 u_2^k \left[P_t^g - S_t^{k(i_{\text{ter}})}\right] \quad (16)$$

where i_{ter} is the present iteration of the algorithm; $V_t^{k(i_{\text{ter}}+1)}$ indicates the updated velocity of the particle; $V_t^{k(i_{\text{ter}})}$ refers to the present velocity of the particle; P_t^k is the best position reached by the particle; P_t^g corresponds to the best position found by all particles; $S_t^{k(i_{\text{ter}})}$ is the present position of the particle; c_1 and c_2 are positive constant parameters; w is a parameter that may be defined by $w = 0.5 + 1/[2ln(i_{\text{ter}}) + 1]$; u_1^k and u_2^k are two random numbers that follow a normal distribution $N(0,1)$.

Later, the position of each particle in all t stages is modified by:

$$S_t^{k(i_{\text{ter}}+1)} = S_t^{k(i_{\text{ter}})} + \text{Round}[V_t^{k(i_{\text{ter}}+1)}] \quad (17)$$

where $S_t^{k(i_{\text{ter}}+1)}$ is the updated position of the particle; and "Round" is a function that returns a rounded integer number, since the updated velocity may have continuous elements and the variables of the TEP problem are discrete.

3.2.2 Ant Colony Optimization

Ants are insects that live in community and aim to establish the shortest route paths from their colonies to the feeding sources. Such result is only possible because the ants lay on the ground a substance known as pheromone during their searches. In the ACO metaheuristic, the ants use probabilistic rules to move around the search space. These rules are based on some knowledge about the problem (heuristic function) and pheromone trails laid on the paths.

In the TEP problem, each ant represents an attempt of finding a sequence, (1), where the paths refer to the eligible branches to receive reinforcements. Among several algorithms found in the literature, the so-called ant colony system (ACS) has achieved better results to solve the TEP problem, as mentioned in [15].

Considering the ACS algorithm, an ant chooses to add reinforcements to the network, at each stage t of the planning horizon, through the following state transition rule:

$$c = \begin{cases} \underset{(i,j)\in T_t^k}{\arg\max}\left\{ \left[\tau(i,j)_t\right] \left[\eta(i,j)_t\right]^{\beta} \right\}, & \text{if } \quad q \le q_0 \\ C, & \text{otherwise} \end{cases} \tag{18}$$

where c indicates the selected reinforcement; $\tau(i,j)_t$ corresponds to the pheromone trail laid on the branch that connects buses i and j; $\eta(i,j)_t$ represents the value of the heuristic function associated with branch (i,j); β defines the importance of the heuristic function; arg max is a function that selects the maximum value for the result $[\tau(i,j)_t][\eta(i,j)_t]^{\beta}$; T_t^k is the set of branches that did not achieve the maximum number of reinforcements permitted; q refers to a random number that follows a uniform distribution (0,1); q_0 is a constant parameter ($0 \le q_0 \le 1$); C represents a random variable that follows a discrete distribution given by (19), where $p_t^k(i,j)$ is the probability of choosing branch (i,j) to receive a reinforcement, as follows:

$$p_t^k(i,j) = \begin{cases} \dfrac{[\tau(i,j)_t][\eta(i,j)_t]^{\beta}}{\sum\limits_{(t,u)\in T_t^k} \{[\tau(t,u)_t][\eta(t,u)_t]^{\beta}\}}, & \text{if } (i,j) \in T_t^k \\ 0, & \text{otherwise} \end{cases} \tag{19}$$

Regarding the pheromone trails, two rules are used to update them. The local update rule is carried out while the sequences are being built up by the ants. The objective is to reduce the trails over the most relevant branches avoiding the same reinforcements to be selected by all ants. On the other hand, the global update rule is applied only when all ants have already finished their searches. This rule uses the information of the best sequence found by the ants to update the pheromone trails to the next algorithm iteration, i.e., the search of a new group of ants. The objective is to increase the pheromone trails over the branches that received reinforcements and reduce the trails over the least important branches, considering the configuration of the best sequence. More details about equations and how to use the pheromone update rules can be found in [15].

3.3 Tabu Search

The Tabu search metaheuristic is an optimization technique that uses a flexible memory of the previously visited states. This memory guides the process in such a way that the search is not interrupted or disturbed when there are no movements that improve the current solution. The metaheuristic avoids that recently visited local optima have an attractive effect on the search trajectory, ensuring a more intelligent exploration as compared to traditional local search methods.

The basic concepts of the TS approach are: neighborhood, movement, Tabu list, aspiration criterion, intensification, and diversification.

Bearing in mind the TEP problem [6, 7, 15], the neighborhood of a sequence k, (1), is defined as the set of sequences obtained through elementary modifications (movements) performed in the present sequence. A movement corresponds to an addition or removal of just one reinforcement in any stage t of the planning horizon, as long as (3) is respected.

The process of defining the neighborhood and selecting the best neighbor to the next iteration is named intensification. This procedure is applied even when the best neighbor is worse than the current sequence. This is important, but not sufficient to avoid a premature imprisonment in local optima. In addition, a Tabu list is used, which is made up of rules that prohibit opposite movements previously performed, also preventing the repetition of cycles. The size of the tabu list is an important parameter that defines how many iterations a reverse movement will stay in the list. In the multi-stage TEP problem, each constraint of the tabu list must inform the movement type (addition or removal), the location (which system branch), and what stage t of the planning horizon.

A tabu constraint may be ignored when the visited solution is the best one found until that point. This is determined by the aspiration criterion that allows a solution prohibited by the tabu list to be exceptionally visited.

Finally, a diversification procedure should be employed from time to time restarting the intensification process with new initial sequences, and helping the metaheuristic to avoid the imprisonment in local optima. Diversification allows the algorithm to make a broader search, visiting regions not yet explored and with a greater probability of involving the whole universe of possible sequences for the problem.

3.4 Final Remarks

Several works [6, 7, 9–12, 15, 16, 18] discuss the importance of using a procedure to build initial good-quality sequences, which contributes to a better performance of the optimization approaches based on metaheuristics. In these references, it is shown that there is a higher chance of finding best plans when metaheuristics consider sequences of good quality as starting points for the search.

A strategy is to use the heuristic function given by (6) as the basic knowledge to build initial good-quality sequences. Among the metaheuristics described in the previous sections, the only one that does not need initial good-quality sequences is the ACO. However, the same heuristic function is essential to help in the building process of sequences, as it can be seen by the state transition rule, (18) and (19). Examples of procedures to build these initial sequences, bearing in mind the multi-stage TEP problem, are presented in [15, 16], which is called *Intelligent Initialization*.

4 Results

A case study on the Garver system [25] is presented in this section. The objective is to make a comparison among the best sequences found by an optimization approach, considering or not the "N-1" deterministic security criterion. The comparison also takes into account the LOLC index that is included in the objective function, (10), together with the investments and operational costs associated with the ohmic losses. A discussion about how to ensure security criteria using deterministic and probabilistic approaches is presented. The optimization program was developed using the MATLAB environment, while the pseudo-chronological Monte Carlo simulation was built using a FORTRAN compiler. All results refer to a *Pentium Core 2 Duo* processor (2.66 GHz).

4.1 Garver System

The configuration of the Garver system before expansion is given in Fig. 1. The system has 6 buses and 6 transmission lines. The installed capacity is 1,110 MW and the load peak is 760 MW. The generation and load data are presented in Table 1. In the reference year, the objective is to connect the generation bus #6 to the system and to satisfy the operational constraints. The system expansion horizon is divided into 6 stages and for each stage the load and generating capacities are increased by 5%. Therefore, the installed capacity and load will be of about 1,417 MW and 970 MW, respectively, at the end of the period of analysis. The connection between any two buses in the system is allowed with a limit of four parallel lines in each right-of-way. The existing and new transmission lines data are shown in Table 2. An investment cost of 1.00×10^6 US$/mile is considered for any new transmission line added to the system.

To calculate the present value of the objective function, a discount rate (e) of 10% is used. In relation to the operational costs associated with the ohmic losses, the following parameters are used: $C_{\mathrm{kWh}} = 0.10$ US$/kWh and LF = 0.6144.

Since transmission is the major concern of this work, a failure rate of 0.0781 per year per mile and a repair time of 10 h are assumed for the transmission lines, while the failure rates of the generating units are made nil. However, their capacities are taken into account in the transmission reliability evaluation that is performed in an external process using the pseudo-chronological Monte Carlo simulation [21], i.e., after the identification of the best sequences by the optimization approach. To obtain the LOLC index, a unitary interruption cost of 1.50 US$/kWh and the peak load are considered in the reliability studies.

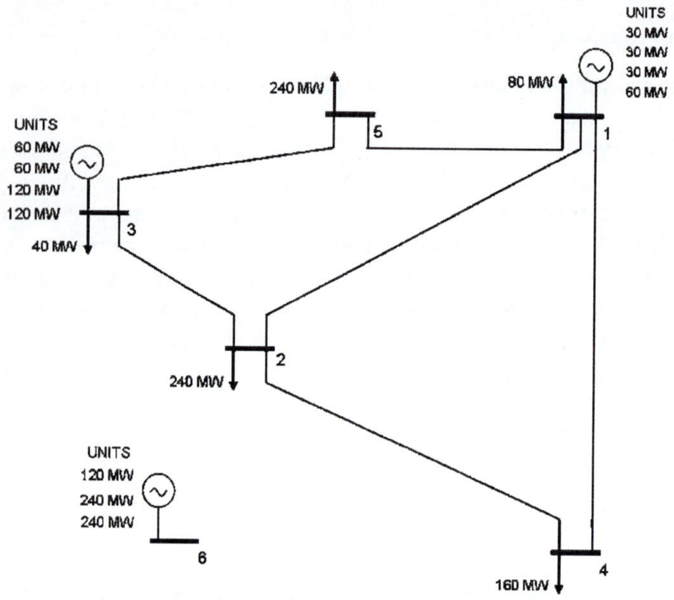

Fig. 1 Garver system

Table 1 Garver system—generation and load data

Bus	Generation (MW)	Load (MW)
1	150	80
2	–	240
3	360	40
4	–	160
5	–	240
6	600	–
Total	1,110	760

4.2 Optimization Approach

Among several optimization approaches described in Sect. 3, the metaheuristic evolution strategies (ES) was chosen to obtain the 10 best sequences that minimize the investment and operational costs related to ohmic losses, (7). Considering the ES metaheuristic, the following parameters were selected: a parent population of 50 individuals (sequences), $\mu = 50$; mutation magnitude applied to branches in the last stage y of planning horizon, $\sigma_y = 0.3$; mutation magnitude for the other t stages, $\sigma_t = 0.6$; an offspring population of 50 individuals, $\lambda = 50$. The algorithm is interrupted after a maximum number of 100 generations.

Table 2 Transmission line data

Terminals	R (pu)	X (pu)	Capacity (MW)	Length (miles)
1-2	0.10	0.40	100	40
1-3	0.09	0.38	100	38
1-4	0.15	0.60	80	60
1-5	0.05	0.20	100	20
1-6	0.17	0.68	70	68
2-3	0.05	0.20	100	20
2-4	0.10	0.40	100	40
2-5	0.08	0.31	100	31
2-6	0.08	0.30	100	30
3-4	0.15	0.59	82	59
3-5	0.05	0.20	100	20
3-6	0.12	0.48	100	48
4-5	0.16	0.63	75	63
4-6	0.08	0.30	100	30
5-6	0.15	0.61	78	61

The ES metaheuristic is applied to solve the multi-stage TEP problem considering two different conditions. First, the final plans must only satisfy the intact network criterion. In the second condition, the final plans must also ensure the security criterion based on the "N-1" determinist approach. The 10 best sequences found, for each condition, are shown in Tables 3 and 4, respectively. The sequences are ordered by the summation of the present values of the investments and operational costs. For each condition, 10 cases were simulated using different seeds randomly chosen. In all simulated cases, the quality of the obtained sequences was about the same, based on the present value costs. The mean CPU time spent in each case considering the first condition was 24.7 min. Regarding the second condition, each case spent on average 224.7 min, which represent the factor of 9.1 of extra computational effort. This happens due to the high number of

Table 3 Present value of the 10 best sequences without "N-1" deterministic approach

Sequence	Invest. (10^6 US$)	Oper. losses (10^6 US$)	LOLC (10^6 US$)	Total (10^6 US$)
A	156.53	–	43.25	199.78
B	158.18	1.91	35.31	195.41
C	160.00	3.76	29.50	193.26
D	171.56	−4.53	36.97	203.99
E	173.06	−5.55	35.00	202.51
F	171.56	−3.93	35.70	203.32
G	168.95	−1.23	39.07	206.79
H	170.19	−2.38	37.17	204.98
I	170.19	−2.33	35.51	203.37
J	168.95	−0.90	39.70	207.75

Table 4 Present value of the 10 best sequences with "N-1" deterministic approach

Sequence	Invest. (10^6 US$)	Oper. Losses (10^6 US$)	LOLC (10^6 US$)	Total (10^6 US$)
K	232.98	11.18	0.08	244.23
L	234.79	9.53	0.06	244.39
M	242.79	2.27	0.14	245.20
N	264.52	−18.68	0.09	245.94
O	244.29	2.19	0.13	246.62
P	266.18	−18.64	0.09	247.63
Q	246.34	1.51	0.09	247.94
R	244.84	3.20	0.10	248.14
S	245.94	2.11	0.12	248.18
T	266.34	−18.25	0.09	248.18

additional analyses needed to evaluate if each candidate plan ensures the security criterion "N-1" during the ES metaheuristic search.

Since the operational costs (i.e., ohmic losses) are directly related to the system load and are similar to any expansion plan, a reference value was adopted for all sequences found by the ES metaheuristic. The reference value corresponds to 118.29×10^6 US$ and it was obtained for Sequence A, shown in Table 3. Therefore, for all others sequences presented in Tables 3 and 4, the informed operational costs refer to the difference between the values obtained for each sequence and the reference value. For this reason, the operational costs are nil for Sequence A, and some sequences present a negative value, which means that their operational costs are lower than the reference value.

Comparing the investments and operational costs of the sequences between Tables 3 and 4, one can observe that, on average, an additional investment of 81.00×10^6 US$ is needed to ensure the "N-1" security criterion.

4.3 Reliability Studies

In this section, a comparison among the best sequences shown in Tables 3 and 4 is presented. The LOLC index is also taken into account, which is calculated in an external loop, after the application of the ES metaheuristic, by a pseudo-chronological Monte Carlo simulation. The obtained LOLC indices refer to the reliability evaluation of the transmission system for the peak load condition and are shown in Tables 3 and 4.

The CPU time spent to estimate the reliability indices of the sequences in Table 3 was 17 min. Regarding the sequences in Table 4, 415 min was needed to estimate their reliability indices. Since the sequences shown in Table 4 ensure the security criterion "N-1", which corresponds to a more reliable system, the Monte Carlo simulation needs to evaluate a huge number of states to achieve the convergence of the reliability indices.

Table 5 LOLE index comparison

Stages	Sequence C LOLE (h/year)	Sequence K LOLE (h/year)
5	202.47	0.63
4	157.04	0.38
3	143.21	0.38
2	95.28	0.38
1	94.74	1.03
0	47.82	0.72

Considering the results presented in Table 3, one can observe that if the objective is to minimize the present costs including the LOLC, (10), Sequence C becomes the winner with a total cost of 193.26×10^6 US$. In this condition, the LOLC index is decisive to correctly classify the best sequences, since it represents on average 18.2% of the total cost. If the "N-1" security criterion is considered (Table 4), the LOLC index becomes worthless since it corresponds to only 0.04% of the total cost. So, Sequence K continues being the winner after LOLC costs have been taken into consideration. Regarding the total cost, the sequences in Table 4 are, on average, 44.53×10^6 US$ more expensive than the sequences in Table 3 that only ensure the intact network criterion.

Another important reliability measure generally used to measure the quality of the system is the LOLE index. A comparison in terms of LOLE, between the winner Sequences C and K, is presented in Table 5 considering all stages along the planning horizon. There is no doubt that an adequate quality level of energy supply needs to be ensured to customers, but one has to wonder if the consideration of the "N-1" security criterion is really the best alternative. Comparing the LOLE indices among the sequences presented in Table 5, surely one would have a positive answer for this question. It is unacceptable to have, for example, a transmission system with a LOLE of 202.47 h/year as it can be seen in stage 5 of Sequence C or even 47.82 h/year in the reference year. The LOLE index is far better for Sequence K along the planning horizon. Although not shown, the same conclusion is reached comparing any sequences between Tables 3 and 4.

However, the reliability evaluation was performed considering the peak load model during the entire year. It would be fairer to make such comparison considering an hourly load model. For this purpose, the hourly load curve of the IEEE-RTS [27] is adopted and the new LOLC indices of the sequences in Table 3 are presented in Table 6. As the LOLC indices have a very low value for all sequences in Table 4, even for the peak load model, the reliability indices, including the LOLC, become insignificant under the hourly load model. For example, the LOLE indices of Sequence K are reduced to less than 1 min/year. For this reason, the results considering the sequences that ensure the "N-1" security criterion are not shown under the hourly load model.

Table 6 Present value of the 10 best sequences without "N-1" deterministic approach—hourly load model

Sequence	Invest. (10^6 US$)	Oper. losses (10^6 US$)	LOLC (10^6 US$)	Total (10^6 US$)
A	156.53	–	0.56	157.08
B	158.18	1.91	0.41	160.50
C	160.00	3.76	0.33	164.10
D	171.56	−4.53	0.49	167.52
E	173.06	−5.55	0.46	167.97
F	171.56	−3.93	0.46	168.09
G	168.95	−1.23	0.40	168.12
H	170.19	−2.38	0.48	168.28
I	170.19	−2.33	0.32	168.18
J	168.95	−0.90	0.51	168.55

Table 7 Expansion plan—Sequence A

Stage	Terminals			LOLE (h/year)	Invest. (10^6 US$)	LOLC (10^6 US$)
	2-6	3-5	4-6			
5	–	–	–	10.24	–	0.29
4	–	–	–	5.83	–	0.13
3	–	–	–	2.62	–	0.05
2	–	1	–	1.04	20.00	0.02
1	–	–	–	6.65	–	0.18
0	2	1	2	3.39	140.00	0.08
Present value cost (10^6 US$)					156.53	0.56

Under the hourly load model, the LOLC indices of the sequences that ensure only the intact network criterion are reduced dramatically. The new values represent on average 1.2% of the LOLC indices calculated for the peak load model. In Table 7, the expansion plan and the new LOLE indices of winner Sequence A are presented, considering all stages of the planning horizon. The added reinforcements and the costs at a particular stage of the planning horizon are also shown. As this sequence is the reference for the operational costs, their respective values are nil.

Adopting a more realistic situation, i.e., an hourly load model, the LOLE indices of the sequences that ensure only the intact network criterion are more reasonable, as it is shown in Table 7 for Sequence A. The worst situation, a LOLE of 10.24 h/year, appears in stage 5. Now, under this load model, it is not obvious if the consideration of the "N-1" security criterion is the best alternative.

Generally, a tariff is used to recover all the costs involved in the TEP, such as investments, operational, and possible customer premiums due to load interruption damages. Bearing in mind Sequence A (Table 7), the tariff related to interruption costs (LOLC) represents only 0.36% of the total costs to be recovered. This amount can be interpreted as a kind of energy insurance. Although all sequences that ensure the "N-1" security criterion do not need to recover a tariff related to load interruption, since the LOLC index is almost nil considering the hourly load model, on average, there is an additional investment cost of 81.00×10^6 US$ that needs to be recovered by the tariff. One wonders if it would be better to pay only 0.56×10^6 US$ related to the load interruption costs and disregard the "N-1" security criterion. Companies do not reach a consensus about reliability index values that represent an acceptable quality level for their power networks. Although it seems that the "N-1" security criterion is not the best choice, since there is a high tariff to be recovered, this has been the strategy adopted by most transmission companies around the world.

An alternative to the "N-1" deterministic approach would be the use of a probabilistic one during the optimization process, where some criterion based on reliability indices, e.g., LOLE, could be considered to select the best sequences. This would guarantee an acceptable LOLE index and a lower investment cost than those obtained by the "N-1" deterministic approach. However, as already mentioned in Sect. 2.3.2, this may be impracticable due to the high computational effort required.

As a conclusion, the best strategy appears to be the acceptance of the "N-1" security criterion, but only in the most vital areas of the system. In real systems, this criterion is guaranteed only in meshed network areas and main branches or interconnections. A huge amount of money would be needed if the whole system had to ensure the "N-1" security criterion including sub-transmission areas. After the optimization process, a probabilistic approach could be used to evaluate the best sequences considering interruption costs and an hourly load model. Finally, expansion plans with a good quality network and cheaper than that ones achieved considering the "N-1" security criterion for the entire network could be selected.

5 Large Networks

In this section, a methodology to solve large multi-stage TEP problems considering security criteria and the treatment of external uncertainties is proposed. A case study in a real transmission network is presented. The system is composed of 33 buses and 55 circuits. Although this represents a small network, the proposed methodology can be easily applied to a large system. The installed capacity is 465 MW and the peak load is 248.4 MW for the reference year. A simplified diagram of the system, which includes one of the best expansion plans for the horizon year, is shown in Fig. 3 at the end of the chapter. The objective is to obtain the best expansion strategies to connect bus "1832" to the 138 kV system along the planning horizon. The system expansion horizon is divided in 8 stages and for each stage the load is increased on average by 7.7%, which results in a peak load of 343.7 MW in the final year. The "N-1" security criterion is not considered in the analysis, but the reliability aspects are. The proposed methodology is described in the following sections.

5.1 Equivalent System

Bearing in mind large multi-stage TEP problems, it is interesting to obtain an equivalent system through a reduction of the complete system. This system reduction provides an important computational gain for the optimization approaches, such as the metaheuristics, and for the system reliability evaluation. To determine the equivalent system, the internal and external systems need to be defined. The first is composed by the buses located at the interest area, i.e., where the expansion studies will be performed. The second system is composed of the region that will be reduced to the boundary buses with the respective equivalent power injections through the Ward equations [28]. Considering the power injections, the negative ones are represented as equivalent loads, while the positive injections refer to equivalent generations. In the last case, an increase of 10% is used to simulate the maximum generation capacity on each boundary bus, in order to respond to eventual contingencies occurred in the internal system. Regarding the equivalent circuits, an unlimited capacity is assumed.

5.2 Optimization Approach

After the definition of the equivalent system, an optimization approach needs to be applied along the planning horizon, in order to obtain a set of best expansion plans that minimizes the investments and operational costs. Such tools aim at helping the planner in creating an initial set of expansion plans, which will be evaluated later considering uncertainties and an AC power flow model. Considering the real transmission network (Fig. 3), the connection between all 138 kV buses is permitted since the distance between the buses are less than 60 miles and a limit of three parallel lines in each right-of-way is not violated. As in Sect. 4, the ES metaheuristic is selected as the optimization approach, where the same parameters are used for this real system. In relation to the operational costs associated with the ohmic losses, the following parameters are used: $C_{\mathrm{kWh}} = 0.10$ US$/kWh and LF $= 0.5$.

5.3 Reliability Studies

The next step is to perform reliability studies considering the best sequences selected by the optimization approach. A pseudo-chronological Monte Carlo simulation is used to estimate the reliability indices [21]. Thus, the final objective is to minimize the total costs, including investment, operational, and interruption costs referred to the LOLC index. To obtain the LOLC index, a single unit interruption cost of 1.50 US$/kWh and the peak load are considered in the reliability study.

Table 8 Present value of the 5 best sequences

Sequence	Invest. (10^6 US$)	Oper. losses (10^6 US$)	LOLC (10^6 US$)	Total (10^6 US$)
A	33.15	–	12.15	45.30
B	34.10	0.77	12.19	47.06
C	32.04	3.25	22.19	57.48
D	34.80	0.54	10.02	45.36
E	32.08	3.26	21.96	57.30

In Table 8, the present value costs of the five best sequences found by the ES metaheuristic are shown. The operational costs, referred to ohmic losses, correspond to the differences in relation to the value calculated for Sequence A, which is the best one found by the ES metaheuristic. The winner sequence until this moment, Sequence A, has the following reinforcements: a line between buses 1832 and 4762 (added in the reference year) and two more lines added in the second stage of the planning horizon (1821–1828 and 1828–1832). Although all sequences presented in Table 8 have three reinforcements, their respective locations promote significant variations in the LOLC index that vary from 10.02×10^6 to 22.19×10^6 US$. Therefore, the reliability evaluation can be decisive in selection of the best plans.

5.4 AC Load Flow Analysis

The sequences obtained by the optimization approach also need to be evaluated considering an AC load flow model. In this analysis, the behavior of the transmission loads and bus voltage magnitudes in the internal system are observed. Since the optimization approach is based on a DC flow, the AC model is used to decide if the best sequences have problems related to the bus voltages and overloads in some circuits due to reactive power flows. With this analysis, one has a better selection of the expansion plans.

The three best sequences presented in Table 8 (A, B and D), considering the total costs in present values, are evaluated by the AC model. As it can be seen from Table 9, the three sequences present a voltage mean value that satisfies the limits adopted for the system (0.95–1.05 pu). Furthermore, on average, the transmission load corresponds to about 34% of the total transmission capacity. Only the transmission circuits and buses of the internal system were considered in this

Table 9 Mean values of bus voltages and transmission loadings

Sequence	Bus voltages (pu)	Transmission loadings (%)
A	1.007	34.23
B	1.006	34.22
D	1.006	33.58

evaluation. No voltage or transmission capacity violation was observed. Therefore, the decision must mainly take into account the present values of the total costs.

5.5 External Uncertainties

The best sequences considering the total costs and the AC load flow analysis are finally evaluated taking into account the external uncertainties. In the case study presented in this section, the scenario technique is used to treat the uncertainties related to the load growth. Uncertainties in other parameters, such as discount rate and unitary interruption cost could also be treated using the scenario techniques.

The expansion study under load growth uncertainties considers three distinct scenarios. Each scenario is defined by the annual load growth of the internal system, as follows:

- Low scenario: mean annual load growth of 5.5%;
- Normal scenario: mean annual load growth of 7.7% (Table 8);
- High scenario: mean annual load growth of 10.1%.

Figure 2 shows the load growth of the internal system for all scenarios. The maximum generation capacity of the internal system is kept constant, while the generation capacities in the boundary buses are defined by the equivalent models of the external area in each scenario.

The procedure adopted to obtain the best sequences in the normal scenario, Table 8, is repeated to find the best ones in the low and high scenarios. Table 10 shows the costs in present values of the three best sequences in the low and high scenarios. It is important to mention that the operational costs of the sequences have as reference the operational cost of the best sequence in each scenario. Afterwards, the best sequences (strategies) of each scenario must be evaluated under the other scenarios. If necessary, some modifications, such as addition or removal of reinforcements, may be performed to adapt them to the new conditions.

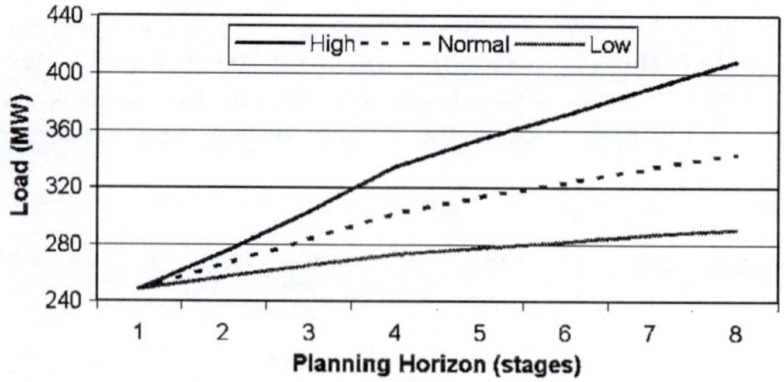

Fig. 2 Load growth of the internal system—high, normal, and low scenarios

Table 10 Present value of the 3 best sequences—low and high scenarios

Sequence	Scenario	Invest. (10^6 US$)	Oper. Losses (10^6 US$)	LOLC (10^6 US$)	Total (10^6 US$)
F	Low	32.54	–	10.90	43.43
G	Low	31.83	0.20	12.78	44.82
H	Low	31.43	2.42	12.50	46.36
I	High	41.30	–	16.64	57.94
J	High	40.90	1.74	20.83	63.47
K	High	40.20	2.02	23.25	65.47

Table 11 Total and expected costs

Strategies	Total cost (10^6 US$)			Expected cost (10^6 US$)
	Low	Normal	High	
F	43.43	60.73	73.52	59.60
G	44.82	57.58	73.92	58.48
H	46.36	56.25	64.48	55.84
A	45.25	45.30	62.94	49.70
B	46.74	45.36	63.47	50.23
D	48.10	47.06	65.47	51.92
I	44.84	48.57	57.94	49.98
J	46.74	45.36	63.47	50.23
K	48.10	47.06	65.47	51.92
Probability	0.25	0.50	0.25	

The decision criteria Minimum Expected Cost and MinMax Regret are used to indicate the best strategies observing all scenarios. Through these criteria, it is possible to evaluate the flexibility and robustness of the selected strategies when submitted to different scenarios.

The results for the Minimum Expected Cost decision criterion are presented in Table 11. The probabilities 0.25, 0.50, and 0.25 are specified for the occurrence of the low, normal, and high scenarios, respectively. As it can be seen, Strategy A has the lowest expected cost being, therefore, the winner for the considered scenarios. Furthermore, the winners strategies of each scenario are generated in the own scenario. This proves the efficiency of the ES metaheuristic to find the best sequences in each scenario.

Table 12 presents the MinMax Regret decision criterion, where one can see the regrets for the low, normal, and high scenarios and the maximum regret for each strategy. Contrary to the Minimum Expected Cost decision criterion, Strategy I is indicated as the winner when the MinMax Regret is adopted. From Tables 11 and 12, one concludes that Sequence I may be considered the best one since it has an expected cost a little higher than Sequence A (0.28×10^6 US$) and the minimum maximum regret (1.73×10^6 US$ lower than Sequence A). The final choice of a sequence depends on the risk adopted by the utility. Figure 3 shows the single line

Table 12 Regret matrix

Strategies	Regret (10^6 US$)			Maximum regret (10^6 US$)
	Low	Normal	High	
F	0.00	15.43	15.58	15.58
G	1.39	12.29	15.98	15.98
H	2.92	10.96	6.54	10.96
A	1.82	0.00	5.00	5.00
B	3.31	0.05	5.53	5.53
D	4.66	1.75	7.53	7.53
I	1.41	3.27	0.00	3.27
J	3.31	0.05	5.53	5.53
K	4.66	1.75	7.53	7.53

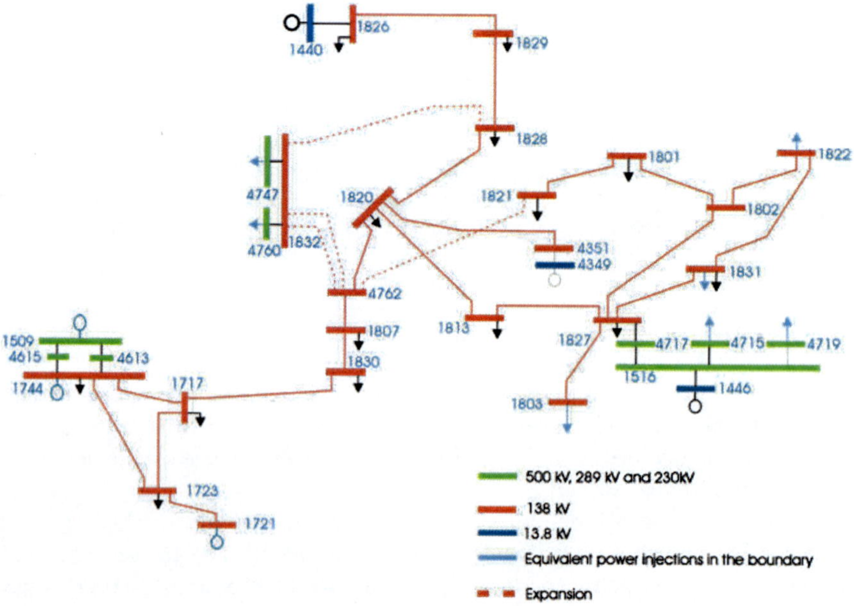

Fig. 3 Single line diagram of Sequence I

diagram of the transmission network studied, which represents the configuration of Sequence I in the horizon year and normal load scenario.

6 Final Remarks

This chapter proposes a new methodology for solving a multi-stage TEP problem, considering security criteria and the treatment of external uncertainties.

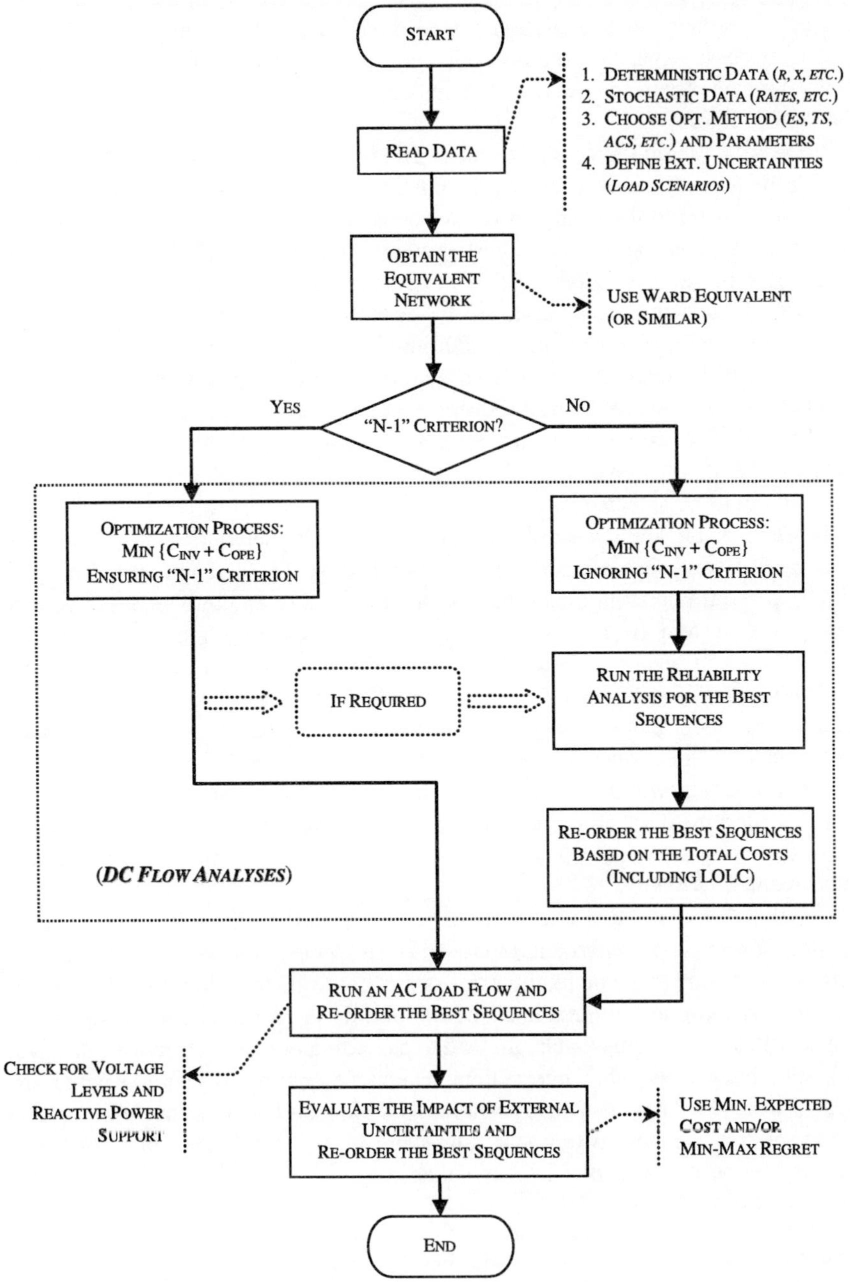

Fig. 4 Simplified flow chart

In Sect. 2, a detailed formulation of the multi-stage TEP problem and how to obtain the best expansion plans using optimization approaches were presented. Several metaheuristics were described in Sect. 3, which could be selected to solve the TEP problem.

In Sect. 4, a reliability study and a discussion on how to consider security criteria using deterministic and probabilistic approaches are included. Although the "N-1" deterministic approach has been adopted by most electric power companies around the word, the best strategy appears to ensure this security criterion only in the most vital areas of the system network. Vital areas have to be carefully defined but, in general, are those with higher voltage levels, interconnections, special equipment, etc. In addition, a probabilistic approach, such as the pseudo-chronological Monte Carlo simulation, could be used after the optimization process to evaluate the best sequences bearing in mind the interruption costs. At the end, the expansion plans with high-quality network performance and cheaper than that achieved considering the "N-1" security criterion for the entire network could be selected.

In Sect. 5, an application for a real transmission network is provided considering the treatment of external uncertainties, as the load growths. The proposed procedure consists of, in the first step, obtaining an equivalent network to reduce the system to the interest area; second, selecting the best sequences through an optimization approach; third, performing reliability studies; fourth, evaluating the best sequences using an AC power flow model; and finally, treating the uncertainties using decision-making models, such as the Minimum Expected Cost and MinMax Regret. Figure 4 shows a simplified flow chart of the proposed algorithm considering both situations ensuring and ignoring the "N-1" criterion. The proposed methodology proved to be relatively efficient from the computational point of view and robust in terms of transmission reinforcement strategy.

Finally, there will be new challenges related with the TEP subject considering the new concepts like smart and micro grids. In theory, they will incorporate new networking technology, including sensors and controls, to monitor electricity use in real time and make automatic changes, not only to reduce energy waste, but to instantly detect problems that could lead to cascading outages. Moreover, the grids will start integrating more intermittent renewable energy sources like wind and solar. These new aspects will tremendously influence the way transmission networks will be planned in the near future and increase the value of information assessed by power system reliability studies.

Acknowledgments The authors would like to thank Dr. Cleber E. Sacramento from CEMIG (Compania Energética de Minas Gerais), Brazil, for providing data and discussions on planning strategies. The authors would like to extend their thanks to Mr. Larry Lee and Dr. Gomaa Hamoud from Hydro One, Canada, to Dr. George Anders from Kinectrics, Canada, to Prof. Leonardo M. Honório from UNIFEI, Brazil, and also to Prof. Leonidas Chaves de Resende from UFSJ, Brazil, for discussions on transmission expansion planning and optimization issues.

References

1. Billinton R, Salvaderi L, McCalley JD, Chao H, Seitz Th, Allan RN, Odom J, Fallon C (1997) Reliability issues in today's electric power utility environment. IEEE Trans Power Syst 12:1708–1714
2. Chowdhury AA, Koval DO (2004) Value-based system facility planning. IEEE Power Energy Mag 2:58–67
3. Latorre G, Cruz RD, Areiza JM, Villegas A (2003) Classification of publications and models on transmission expansion planning. IEEE Trans Power Syst 18:938–946
4. Xu Z, Dong ZY, Wong KP (2006) Transmission planning in a deregulated environment. IEE Proc Gener Transm Distrib 153:326–334
5. Gallego RA, Alves AB, Monticelli A, Romero R (1997) Parallel simulated annealing applied to long term transmission network expansion planning. IEEE Trans Power Syst 12:181–187
6. Gallego RA, Romero R, Monticelli A (2000) Tabu search algorithm for network synthesis. IEEE Trans Power Syst 15:490–495
7. Leite da Silva AM, Manso LAF, Resende LC, Rezende LS (2008) Tabu search applied to transmission expansion planning considering losses and interruption costs. In: Proceedings of 10th PMAPS, Puerto Rico
8. Gil HA, da Silva EL (2001) A reliable approach for solving the transmission network expansion planning problem using genetic algorithms. Electr Power Syst Res 58:45–51
9. Escobar AH, Gallego RA, Romero R (2004) Multistage and coordinated planning of the expansion of transmission systems. IEEE Trans Power Syst 19:735–744
10. Binato S, Oliveira GC, Araújo JJ (2001) A greedy randomized search procedure for transmission expansion planning. IEEE Trans Power Syst 16:247–253
11. Faria H Jr, Binato S, Resende MGC, Falcão DM (2005) Power transmission network design by greedy randomized adaptive path relinking. IEEE Trans Power Syst 20:43–49
12. Leite da Silva AM, Sales WS, Resende LC, Manso LAF, Sacramento CE, Rezende LS (2006) Evolution strategies to transmission expansion planning considering unreliability costs. In: Proceedings of 9th PMAPS, Stockholm
13. Dong ZY, Lu M, Lu Z, Wong KP (2006) A differential evolution based method for power system planning. In: IEEE congress on evolutionary computation 2699-2706, Vancouver
14. Jin YX, Cheng HZ, Yan JY, Zhang L (2007) New discrete method for particle swarm optimization and its application in transmission network expansion planning. Electr Power Syst Res 77:227–233
15. Leite da Silva AM, Sacramento CE, Manso LAF, Rezende LS, Resende LC, Sales WS (2008) Metaheuristic-based optimization methods for transmission expansion planning considering unreliability costs. In: Castronuovo ED (ed) Optimization advances in electric power systems, 1st edn. Nova Publishers, USA
16. Rezende LS, Leite da Silva AM, Honorio LM (2009) Artificial immune system applied to the multi-stage transmission expansion planning. In: Proceedings of 8th ICARIS. LNCS (To be published)
17. Lee KY, El-Sharkawi MA (2008) Modern heuristic optimization techniques: theory and applications to power systems. IEEE Press Series on Power Engineering, Wiley
18. Silva IJ, Rider MJ, Romero R, Garcia AV, Murari CA (2005) Transmission network expansion planning with security constraints. IEEE Proc Gener Transm Distrib 152:828–836
19. Tor OB, Guven AN, Shahidehpour M (2008) Congestion-driven transmission planning considering the impact of generator expansion. IEEE Trans Power Syst 23:781–789
20. Manso LAF, Leite da Silva AM (2004) Probabilistic criteria for power system expansion planning. Electr Power Syst Res 69:51–58
21. Leite da Silva AM, Manso LAF, Mello JCO, Billinton R (2000) Pseudo-chronological simulation for composite reliability analysis with time varying loads. IEEE Trans Power Syst 15:73–80

22. CIGRE W G 37.10 (1993) Dealing with uncertainty in system planning—has flexibility proved to be an adequate answer? ELECTRA 151:53–65
23. CIGRE W G 37.10 (1995) Methods for planning under uncertainty—towards flexibility in power system development. ELECTRA 161:143–164
24. Buygi MO, Shanechi HM, Balzer G, Shahidehpour M, Pariz N (2006) Network planning in unbundled power systems. IEEE Trans Power Syst 21:1379–1387
25. Garver LL (1970) Transmission network estimation using linear programming. IEEE Trans PAS 89:1688–1697
26. Castro LN, Zubben FJV (2002) Learning and optimization using the clonal selection principle. IEEE Trans Evol Comput 6:239–251
27. IEEE APM Subcommittee (1979) IEEE reliability test system. IEEE Trans PAS 99:2047–2054
28. Ward JB (1949) Equivalent circuits for power flow studies. AIEE Trans 98:498–508

The Economic Evaluation of System Security Criterion in a Competitive Market Environment

Teoman Güler, George Gross, Eugene Litvinov and Ron Coutu

1 Introduction

In the restructured environment, the improvement of the economic efficiency of electricity markets has been the focus of recent efforts [1, 2]. Central to these efforts is the better understanding of the nature of the tight coupling between market and system operations. An important aspect of this coupling is the dependence of the market outcomes on the way the system is operated. A key driver in system operations is the security criterion, with which compliance must be ensured. The focus of this work is the dependence of market performance on system security. In this work, we propose an approach to quantify the market performance as a function of a specified security criterion for both single- and multi-settlement environments. We illustrate the application of the proposed approach on the large-scale ISO-NE system.

System security is defined as the ability of the interconnected system to provide electricity with the appropriate quality under normal and contingency conditions [3]. The security criterion consists of the set of postulated contingencies and the associated preventive and/or corrective control actions [4]. For a given operating

T. Güler (✉) and G. Gross
University of Illinois at Urbana Champaign, 1406 W. Green Urbana, Urbana, IL 61801, USA
e-mail: teoman.guler@ieee.org

G. Gross
e-mail: gross@illinois.edu

E. Litvinov and R. Coutu
ISO New England Inc, 1 Sullivan Road, Holyoke, MA 01040, USA
e-mail: elitvinov@iso-ne.com

R. Coutu
e-mail: rcoutu@iso-ne.com

G. Anders and A. Vaccaro (eds.), *Innovations in Power Systems Reliability*, Springer Series in Reliability Engineering, DOI: 10.1007/978-0-85729-088-5_7,

state, security assessment entails the verification that no violation occurs for any of the postulated contingencies taking fully into account the deployment of the associated security control actions. As these actions affect the market outcomes, a key step in the efforts to improve market performance is the assessment of these impacts of complying with the security criterion in monetary terms. Such studies are, typically, not performed by today's large regional transmission organizations, or RTOs. Consequently, there is a need for an appropriate methodology to quantitatively measure the market performance impacts of complying with security criterion.

Under single-settlement systems, a market trades the electricity commodity that is physically produced and consumed. However, the design and implementation of electricity markets in many jurisdictions involves two or more inter-related markets that are cleared at different points in time. The sequence of markets trades the commodity that is physically produced and consumed in real time. Each market, be it a day-ahead hourly market (DAM) or one of the real-time markets (RTMs) associated with that hour, trades the MWh commodity at different prices that reflect the information on the system and market conditions available at the time the MWh commodity is cleared. As these conditions are subject to continuous changes in real time, real-time markets (RTMs) are cleared at a high frequency, typically every 5 min. On the other hand, markets run ahead of real-time system operations have a lower clearing frequency, e.g. hourly clearing in the day-ahead markets, reflecting the lower resolution of the imperfect information on the real-time conditions in the next day. The hour h DAM is cleared on the forecasts of the real-time conditions for that hour the next day. The hourly clearing influences each RTM in the near-real-time during that hour. The RTM clearing determines the volume of deviations from the hour h DAM value and the associated price. Such a market design with different lead times and clearing frequencies is commonly referred to as a multi-settlement system [5–7].

The RTO or the independent system operator (ISO) manages system operations through market forces. The various inter-relationships between the physical network security management and the clearing of a single market or the sequence of markets imply the strong interdependence between system security and market outcomes. Although the DAMs are financial markets in contrast to the purely physical RTMs, both markets are cleared using the same approach. A key difference between these markets is the nature of the participants. While financial entities may participate in the DAMs, only players with physical resources or loads can participate in the RTMs. In addition, while the demand may be price responsive in the DAMs, the demand is, typically, fixed in the RTMs. As the DAM outcomes impact the associated RTMs, financial entities may very well impact the system operations in real-time.

The economic efficiency of electricity markets under single-settlement systems is analyzed in various contexts by both empirical and analytical means. The empirical studies investigate the adverse impacts of market participants' behaviors on the performance of electricity markets [8–11]. The analytical studies, on the other hand, focus on the impacts of constrained system operations on markets to determine the unavoidable losses in the economic efficiency of electricity markets [12, 13]. The interactions between the system security criterion and the associated economics are investigated in terms of the marginal costing—used to determine the security prices—and the evaluation of expected system security costs. The security prices, determined in this way, explicitly incorporate the willingness of the market participants to provide security control capability into the market clearing process [14–16]. The expected system security costs are evaluated taking explicitly into account the random nature of the outages and the costs of the required security control actions to deal with them [17, 18]. A key result of [18] is that the security criterion may be set by the cost/benefit analysis taking into account the expected costs of operating the system and the expected outage costs. Such an approach may be viewed as the application of the notion of "value of reliability" introduced in [19] which was used for operational planning purposes [20]. However, there is a clear need, in the restructured environment, to quantify the market performance as a function of system security in a way that appropriately reflects the RTO operations. This quantification further requires the consideration of different market and system conditions that may exist within a period in order to capture the range of impacts under such conditions.

The economic benefits of multi-settlement systems have been previously analyzed [7, 21–23]. The study in [7] discusses the role of the DAMs in terms of providing incentives for accurate forecasts in real-time operations and for facilitating trades through ex-ante price discovery. The duopoly model in the simple two-node network shows that for small probabilities of congestion, multi-settlement systems are welfare-enhancing when compared to single-settlement market designs [21]. The analysis of the impacts of congestion in a multi-settlement environment makes clear that the welfare-enhancing role of multi-settlement system is highly sensitive to the presence of real-time congestion [22]. Further evidence of the welfare-enhancing impacts of multi-settlement systems comes from the empirical analysis of the PJM and NY-ISO markets [23]. We note that when the market design misaligns system and market operations, as was the case in the California prior to the 2000–2001 crisis, the market participants may manipulate markets resulting in decreased overall market performance [24]. Much of the analysis of the role of financial entities has a focus on the monetary impacts of such participants [25–27]. As the participation of financial entities not only increases market liquidity but may also result in price convergence, multi-settlement systems are viewed as improving the economic efficiency of electricity markets when compared to the single-settlement markets [25–27]. The ISO-NE study [25] identifies that after the introduction of "virtual bidding", the actual offer/bid mechanism that enables financial entity participation in the DAMs, the RTM and DAM prices tend to converge. The convergence implies an improvement in market performance as temporal arbitrage opportunities diminish. Under speculative trading, financial entity participation may lead to increased market efficiency and decreased average electricity price [26]. The inter-relationship between the price convergence and the profits of the financial entities is a "self-correcting" feature of multi-settlement systems [27]. As such, multi-settlement systems provide appropriate price signals to the financial entities, which they may use to make profitable trades that may lead to price convergence. On the other hand, the discussion of the impacts of the behavior of the financial entities driven by such price signals on the real-time system security is absent in the literature.

The provided literature review pinpoints the specific needs in better understanding and quantifying of the tight coupling between system and market operations. An important need is an integrated analysis approach to quantify the interdependence between the market performance and the way the power systems are operated to ensure security that appropriately reflects the actual RTO operations. Furthermore, such a need has to take into account the multi-settlement environment. Specifically, there is a clear need to analyze whether the behavior of financial entities driven by price signals can impact, in some measurable way, system security. We address these and other related issues in the analysis and economics of power system security in the competitive environment by proposing a set of appropriate approaches and tools that are effective for the analysis and quantification of a wide range of issues for large-scale networks that we encounter in actual power systems. We illustrate a number of representative applications on the large-scale ISO-NE system and discuss the insights we obtain from our studies.

The highly challenging task of security management becomes even more complex in the competitive market environment. We use the insights we developed into the tight coupling between market and system operations under restructuring to characterize analytically the inter-relationships between the way the power systems are operated and the performance of the electricity markets. Such characterization leads us to the development of a systematic approach that quantifies the market performance as a function of security criterion under diverse system and market conditions for single-settlement systems. This approach permits the quantification of the market performance impacts arising from a change from a given to another security criterion. The approach provides, for the first time, an economic justification for the RTO decision to modify the security criterion. Furthermore, the approach can be used in the cost/benefit analysis of network improvements to mitigate the market performance impacts of a set of contingencies or their associated security control actions. Another application of the approach is to the assessment of the impacts of specific behavioral changes in market participants on system security.

We extend this approach to quantitatively characterize the linkages between the real-time system operations and the DAMs and their associated RTMs for a multi-settlement environment. We explicitly show with the extended approach that the auction surplus attained in the multi-settlement system is equivalent to the sum of the auction surpluses attained in each RTM. Therefore, the mere presence of the DAMs results in surplus transfers among market participants. Furthermore, the extended approach provides a very useful tool to analyze the nature of the DAM-RTM price deviations and the impacts of financial entities on real-time system security.

We illustrate the application of the approaches on the large-scale ISO-NE system using the historical 2005–2006 data—the system model and the bids/offers submitted—and the actual market clearing methodology. The large-scale ISO-NE DAM application study provides important insights into the role of price-responsive demand and selected security control actions by demonstrating that the economic efficiency of the electricity markets need not decrease when a power system is operated under a stricter criterion, as long as there is effective price-responsive demand and appropriate utilization of the corrective control capabilities of the resources. The ISO-NE multi-settlement application study bear out the well known fact that the participation of financial entities leads to the convergence of the DAM and the associated RTM prices. Moreover, the study also illustrate that such participation leads to improved forecasts of the real-time system operations, and consequently results in improving the assurance of system security.

This chapter contains six additional sections. The market performance quantification for a system snapshot is described in Sect. 2. We devote Sect. 3 to extension of the quantification to the multi-settlement environment. In Sect. 4, we describe the proposed approach and discuss its possible applications. In Sect. 5, we apply the proposed approach to the ISO-NE system and present the study results in detail. We conclude in Sect. 6 with a summary of the work.

2 Market Assessment for a System Snapshot

We introduce specific assumptions on unit commitment decisions, ancillary services and the market participants' behaviors so as to allow the side-by-side comparison of different security criteria impacts for a given system. We assume that the unit commitment decisions fully reflect the requirements of the security criterion under consideration. In particular, this assumption ensures the feasibility of meeting the system fixed demands under such a criterion. As the focus of this investigation is limited to energy only markets, we assume that the ancillary services provision and acquisition requirements under the RTO framework do not impose any additional constraints on the system. For the purposes of this study, we furthermore assume that the bidding behavior of each market participant is independent of the security criterion in force. Since we replicate the RTO actions, we ensure compliance with the security criterion in force, but implicitly ignore the probability of any contingency in the studies.

We associate a security criterion C with a specific contingency list and a specified control action (preventive or corrective) for every contingency on that list. A preventive control action associated with a postulated contingency entails the modification of the pre-contingency—base case—state, to eliminate any potential violation, if that contingency were to occur. On the other hand, an associated corrective control action may involve the modification of both the pre- and the post-contingency states. The modification of the pre-contingency state involves steering the operating point into a state in which the RTO is able to modify the resources' dispatch, including those of both load and generation, to alter the post-contingency state only after the contingency actually occurs.

As such, there may be no change in resource utilization if the contingency fails to happen. For some contingencies, such as a generator outage or a sudden load change, the RTO may take only corrective control actions.

We consider a power system consisting of $(N+1)$ nodes and denote by $\mathbf{N} = \{0, 1, \ldots, N\}$ the set of buses, with the slack bus at bus *0.* The security criterion C has the contingency list $\mathbf{J}_C$ and the specified control action for each contingency. Let $\mathbf{S}$ ($\mathcal{B}$) denote the collection of sellers (buyers). Each seller (buyer) submits its price and quantity offer (bid), indicating the willingness to sell (buy) the amount of energy for the duration represented by the snapshot to (from) the RTO. We note that the offers and bids need not necessarily be the true marginal costs and benefits of the participants [28–30]. We represent a bilateral transaction ω_w by $\omega_w = \{m_w, n_w, \bar{t}_w\}$. Here, m_w denotes the *from* node, n_w the *to* node, and $\bar{t}_w$ the desired transaction amount. The set of all the bilateral transactions is denoted by $\boldsymbol{W} = \{\omega_1, \ldots, \omega_W\}$. Each transaction submits a willingness to pay function, which states a willingness to pay maximum transmission usage fees for receiving the requested transmission services as a function of the transaction amount delivered [31]. The RTO weighs the willingness to pay of the bilateral transactions with that of the individual market participants to determine the amount of transmission service provision to each player. For this purpose for a given snapshot of the system, the RTO solves a security-constrained OPF, or SCOPF, problem with the objective to maximize the auction surplus under the security criterion C whose contingency index set is denoted by $\mathbf{J}_C$. We state the SCOPF problem as

$$\max \mathbf{S} \triangleq \sum_{i=0}^{N} \left(\sum_{b \in \mathcal{B} \text{ at node } i} \beta_b(p_b) - \sum_{s \in \mathbf{S} \text{ at node } i} \beta_s(p_s) \right) + \sum_{w=1}^{W} \alpha_w(t_w) \tag{1}$$

subject to

$$\underline{g}^{(0)}\left(\underline{p}_s^{(0)}, \underline{p}_b^{(0)}, \underline{t}^{(0)}, \underline{\chi}^{(0)}, \underline{\gamma}^{(0)}\right) = \underline{0} \leftrightarrow \underline{\lambda}^{(0)} \tag{2}$$

$$\underline{h}^{(0)}\left(\underline{p}_s^{(0)}, \underline{p}_b^{(0)}, \underline{t}^{(0)}, \underline{\chi}^{(0)}, \underline{\gamma}^{(0)}\right) \leq \underline{0} \leftrightarrow \underline{\mu}_h^{(0)} \tag{3}$$

and for every $j \in \boldsymbol{J}_C$.

$$\underline{g}^{(j)}\left(\underline{p}_s^{(j)}, \underline{p}_b^{(j)}, \underline{t}^{(j)}, \underline{\chi}^{(j)}, \underline{\gamma}^{(j)}\right) = \underline{0} \leftrightarrow \underline{\lambda}^{(j)} \tag{4}$$

$$\underline{h}^{(j)}\left(\underline{p}_s^{(j)}, \underline{p}_b^{(j)}, \underline{t}^{(j)}, \underline{\chi}^{(j)}, \underline{\gamma}^{(j)}\right) \leq \underline{0} \leftrightarrow \underline{\mu}_h^{(j)} \tag{5}$$

$$\left|\underline{p}_s^{(j)} - \underline{p}_s^{(0)}\right| \leq \underline{\Delta p}_s^{(j)} \leftrightarrow \underline{\mu}_s^{(j)} \tag{6}$$

$$\left|\underline{p}_b^{(j)} - \underline{p}_b^{(0)}\right| \leq \underline{\Delta p}_b^{(j)} \leftrightarrow \underline{\mu}_b^{(j)} \tag{7}$$

$$\left|\underline{t}^{(j)} - \underline{t}^{(0)}\right| \leq \underline{\Delta t}^{(j)} \leftrightarrow \underline{\mu}_t^{(j)} \tag{8}$$

Here, we consider a lossless system and use the superscript (j) to denote the contingency cases with the base case denoted by (0). The vector associated with the right-hand side of a constraint is the dual variable of that constraint. The relations in Eqs. 2 and 3 represent the operational constraints for the base case, while those in Eqs. 4–8 represent the operational constraints for the contingency cases. The $|\boldsymbol{J}_C|+1$ equality constraints in Eqs. 2 and 4 state the nodal power balance equations for the base case and for each postulated contingency case, respectively. The base case (Eq. 3) and contingency case (Eq. 5) inequality constraints state the system components' operational limits, as well as, the so-called generic limitations representing the physical, engineering and policy considerations. The range of the decision variables of the security control action for each contingency $j \in \boldsymbol{J}_C$. is given in Eqs. 6–8 together with the limiting values of these ranges. The preventive control actions have a zero range in contrast to the corrective actions whose non-zero range reflects the additional flexibility to address the onset of the contingency. Note that, in the SCOPF, we explicitly take into account the costs of modifying the pre-contingency state but ignore any costs related to the post-contingency state modification. We denote the security-constrained market problem Eqs. 1–8 by $\mathcal{M}(\mathcal{S},\mathcal{B},\mathcal{W};\mathcal{C})$.

We distinguish between fixed demand buyers and those with price-responsive demand. The fixed demand bid is a special case of the price-sensitive bid in which a specified quantity is submitted with no price information. Such a bid indicates an unlimited willingness to pay for the electricity purchases to meet the fixed quantity bid, i.e., the buyer is willing to pay any price to obtain the electricity. There are, however, difficulties in determining the appropriate value of the benefits of such fixed demand buyers. In order to include these buyers' benefits in the objective function of the $\mathcal{M}(\mathcal{S},\mathcal{B},\mathcal{W};\mathcal{C})$, we use a constant τ per MWh benefit value for the fixed demand, where τ is set to have a high value to indicate the payments that may be incurred due to outages, say as much as 10,000 \$/MWh [32].

The market performance of the snapshot system under the specified security criterion C is quantified from the market clearing given by the solution of $\mathcal{M}(\mathcal{S},\mathcal{B},\mathcal{W};\mathcal{C})$. We quantify the market performance in dollar terms on a system-wide basis, as well as for each individual market participant. We use the optimal auction surplus attained under C as the measure of the overall economic performance of the market.[1] As market participants are not obligated to reveal their actual costs and benefits, we use the bids and the offers to evaluate the bid/offer

[1] In a highly competitive market environment with a uniform price auction mechanism, the market participants tend to reveal their true marginal costs and benefits. Under such conditions, the auction surplus becomes a good proxy for the social welfare and, therefore, an appropriate approximation of the economic efficiency of the markets.

surplus of each participant at the optimum of Eqs. 1–8. We denote the values of the optimal variables by the superscript *. The seller s at node i has an offer surplus of

$$S_s = \lambda_i^* p_s^* - \beta_s(p_s^*) \tag{9}$$

where, λ_i^* is the locational marginal price (LMP) at node i, i.e. the price at which each MWh at node i is bought and sold. The bid surplus of the buyer b with demand at node i is similarly given by

$$S_b = \beta_b(p_b^*) - \lambda_i^* p_b^* \tag{10}$$

When the grid becomes constrained, the LMP at each node may change: in fact, for a lossless system, the non-zero LMP difference provides a measure of the congestion impacts. Absent congestion, the revenues collected in a lossless system from the buyers exactly equal the payments made to all the sellers. When congestion occurs, however, the two quantities are no longer equal. The difference between the revenues and the payments

$$K = \sum_{i=I}^{N} \sum_{b \in \mathcal{B} \text{ at node } i} \lambda_i^* p_b^* - \sum_{i=I}^{N} \sum_{s \in \mathcal{S} \text{ at node } i} \lambda_i^* p_s^* + \sum_{w=1}^{W} \left(\lambda_{m_w}^* - \lambda_{n_w}^*\right) t_w^* \tag{11}$$

is the congestion rents collected by the RTO, with the last term in Eq. 11 being the payments by the bilateral transactions.

We use the total dispatched load to evaluate the total cleared demand quantity under C

$$P_B = \sum_{b \in \mathcal{B}} p_b^* + \sum_{w=1,}^{W} t_w^* \tag{12}$$

In the next section, we extend the snapshot quantification approach for a multi-settlement environment.

3 Market Performance Quantification of a DAM and of its associated RTMs

In a multi-settlement environment, we deal with inter-related electricity markets. The actual system conditions during the hour h may differ from those used to determine the DAM hour h outcomes. The RTO uses market forces to manage such deviations and runs the RTMs, typically, every 5–10 min. As such, we may refer to RTMs as balancing energy markets. We associate with the DAM $\mathcal{D}_h$ for the hour h, the M RTMs $\mathcal{R}|_{(h,1)}\mathcal{R}|_{(h,2)}, \ldots, \mathcal{R}|_{(h,M)}$.

The DAMs are 24 separate hourly energy markets, one for each hour of the next day. Their financial nature makes possible the participation of financial entities, in addition to the players with physical resources. We use a snapshot to represent the

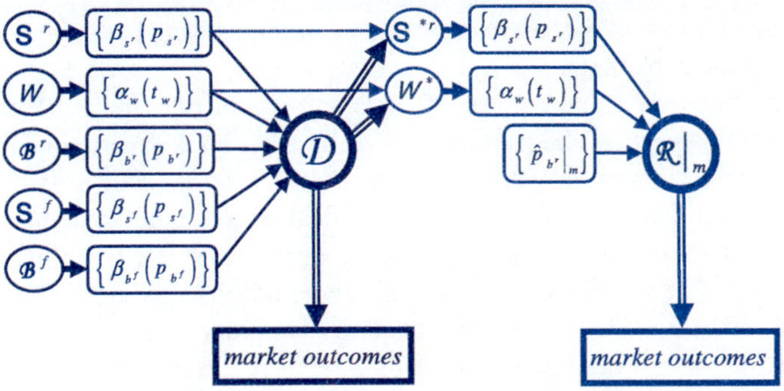

Fig. 1 Information flow in $\mathcal{D}$ for a given hour and an associated $\mathcal{R}|_m$

system for the hour h DAM and an "updated" snapshot for each $\mathcal{R}_{(h,m)}$, $m = 1, \ldots, M$. In what follows, we suppress the hour h notation so as to simplify the notation. We analyze the hour h DAM $\mathcal{D}$ operated in compliance with the security criterion C using $\mathcal{M}(\mathcal{S},\mathcal{B},\mathcal{W};\mathcal{C})$. The problem statement explicitly takes into account all the entities that constitute the set of sellers and the set of buyers in hour h – both financial and physical players. We use the superscript $r(f)$ to denote the participants with physical resources (financial players). Therefore, the set of sellers $\mathcal{S}$ (buyers $\mathcal{B}$) is given by $\mathcal{S} = \mathcal{S}^r \cup \mathcal{S}^f \left(\mathcal{B} = \mathcal{B}^r \cup \mathcal{B}^f\right)$. We denote the subset with non-zero cleared quantities in the DAM by $\mathcal{S}^{*r} \subseteq \mathcal{S}^r$ and the subset of transactions that receive transmission services by $\mathcal{W}^* \subseteq \mathcal{W}$. Even though a physical buyer b^r may have $p^*_{b^r} = 0$ in $\mathcal{D}$, b^r participates in the RTM to meet his fixed demand.

Each RTM is designed to be a purely physical market restricting participation to only those players with actual loads and physical generation assets who have non-zero outcomes in the DAM. For each $\mathcal{R}|_m$, the RTO uses the offers of the physical sellers in $\mathcal{S}^{*r}$, the willingness to pay of the bilateral transactions cleared in $\mathcal{W}^*$ and the real-time fixed demand of the physical buyers in $\mathcal{B}^r$. We use the identical system snapshot approach for $\mathcal{R}|_m$ and so we formulate and solve the market problem $\mathcal{M}(\mathcal{S}^{*r},\mathcal{B}^r,\mathcal{W}^*;\mathcal{C})$ for $\mathcal{R}|_m$.[2]

The metrics of interest—the auction surplus, the market participants' bid/offer surpluses and the congestion rents collected—are evaluated using the relations in Eq. 1 and Eqs. 9–11 for $\mathcal{M}(\mathcal{S}^{*r},\mathcal{B}^r,\mathcal{W}^*;\mathcal{C})$ for the subperiod m. We depict the inter-relationships between $\mathcal{D}$ and an associated $\mathcal{R}|_m$ in Fig. 1. We use the notation "^" to denote the optimal values attained in the clearing of $\mathcal{R}|_m$. The figure clearly

---

[2] Note that it is possible for a physical seller, not cleared at the DAMs, to participate in the RTMs under certain conditions. For example, the RTO may commit additional units after the clearing of the DAMs to ensure the system security in the real time. Furthermore, an RTO may also allow "self-commitment" of those resources. We do not consider these situations in the paper.

indicates the players who participate in each market, as well as the inputs and the outcomes of these markets.

The outcomes of $\mathcal{D}$ and those of $\mathcal{R}|_m$ are inputs into the settlement—the mechanism that specifies the payments to or by each market participant after the fact. We consider a system where the same MWh may be sold in two different markets—$\mathcal{D}$ and a specific $\mathcal{R}|_m$ —and so we deal with a multi-settlement system. Each of the M subperiods of the hour h has a duration of $1/M$ of an hour and we consider the multi-settlement for such a subperiod. A physical seller $s^r \in S^{*r}$, located at node i, who has cleared $p^*_{s^r}$ in $\mathcal{D}$ receives revenues of $1/M(\lambda^*_i p^*_{s^r})$ over that subperiod. As his real-time production $\hat{p}_{s^r}|_m$ may deviate from $p^*_{s^r}$, there is an adjustment to account for the production deviation $1/M(-p^*_{s^r} + \hat{p}_{s^r}|_m)$, which is paid at the $\mathcal{R}|_m$ LMP $\hat{\lambda}_i|_m$. The subperiod m revenues of s^r are

$$\eta_{s^r}|_m = \frac{1}{M}\left\{\lambda^*_i p^*_{s^r} + \hat{\lambda}_i\Big|_m (-p^*_{s^r} + \hat{p}_{s^r}|_m)\right\} \tag{13}$$

We note that if the seller s^r production in real time does not deviate from its DAM value, i.e. $p^*_{s^r} = \hat{p}_{s^r}|_m$, then the revenues $\eta_{s^r}|_m$ are simply the DAM revenues. As such, the $\mathcal{R}|_m$ LMP $\hat{\lambda}_i|_m$ has no impact on the subperiod m of s^r.

A financial seller s^f, located at node i, has revenues of $1/M\left(\lambda^*_i p^*_{s^f}\right)$ for his DAM "production". As s^f cannot participate in $\mathcal{R}|_m$, his real-time production $\hat{p}_{s^f}|_m = 0$, resulting in a deviation of $-p^*_{s^f}$. As a result, the RTM produces an adjustment of $-\hat{\lambda}_i|_m p^*_{s^r}$ to the DAM revenues of s^f. We may view the financial seller s^f as selling $p^*_{s^f}$ in the DAM at λ^*_i and buying back the same amount in the RTM at $\hat{\lambda}_i|_m$. The revenues of the seller s^f in subperiod m are

$$\eta_{s^f}|_m = \frac{1}{M}\left\{p^*_{s^r}\left(\lambda^*_i - \hat{\lambda}_i\Big|_m\right)\right\} \tag{14}$$

We note that as long as the DAM LMP λ^*_i is above the RTM LMP $\hat{\lambda}_i|_m$ the financial seller s^f has positive revenues.

In an analogous manner, the physical buyer b^r located at node i makes payments in the subperiod m of

$$\gamma_{b^r}|_m = \frac{1}{M}\left\{-\lambda^*_i p^*_{b^r} - \hat{\lambda}_i\Big|_m (-p^*_{b^r} + \hat{p}_{b^r}|_m)\right\} \tag{15}$$

The buyer b^r pays λ^*_i for the portion $p^*_{b^r}$ cleared in $\mathcal{D}$ and $\hat{\lambda}_i|_m$ for the remainder of his real-time demand. The payments of a financial buyer b^f in the subperiod m are

$$\gamma_{b^f}|_m = \frac{1}{M}\left\{p^*_{b^f}\left(-\lambda^*_i + \hat{\lambda}_i\Big|_m\right)\right\} \tag{16}$$

We use the same reasoning to determine the payments by the bilateral transaction $w \in W^*$ in the subperiod m to be

$$\gamma_w|_m = \frac{1}{M}\left\{\left(\lambda^*_{m_w} - \lambda^*_{n_w}\right)t^*_w + \left(\hat{\lambda}_{m_w}\Big|_m - \hat{\lambda}_{n_w}\Big|_m\right)\left(-t^*_w + \hat{t}_w|_m\right)\right\} \tag{17}$$

We note that if the bilateral transaction w does not deviate from the DAM clearing outcomes in the subperiod, then his payments are independent on the RTM outcomes.

The RTO makes the payments in Eqs. 13 and 14, to the sellers and receives from the buyers and the bilateral transactions the payments in Eqs. 15–17. The difference between these payments is the subperiod m congestion rents collected by the RTO

$$\begin{aligned} K_{\sum}\Big|_m = & \frac{1}{M}\left\{\sum_{b^r \in \mathcal{B}^r} \gamma_{b^r}|_m + \sum_{b^f \in \mathcal{B}^f} \gamma_{b^f}|_m + \sum_{w \in W^*} \gamma_w|_m\right\} \\ & - \frac{1}{M}\left\{\sum_{s^r \in \mathcal{S}^{*r}} \eta_{s^r}|_m + \sum_{s^f \in \mathcal{S}^f} \eta_{s^f}|_m\right\}. \end{aligned} \tag{18}$$

We use the results in Eqs. 13–18 for the quantification of the performance of the multi-settlement system in the subperiod m. The output of the seller $s^r \in \mathcal{S}^{*r}$ is produced in the real time, i.e. in the subperiod m, and is offered for sale for $1/M\beta_{s^r}(\hat{p}_{s^r}|_m)$. The offer surplus of the seller s^r in the subperiod m is expressed in terms of the difference between the revenues and the offer, i.e.

$$S_{s^r}|_m = \eta_{s^r}|_m - \frac{1}{M}\left\{\beta_{s^r}\left(\hat{p}_{s^r}|_m\right)\right\} \tag{19}$$

The fact that the financial seller s^f has no real-time production implies that the offer surplus of s^f equals his revenues

$$S_{s^f}|_m = \eta_{s^f}|_m \tag{20}$$

The physical buyer b^r consumes the energy in the subperiod resulting in the bid surplus given by the difference between the b^r willingness to pay in real time and the actual payments:

$$S_{b^r}|_m = \frac{1}{M}\left\{\hat{\beta}_{b^r}\left(\hat{p}_{b^r}|_m\right)\right\} - \gamma_{b^r}|_m \tag{21}$$

We note that the real time $\hat{\beta}_{b^r}$ may differ from β_{b^r} due to the fact that the real-time demand is viewed as fixed.

The financial buyer b^f cannot consume in the real time and so has the bid surplus

$$S_{b^f}|_m = -\gamma_{b^f}|_m \tag{22}$$

The bilateral transaction $w \in W^*$ receives the actual transmission service in the real time resulting in a surplus of

$$S_w|_m = \frac{1}{M}\{\alpha_w(\hat{t}_w|_m)\} - \gamma_w|_m \tag{23}$$

We make use of the market participants' bid/offer surpluses, including those of the bilateral transactions and the congestion rents collected by the RTO, to evaluate the total auction surplus attained in the multi-settlement system:

$$S_{\sum}\Big|_m = \sum_{s^r \in \mathcal{S}^{*r}} S_{s^r}|_m + \sum_{b^r \in \mathcal{B}^r} S_{b^r}|_m + \sum_{s^f \in \mathcal{S}^f} S_{s^f}|_m + \sum_{b^f \in \mathcal{B}^f} S_{b^f}|_m + \sum_{w \in W^*} S_w|_m + K_{\sum}\Big|_m. \tag{24}$$

We substitute Eqs. 18–23 into Eq. 24 to simplify and get

$$S_{\sum}\Big|_m = \frac{1}{M}\left\{\sum_{b^r \in \mathcal{B}^r} \hat{\beta}_{b^r}(\hat{p}_{b^r}|_m) - \sum_{s^r \in \mathcal{S}^{*r}} \beta_{s^r}(\hat{p}_{s^r}|_m)\right\} + \frac{1}{M}\sum_{w \in W^*} \alpha_w(\hat{t}_w|_m). \tag{25}$$

Now, the auction surplus attained in the RTM $\mathcal{R}_m$ is $\hat{S}|_m$ and its value is given by Eq. 1:

$$\hat{S}\Big|_m = \frac{1}{M}\left\{\sum_{b^r \in \mathcal{B}^r} \hat{\beta}_{b^r}(\hat{p}_{b^r}|_m) - \sum_{s^r \in \mathcal{S}^{*r}} \beta_{s^r}(\hat{p}_{s^r}|_m)\right\} + \frac{1}{M}\sum_{w \in W^*} \alpha_w(\hat{t}_w|_m). \tag{26}$$

We conclude that

$$S_{\sum}\Big|_m = \hat{S}\Big|_m. \tag{27}$$

Therefore, the total auction surplus of the multi-settlement system attained in the subperiod m is precisely the auction surplus attained in $\mathcal{R}|_m$. We, furthermore, conclude that the outcomes of $\mathcal{D}$ do not explicitly impact the total auction surplus. $S_{\sum}|_m$, but impact the allocation of the total auction surplus among the market participants.

The performance metrics in Eqs. 18–25 are for the subperiod m of the hour h. We aggregate them for the M subperiods of hour h to evaluate the hourly metrics. In particular, we compute the hour h auction surplus $S_{\sum}$ attained in the multi-settlement system to be

$$S_{\sum} = \sum_{m=1}^{M} S_{\sum}\Big|_m = \sum_{m=1}^{M} \hat{S}\Big|_m \tag{28}$$

The performance quantification of the multi-settlement system clearly makes use of the inter-relationships between the DAM and its associated RTMs, as illustrated in Fig. 2.

We note the clearing of the financial entities in $\mathcal{D}$ impacts the clearing of the physical generation as well as the clearing of the physical loads. As such, the participation of the financial entities impacts the deviations of the physical resources clearing in the real time. Such deviations have implications on the

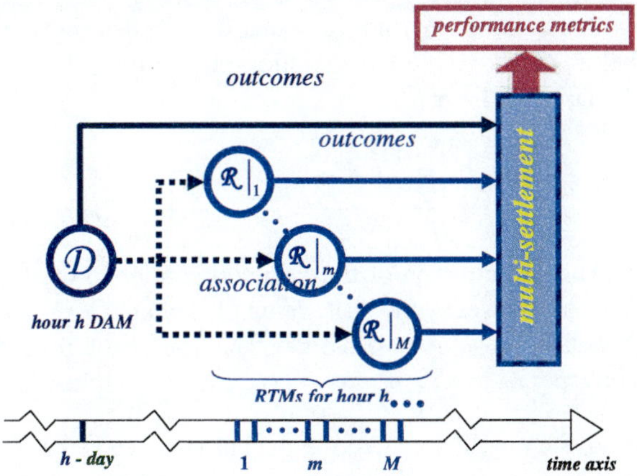

Fig. 2 Interactions between $\mathcal{D}$ and $\mathcal{R}|_m$ and the performance quantification in a multi-settlement environment

market and the system operations. In particular, they impact the ability of the RTO to ensure real-time system security.

We also note that a single-settlement system is a special case of multi-settlement systems. An example of such a case is that every transaction at the DAM corresponds to physical production and the DAM clearing perfectly forecasts real-time system conditions, as such no net injection deviation at each node in real time. Under such conditions, the clearing of the DAMs or their associated RTMs represents a single-settlement system.

In the next section, we describe the proposed approach to quantitatively assess the impacts of operating a system under a specified criterion C on the market performance in a multi-settlement environment. Also, we quantify the impacts of financial entities on the ability of the RTO to meet system security C in the near-real-time.

4 Proposed Approach

The maintenance of secure power system operations is a task that strongly depends on the outcomes of the DAMs. We may view the DAM physical generation and consumption as a rough guess of the actual outcomes in the associated RTMs. As the system and the market conditions in the near to real time may change from those forecast and cleared in the DAM, the RTMs are run to manage the resulting deviations.

The actual physical demand in the M RTMs gives rise to the physical demand deviation in hour h:

$$\delta \hat{p}_{\mathcal{B}^r} = \frac{1}{M}\left[\sum_{m=1}^{M}\left(\sum_{b^r \in \mathcal{B}^r} \hat{p}_{b^r}|_m + \sum_{w \in W^*} \hat{t}_w|_m\right)\right] - \left[\sum_{b^r \in \mathcal{B}^r} p^*_{b^r} + \sum_{w \in W^*} t^*_w\right] \tag{29}$$

A non-zero $\delta\hat{p}_{\mathcal{B}^r}$ indicates that the physical buyers' DAM purchases are either below or above the consumption in real time. Similarly, the physical generation deviation in hour h is

$$\delta\hat{p}_{\mathcal{S}^{*r}} = \frac{1}{M}\left[\sum_{m=1}^{M}\left(\sum_{s^r\in\mathcal{S}^{*r}}\hat{p}_{s^r}|_m + \sum_{w\in\mathcal{W}^*}\hat{t}_w|_m\right)\right] - \left[\sum_{s^r\in\mathcal{S}^{*r}} p^*_{s^r} + \sum_{w\in\mathcal{W}^*} t^*_w\right]. \tag{30}$$

The participation of the financial entities in the DAM gives rise to the lack of balance between physical demand deviation $\delta\hat{p}_{\mathcal{B}^r}$ and the physical generation deviation $\delta\hat{p}_{\mathcal{S}^{*r}}$. A positive net injection of the financial participants in the DAM corresponds to $\sum_{s^f\in\mathcal{S}^f} p^*_{s^f} > \sum_{b^f\in\mathcal{B}^f} p^*_{b^f}$, which implies that $\sum_{s^r\in\mathcal{S}^{*r}} p^*_{s^r} < \sum_{b^r\in\mathcal{B}^r} p^*_{b^r}$. In this case, the physical generation deviation exceeds the physical demand deviations so that $\delta\hat{p}_{\mathcal{S}^{*r}} > \delta\hat{p}_{\mathcal{B}^r}$. Therefore, more generation is required in real time than cleared in $\mathcal{D}$ leading to the deviations in the physical sellers' outcomes. In case of $\sum_{s^f\in\mathcal{S}^f} p^*_{s^f} < \sum_{b^f\in\mathcal{B}^f} p^*_{b^f}$ —a negative net injection of the financial entities—some of the physical generation serves the demand of the financial buyers in $\mathcal{D}$ and so $\sum_{s^r\in\mathcal{S}^{*r}} p^*_{s^r} > \sum_{b^r\in\mathcal{B}^r} p^*_{b^r}$. In this case, $\delta\hat{p}_{\mathcal{S}^{*r}} < \delta\hat{p}_{\mathcal{B}^r}$. Whenever there is zero net injection by the financial entities, the physical generation deviation and the physical demand deviation are in exact balance. The absence of financial entity participation is a special case of this zero net injection. While the injection/withdrawal deviation metrics of Eqs. 29 and Eqs. 30 provide system-wide aggregate measures, we can also introduce analogous metrics for zonal, as well as, nodal measures in order to meet the requirements at the different levels of granularity.

We use the auction surplus in Eq. 28, the total congestion rents in Eq. 18 and each market participants' bid/offer surplus metric in Eqs. 19–23 to evaluate the overall economic performance of the multi-settlement system and that of each market participant, respectively. In addition, we need appropriate metrics to analyze the combined impacts of the DAM-RTM clearing outcomes.

As market and system conditions may change, the price of the MWh commodity in each $\mathcal{R}|_m$ at a specified node may deviate from that in $\mathcal{D}$. The hour h price deviation at node i is

$$\delta\lambda_i = \frac{1}{M}\sum_{m=1}^{M}\left[\lambda^*_i - \hat{\lambda}_i\Big|_m\right]. \tag{31}$$

Whenever $\delta\lambda_i \neq 0$ over a nontrivial subset of hours, arbitrage opportunities exist, implying market inefficiency [26]. A financial entity can participate in the market to take advantage of price arbitrage opportunities at such a node. As more and more financial entities eye such opportunities, leading to their participation in the markets to arbitrage the price deviation, the arbitrage opportunities begin disappearing. As such, $\delta\lambda_i \to 0$, leading to the improved economic efficiency of the markets. Thus, price convergence is a desirable outcome in multi-settlement systems. We also note that $\delta\lambda_i - \delta\lambda_j$, $i \neq j$, quantifies how well the $\mathcal{D}$ outcomes forecast the nodal price difference between nodes i and j in real time taking into account the actual system congestion and losses.

The price deviation $\delta\lambda_i$ also impacts the surplus of each market participant. The output of the seller $s^r \in \mathcal{S}^{*r}$, located at node i, is produced in the real time. Therefore, his offer is, unlike his revenues in Eq. 13, independent of the $\mathcal{D}$ outcomes. Therefore, the $\mathcal{D}$ outcomes impact the surplus of the seller s^r in hour h. Using Eq. 19, the s^r offer surplus in hour h is

$$\mathcal{S}_{s^r} = \frac{1}{M}\sum_{m=1}^{M}\left\{\hat{p}_{s^r}|_m \hat{\lambda}_i\Big|_m - \beta_{s^r}\left(\hat{p}_{s^r}|_m\right)\right\} + \delta\mathcal{S}_{s^r} \tag{32}$$

Here, $\delta\mathbf{S}_{s^r}$ is the physical seller offer surplus deviation metric

$$\delta\mathbf{S}_{s^r} = p^*_{s^r}\delta\lambda_i \tag{33}$$

and quantifies the impact of the $\mathcal{D}$ outcomes on the revenues of the seller s^r. A positive (negative) $\delta\mathcal{S}_{s^r}$ implies that s^r captures more (less) revenues for his real-time production than those in $\mathcal{D}$. We consider a specific case to illustrate the nature of $\delta\mathcal{S}_{s^r}$. For a system with s^r, the marginal seller in both $\mathcal{D}$ and an associated $\mathcal{R}|_m$ and with $\hat{\lambda}_i|_m > \lambda^*_i$ and $\hat{p}_{s^r}|_m > p^*_{s^r}$. While $p^*_{s^r}$ is paid at λ^*_i the $\hat{p}_{s^r}|_m - p^*_{s^r}$ is paid at $\hat{\lambda}_i|_m$. Therefore, the portion $p^*_{s^r}$ receives less revenues per MWh than $\hat{p}_{s^r}|_m - p^*_{s^r}$. The fact that s^r participates in $\mathcal{D}$ and sells $p^*_{s^r}$ implies that for this case he receives lower revenues than had he participated in only $\mathcal{R}_m$. As such, s^r is better off clearing a lesser amount than $p^*_{s^r}$ whenever $\hat{\lambda}_i|_m > \lambda^*_i$ so as to increase revenues for its actual production $\hat{p}_{s^r}|_m$. The negative $\delta\mathcal{S}_{s^r}$ is illustrated in Fig. 3. Whenever a financial buyer b^f realizes such a price deviation, then his participation in $\mathcal{D}$ that may result in an increase in λ^*_i. In turn, the revenues of s^r may increase and he may be willing to produce more in $\mathcal{D}$ than in the case of without the financial buyer b^f. We note that the surplus driven offer/bid decisions furthermore impact the

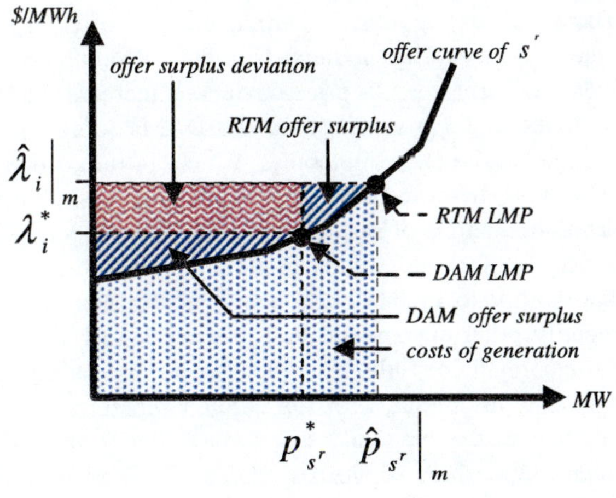

Fig. 3 The effects of $\mathcal{D}$ and $\mathcal{R}|_m$ clearing on the offer surplus deviation of the physical seller s^r located at node i for the case $\hat{\lambda}_i|_m > \lambda^*_i$

deviation of the physical production/consumption. Under the conditions of the example, as the physical seller has the incentive to clear a lesser amount in $\mathcal{D}$ due to the price deviation, the need may arise for additional amounts cleared in the near-real-time. We conclude that such incentives may result in conditions that the 'physical production/consumption in $\mathcal{D}$does not appropriately forecast the real-time conditions, and may lead to "stressed" real-time operations, thereby lessening the ability to ensure secure power system operations.

Once we compute the individual offer surplus deviation of a physical seller, we can determine the offer total surplus deviation of the subset of the physical sellers using

$$\delta S_{\mathcal{S}^{*r}} = \sum_{i=0}^{N} \sum_{s^r \in \mathcal{S}^{*r}} p^*_{s^r} \delta\lambda_i \varepsilon_{is^r}, \varepsilon_{is^r} = \begin{cases} 1, & s^r \text{ is at node } i \\ 0, & \text{otherwise} \end{cases} \tag{34}$$

Similarly, we can evaluate the total physical buyers' bid surplus deviations

$$\delta S_{\mathcal{B}^r} = \sum_{i=0}^{N} \sum_{b^r \in \mathcal{B}^r} p^*_{b^r} \delta\lambda_i \varepsilon_{ib^r}, \varepsilon_{ib^r} = \begin{cases} 1, & b^r \text{ is at node } i \\ 0, & \text{otherwise} \end{cases} \tag{35}$$

A positive $\delta S_{\mathcal{B}^r}$ implies that the physical buyers pay less for their aggregate real-time demand in $\mathcal{D}$ than in the associated RTMs. This happens because the physical buyers benefit from the lower DAM prices that they pay for the portion of the demand cleared in the DAM.

The MW deviation metrics along with the price and the bid/offer surplus deviation metrics capture important aspects of system and market operations in a multi-settlement environment. The physical generation and demand deviation metrics quantify how "close" the real-time system conditions are to those forecasted in the clearing of the DAM. Smaller magnitude deviations imply improved "forecasts" of the system conditions in the DAM, which, in turn, result in the improved ability of the RTO to ensure real-time system security. Therefore, the DAM clearing is strongly inter-related with the real-time system operations. The price and the physical participants' bid/offer surplus deviation metrics, on the other hand, quantify the impacts of the DAM outcomes on the market participants' bid/offer surpluses. As price deviations increase, the financial entity participation becomes more pronounced in the DAMs [26, 27]. Such participation leads to changes in the DAM outcomes, which, in turn, impact how the real-time system conditions are forecasted in the DAM. A desirable market outcome is that the deviation metrics of surplus and of production/consumption tend to zero since the lower the absolute values of these metrics, the "better" the markets perform. The ability to quantify the economic impacts of compliance with a specified security criterion renders these metrics highly appropriate in the preparation of various regulatory filings, as well as in applications to longer-term planning and shorter-term studies with the explicit representation of the financial entities in addition to the physical asset owners. The proposed metrics capture the strong inter-relationships between system and market operations in the multi-settlement

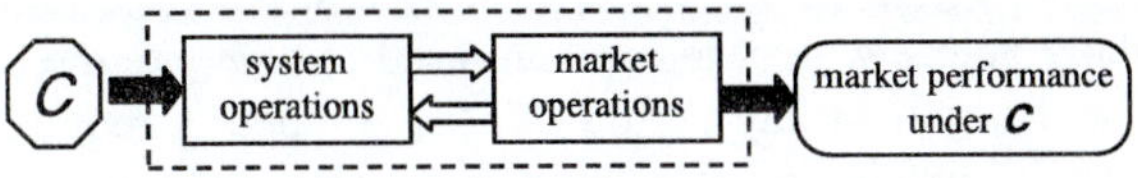

Fig. 4 The market performance quantification under security criterion C

environments. Therefore, they effectively quantify the performance of the multi-settlement systems.

The value of the metrics given in Eqs. 9–35 depends on the specified security criterion C and constitutes the basic building of the approach. We conceptually represent this snapshot assessment framework in Fig. 4.

Under a different security criterion C', the RTO explicitly considers the solution of the problem $\mathcal{M}(S, \mathcal{B}, W; C')$ at each system snapshot, be it a DAM or an RTM. The constraints expressed in Eqs. 2–8 apply to each contingency in the set $J_{C'}$. We measure the impacts on market performance due to the change in the security criterion from C to C' by the change in each metric of interest from one criterion to the other. For example, the change in the auction surplus metric is given by

$$\Delta S_{\Sigma}\Big|_{C\to C'} = S_{\Sigma}\Big|_{C'} - S_{\Sigma}\Big|_{C} \tag{36}$$

and provides a proxy measure for the change in the economic efficiency of the markets in a multi-settlement environment due to a change in the security criterion from C to C'. We deploy analogous expressions for each metric in Eqs. 9–35 to measure the relative change in response to the security criterion change from C to C'. The changes in the bid/offer surpluses, the total dispatched load and the multi-settlement system deviation metrics are all of interest in our assessment. We also need the changes in the physical demand and generation deviation to quantify the impacts on the ability of the RTO to meet system security in real time. For example, to evaluate the impacts on the physical generation deviation, we use

$$\Delta\delta_{S^{*r}}\big|_{C\to C'} = \delta\hat{p}_{S^{*r}}\big|_{C'} - \delta\hat{p}_{S^{*r}}\big|_{C} \tag{37}$$

These metrics effectively capture the multi-settlement system performance for the security criterion change from C to C' for a given hour h. For example, the RTO can quantify the economic impacts of operating the system under a tightened security criterion by including additional contingencies in the postulated contingency list.

Under a specified security criterion, the hourly snapshots corresponding to different system and market conditions may result in markedly different market performance outcomes. Such differences arise for many reasons including changes in the load, the set of selling entities, and the offers/bids submitted. Consequently, these hourly assessments must be carried out over a longer period to appropriately capture the impacts of the different conditions that exist during that period. Conceptually, we need to assess the market performance of the DAM and the associated RTMs at each hour of the study period. The needs are similar in assessing the market performance impacts due to a change in the security criterion. The hourly values of the relative performance metrics are summed to obtain the daily values which, in turn, are used to compute the relative performance metrics for the entire study period. As the computing requirements to clear each market over a study period for a large-scale system may be large, a practical way to reduce them is to perform the assessments for a smaller representative sample of the hours. For this purpose, we require a scheme that systematically selects this smaller subset of representative hours [33].

A key requirement in selecting these hours is the incorporation of the unit commitment decisions which entail inter-temporal effects across the hours of the commitment. To fully capture the inter-temporal effects, all the hours of the unit commitment period need to be considered. Since, for typical market applications, the unit commitment period is a day, this requirement shifts the selection of representative sample of hours to that of days, since all the hours of such days must be included.

A first step in the selection of representative days is the partitioning of the study period into subperiods. Since many operational studies are carried out on a monthly basis, we use a month as a subperiod. For a given month i, we determine the subset of representative days and construct the set D_r^i using the following scheme.

Let $D^i = \{d_q{:}q = 1,\ldots, D\}$ be the set of days in the month i. We denote the day d_q peak-demand load by p_{d_q}. We reorder the set of the demand values $\{p_{d_1},\ldots,p_{d_D}\}$ as $\{\tilde{p}_1,\ldots,\tilde{p}_D\}$ with $\tilde{p}_j \geq \tilde{p}_{j+1}$ where $\tilde{p}_j$ denotes the jth largest value of the month. We construct the ordered daily load curve using the set of points $\{(0,\tilde{p}_1),(1,\tilde{p}_2),\ldots,(D-1,\tilde{p}_D)\}$. This curve has at most D distinct load levels. We normalize the time axis using D as the base value and construct the so-called load duration curve $(LDC)L(\cdot)$ as a piece-wise step function using the set of points $\{(0,\tilde{p}_1),(1/D,\tilde{p}_2),\ldots,((D-1)/D,\tilde{p}_D)\}$. We super-pose the grid with k equally distributed LDC factors

$$0 = \psi_0 < \psi_1 < \cdots < \psi_{k+1} = 1 \tag{38}$$

on the time axis. We determine the load level $\hat{p}_j = L(\psi_j)$ for each ψ_j. We choose k so that the $(k+2)$ load values are distinct and

$$\hat{p}_0 > \hat{p}_1 > \cdots > \hat{p}_{k+1} \tag{39}$$

We use the load levels to subdivide the interval between $\hat{p}_0$ and $\hat{p}_{k+1}$ into $(k+2)$ load tranches

$$P_j = \begin{cases} \left[\hat{p}_{k+1}, \dfrac{\hat{p}_k + \hat{p}_{k+1}}{2}\right] & j = k+1 \\ \left(\dfrac{\hat{p}_{j-1} + \hat{p}_j}{2}, \dfrac{\hat{p}_j + \hat{p}_{j+1}}{2}\right] & j = 1,\ldots,k \\ \left(\dfrac{\hat{p}_0 + \hat{p}_1}{2}, \hat{p}_0\right] & j = 0 \end{cases} \tag{40}$$

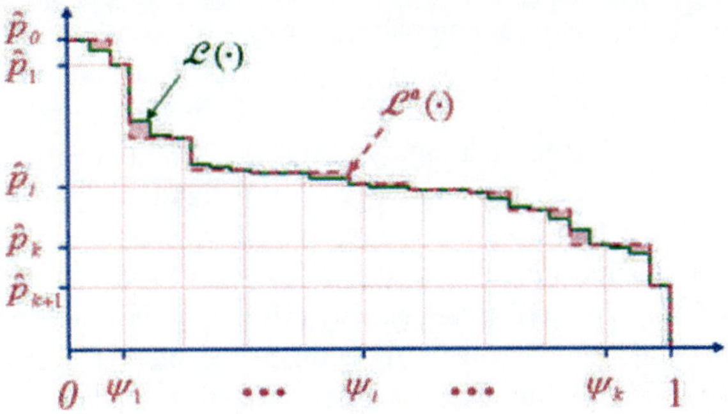

Fig. 5 Construction of $L^a(\cdot)$ from $L(\cdot)$

and determine from the time axis the corresponding duration n_j of each tranche. Note that n_j is an integer multiple of $1/D$ and $\sum_{j=0}^{k+1} n_j = 1$. We define

$$\hat{\psi}_j = \sum_{s=0}^{j-1} n_s, \; j = 1, \ldots, k. \tag{41}$$

We construct $L^a(\cdot)$ from the $(k+2)$ load levels using the set of points $\{(0,\hat{p}_0),(\hat{\psi}_0,\hat{p}_1),\ldots,(\hat{\psi}_k,\hat{p}_{k+1})\}$ and use it to approximate $L(\cdot)$. For each load level $\hat{p}_j$ of $L^a(\cdot)$, we identify the day d_q with $p_{d_q} = \hat{p}_j$. In case of two or more such days, we select the most or more recent day. We construct D_r^i using these $(k+2)$ selected days. We illustrate such construction in Fig. 5.

We measure the "goodness" of the approximation in terms of an error based on the monthly energy. We define the error in the LDC approximation by

$$\varepsilon(k) = \int_0^1 |L^a(x;k) - L(x)|dx \bigg/ \int_0^1 L(x)dx \tag{42}$$

We compare the value of $\varepsilon(k)$ with a specified error tolerance value $\bar{\varepsilon}$. If the error fails to satisfy $\bar{\varepsilon}, k$ is increased until the tolerance check is satisfied and selects the corresponding D_r^i.

We repeat this process for each of the month within the study period and then construct the set of representative days $\boldsymbol{D}_r$ of the study period by the union of the monthly D_r^i. We apply the structure shown in Fig. 4 to each hour of the days in $\boldsymbol{D}$ for each specified criteria. We quantify the hourly relative performance metrics and aggregate them for each day. We use the number of days each day in D_r^i represents and aggregate the daily figures to obtain monthly impacts. The daily figures also serve to evaluate key statistics for each month such as mean, variance and range. The study period impacts then are aggregated form the monthly ones. Thus, we are able to quantify the system and area-wide MW as well as dollar impacts on a daily, monthly and period basis.

The proposed approach provides a useful tool to the RTO to analyze the interdependence between market performance and the system security. The ability to quantify the financial impacts of compliance with a specified security criterion makes the approach highly useful in regulatory proceedings, as well as in longer-term planning and shorter-term investigations with the explicit representation of both the financial and the physical asset owning players. The proposed approach has a wide range of applications such as the justification by the RTO of the decision to modify the security criterion to be used and the cost/benefit analysis of network improvements to mitigate the market performance impacts of a set of specified contingencies. Other applications include the formulation of the control actions for specific contingencies, and the assessment of specific behavioral changes of market participants under various security criteria. The proposed approach furthermore allows us to investigate the role of the DAMs in reallocating the auction surplus among market participants, to analyze the DAM–RTM price deviation issues and to quantify the impacts of the financial entities participation on real-time system security. We devote the following section to present a set of applications to the large-scale ISO New England (ISO-NE) of the proposed approach.

5 Application Studies: The ISO-NE System

In this section, we provide a set of application studies to quantify the economics of secure power system operations for the ISO-NE markets for the case of DAMs only, as a proxy study for single-settlement systems, and for the multi-settlement system. We use the ISO-NE in all the case studies presented since it provides a good, realistic-sized system that is large enough to effectively illustrate the capabilities of the proposed approach to determine practical solutions to a wide range of problems in the various applications. We, first, provide a brief description of the multi-area structure of the ISO-NE system and state the current ISO-NE security criterion. We, next, provide the results of the ISO-NE DAM comparative study and those of the set of studies for the multi-settlement system.

The ISO-NE is the regional transmission organization serving the states of Connecticut, Maine, Massachusetts, New Hampshire, Rhode Island and Vermont. The ISO-NE is a member of the Northeast Power Coordinating Council (NPCC) and is part of the Eastern Inter-connection. An important characteristic of the ISO-NE system is its multi-area structure. Such a structure has major implications for the way the system operations comply with the security criterion. Each area of the ISO-NE multi-area network is characterized as being either an import or an export area. We depict conceptually the multi-area structure of the ISO-NE in Fig. 6. The import areas [34] are

- A^1: Boston/NE Massachusetts
- A^2: Connecticut

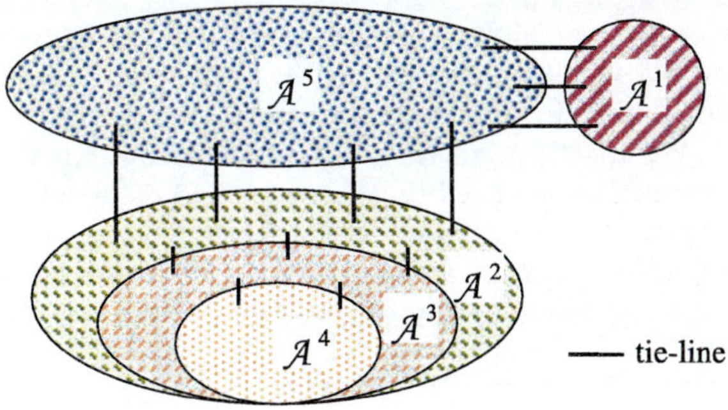

Fig. 6 Multi-area structure of the ISO-NE system

- A^3: SW Connecticut
- A^4: Norwalk/Stamford

We treat rest of the system as a single export area and denote it by A^{5}. Figure 2 illustrates conceptually the multi-area structure of the ISO-NE. A salient feature is the nested structure of the areas $A^{4} \subset A^{3} \subset A^{2}$. From the physical and the economic point of view, the generation of the export area is required to meet the load of the import areas.

The ISO-NE system security criterion takes into account this multi-area structure. This is a modified $(n - 2)$ security criterion. We denote this security criterion by $\mathbf{C}_0$, whose contingency list is $J_{C_0}=J_{n-1} \cup \left(\bigcup_{k=1}^{4} M^k \right)$. Here, $\boldsymbol{J}_{n-1}$ is the set of single element contingencies considered by the ISO-NE and $\boldsymbol{M}^k$ is the set of double tie line contingencies specified for each import area $A^k \in A$, $k=1,\ldots,4$. Each selected tie line pair interconnects the import area A^k to any other area of the system. The set of control actions for the security criterion consists of preventive control actions which are associated with the elements of $\boldsymbol{J}_{n-1}$, and corrective control actions which are associated with the double element tie line contingencies of $\bigcup_{k=1}^{4} M^k$[34].

We next focus on the studies to quantify the economics of secure power system operations in the ISO-NE system. The objective of this study is to analyze whether the economic efficiency of the ISO-NE DAM is adversely impacted by the system operations complying with the security criterion in force.[3] For this purpose, we quantify the market performance as a function of three security criteria and perform comparative assessments. We measure the changes with respect to the outcomes under the current ISO-NE security criterion. We quantify the impacts of the security criterion in force on the market performance using the actual day-ahead data – the system model and the bids/offers submitted—with the actual market

[3] We use the ISO-NE DAM study as a proxy study for single-settlement systems. We assume that the bids and the offers of the financial entities represent physical consumption and production. As such, this study is a special case for multi-settlement systems. .

clearing methodology. The results of this study serve as the reference case, with respect to which we compare the impacts of a tightened and a relaxed security criterion on the market performance.

We select the criterion $\mathbf{C}_0$ as the reference criterion and consider two specific criteria C_1 a modified (n–1) security, and C_2, a modified (n – 2) security. For the criterion C_1, the contingency list $J_{C_1} = J_{n-1}$, and preventive control action is the deployed for each contingency in J_{C_1}. For the criterion C_2, the contingency list $J_{C_2} = J_{C_0}$, but we replace the corrective control actions by the preventive control actions for the contingencies in $\bigcup_{k=1}^{4} M^k$.

The study is performed for the second half of the year 2005. This study period was chosen to allow the use of market and system data that reflects the most up-to-date ISO-NE procedures and rules. The analysis of the load in the selected period shows that the demand levels in the summer months, July and August, are significantly higher than those in the non-summer months—the months from September to December. Furthermore, the range of daily peak demands in the summer months is considerably larger than that in the non-summer months. Due to the maintenance scheduling, the sets of available resources in summer months are different than those in the other months. In addition, the ratings of the system components differ in each summer month from those in the other months. The period under study is further characterized by the existence of two distinct regimes, $\mathcal{R}_1$ and $\mathcal{R}_2$ (pre- and post-October 9, 2005), respectively. The ratio of the hourly price-sensitive bid amounts to the total hourly demand changes markedly from a small value under the regime $\mathcal{R}_1$, to a sizable fraction under the regime $\mathcal{R}_2$. The hourly loads in these two regimes are further distinguished in terms of their peak, base and average values. We present the load characteristics of the regimes $\mathcal{R}_1$ and $\mathcal{R}_2$ in Table 1. The minimum, maximum and the average hourly load values are disaggregated into the fixed and price-responsive components in Table 1. The significant increase in the fraction of price-sensitive demand is due to the bidding behavior change of the large buying entity whose demand corresponds to approximately 25% of the total system demand. This buyer submits, on the average, only 10% of his demand as price-sensitive under the regime $\mathcal{R}_1$. However, the buyer has no fixed demand under regime $\mathcal{R}_2$ as the entire buyers' bids become price-sensitive, as shown in Fig. 7. Due to the size of the buyer's demand, the marked change in his bidding behavior results in a significant portion of the total system demand that is price responsive under the regime $\mathcal{R}_2$.

Table 1 Regimes $\mathcal{R}_1$ and $\mathcal{R}_2$ load characteristics

Regime	$\mathcal{R}_1$		$\mathcal{R}_2$	
Load type	Fixed	Price-sensitive	Fixed	Price-sensitive
Minimum demand	6,232	2,944	5,394	5,139
Maximum demand	21,292	4,845	12,109	9,573
Average	13,075	4,294	8,756	6,964

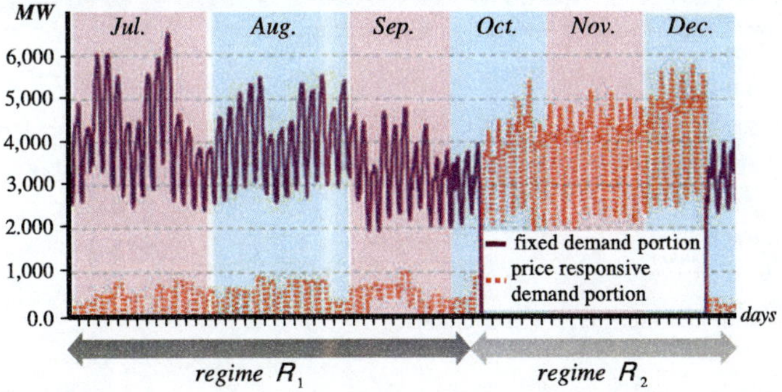

Fig. 7 The bidding behavior change of the larger buying entity

We select the representative days from each month using the scheme introduced in the previous section. We construct the LDC approximation for each summer and non-summer month by 14 and 10 representative days, respectively. Since these approximations provide acceptably small errors, we determine the elements of each D_r^i and construct $\mathcal{D}_r$.

We next discuss the market performance impacts of the change of security criterion $\mathbf{C}_0$ to each of the criteria considered and distinguish those impacts under the two regimes $\mathcal{R}_1$ and $\mathcal{R}_2$

We first focus on the MW impacts. For the reference criterion $\mathbf{C}_0$, we obtain the range and the average values of the total hourly dispatched load P_C under the regimes $\mathcal{R}_1$ and $\mathcal{R}_2$. We compute the changes from the P_C values under the two security criteria and present the results in Table 2. We observe that the price-responsive demand plays an important role in the DAM. For each security criterion, the changes under the regime $\mathcal{R}_2$ are considerably lower than those under the regime $\mathcal{R}_1$. In fact, the changes are more pronounced for the change of the security criterion from $\mathbf{C}_0$ to C_2 than from $\mathbf{C}_0$ to C_1. We hypothesize that the factors that contribute to these distinct outcomes are due to the structure of the system, the effectiveness of the security control actions and the nature of the constraints imposed on the system operations.

The change from the current security criterion to either of the two criteria studied impacts the value of the system transfer capability. The change in the value of the system transfer capability, in turn, affects the ability of the import areas to

Table 2 Total hourly dispatched loads and range of impacts

Metric	Regime	Range (MW)	Average (MW)
$P_{\mathbf{C}_0}$	$\mathcal{R}_1$	(9,177, 25,638)	16,967
	$\mathcal{R}_2$	(8,733, 23,281)	15,421
ΔP_{C^a}	$\mathcal{R}_1$	(0, 452)	141
	$\mathcal{R}_2$	(0, 273)	42
ΔP_{C^b}	$\mathcal{R}_1$	(−818, 0)	−184
	$\mathcal{R}_2$	(−557, 0)	−128

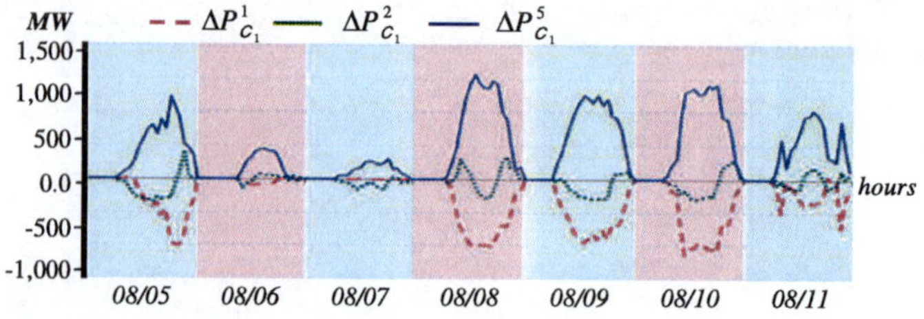

Fig. 8 Area-wide net injection impacts under C_1

bring in energy from the export area. In fact, the analysis of the ISO-NE system during this 2005 study period indicates that the replacement of the security criterion $\mathbf{C}_0$ by the criterion C_1 results in the increased import capabilities of the import areas for each hour. But, the increased capability may not be utilized in every hour. For example, the imports by the stand-alone area $\mathcal{A}^1$ buyers increase their imports from the export area, thereby decreasing their dependence on the less economic $\mathcal{A}^1$ resources. On the other hand, the imports of the nested area $\mathcal{A}^2$, due to the physical constraints of the $\mathcal{A}^2$ network, may not utilize such increased capability in every hour. We measure the changes in the utilization of the increased import capabilities using the relative area-wide net injection metric for the areas $\mathcal{A}^1$, $\mathcal{A}^2$ and $\mathcal{A}^5$. We illustrate the results for the import areas $\mathcal{A}^1$ and $\mathcal{A}^2$, and the export area $\mathcal{A}^5$ for a week in August 2005 in Fig. 8. These plots are typical for the study period, particularly in terms of the more pronounced impacts in the daily peak hours than those in the off-peak hours.

Due to the fact that the system operations under the criterion C_2 are more constraining than those under the criterion $\mathbf{C}_0$, the security change from $\mathbf{C}_0$ to C_2 results in the decreased import capabilities of the import areas for every hour of the study period. In fact, the impacts on the imports of the stand-alone area $\mathcal{A}^1$ are exactly in the opposite direction to those under the criterion change from $\mathbf{C}_0$ to C_1. On the other hand, the imports of the nested area $\mathcal{A}^2$ exhibit similar results to those under the criterion change from $\mathbf{C}_0$ to C_1. We plot these outcomes for the same August week in Fig. 9. We note that the impacts are pronounced in both peak and off-peak hours.

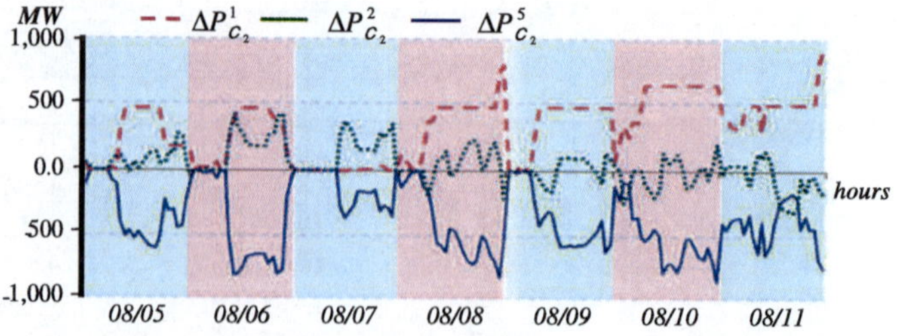

Fig. 9 Area-wide net injection impacts under C_2

We next examine the monetary impacts of the changes in the security criterion as measured by the relative auction surplus metric. We use the daily auction surplus as the basic metric in this investigation. We first normalize the daily auction surplus values using the average value of the daily auction surplus under the reference criterion $\mathbf{C}_0$ as a base value. We use the normalized values to compare the impacts with respect to the values under the reference criterion, as well as, across study periods of different durations. In this way, the comparisons are both consistent and meaningful. We can interpret the results to understand the nature of the impacts and how they relate to the values attained under the reference criterion $\mathbf{C}_0$. For concreteness, we use a value of $\tau = 1{,}000$ \$/MWh/h for evaluating the benefits of the buyers submitting fixed demand. We first consider the economic repercussions of the increased import capabilities arising from the relaxation of the security criterion from $\mathbf{C}_0$ to C_1. Throughout the study period, the increased import capabilities are utilized leading to higher market efficiencies. We may view these improvements as a measure of the "costs" of not violating the constraints due to the double element contingencies in the reference criterion. On the other hand, the decreased import capabilities arising from changing the criterion from $\mathbf{C}_0$ to C_2 may lower the auction surplus. Indeed, such reductions are present throughout the study period. We may interpret these reductions to be a measure of the "costs" of replacing corrective for preventive control actions. The plot of the normalized daily auction surplus values under the reference criterion $\mathbf{C}_0$ is given in Fig. 10 for the set of days $\mathcal{D}^r$. The plots of the changes in auction surplus arising from a change of the security criterion are shown in Fig. 11. In this figure, we also provide the normalized impacts considering a different value of $\tau' = 10{,}000$ \$/MWh/h. Note that, the different values of τ and τ' impact the normalized values but do not affect the nature of the impacts. We provide some of the statistics related to the maximum, the mean and the standard deviation of the values of relative auction surplus metrics under the regimes $\mathcal{R}_1$ and $\mathcal{R}_2$ for each security criterion change in Table 3.

We obtain additional insights into the impacts of the security criterion change on the market participants in each area by studying the disaggregation of the

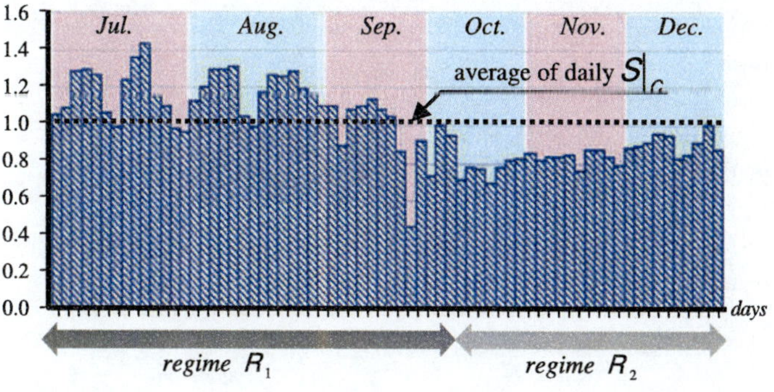

Fig. 10 Normalized daily auction surplus under criterion $\mathbf{C}_0$

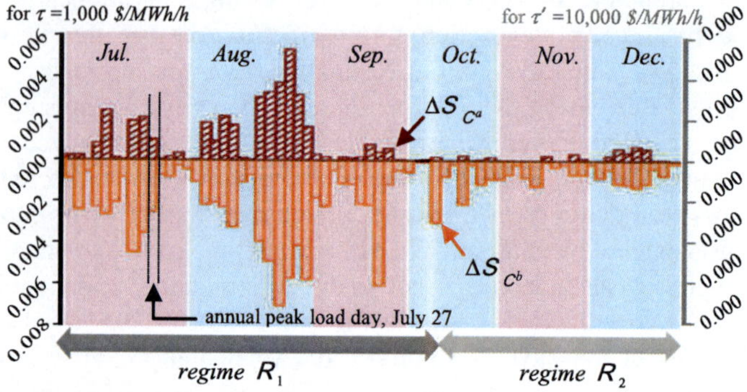

Fig. 11 The normalized daily impacts on auction surplus

Table 3 Statistical analysis of the relative auction surplus metric values under the regimes $\mathcal{R}_1$ and $\mathcal{R}_2$ (basis is $\mathbf{C}_0$)

Criterion	Regime	Maximum	Mean	SD
$\mathbf{C}_1$	$\mathcal{R}_1$	0.00541	0.00098	0.00130
	$\mathcal{R}_2$	0.00070	0.00012	0.00023
$\mathbf{C}_2$	$\mathcal{R}_1$	−0.00715	−0.00224	0.00182
	$\mathcal{R}_2$	−0.00215	−0.00068	0.00051

metrics $\Delta S|_{\mathbf{C}_0 \to \mathbf{C}_1}$ and $\Delta S|_{\mathbf{C}_0 \to \mathbf{C}_2}$. The area by area contribution is in line with the changes in the utilization of the modified import/export capabilities. We plot the changes of the import areas $\mathcal{A}^1$ and $\mathcal{A}^2$, and the export area $\mathcal{A}^5$, contribution to the auction surplus in Fig. 12 (Eq. 13) corresponding to shifting the security criterion from $\mathbf{C}_0$ to $\mathbf{C}_1$ ($\mathbf{C}_2$).

The price-responsive demand that characterizes regime $\mathcal{R}_2$ plays an important role in the nature of the results. In general, as the willingness to pay of the buyers increases, the absolute value of the relative auction surplus metric increases, attaining its highest value for fixed demand for each security criterion considered.

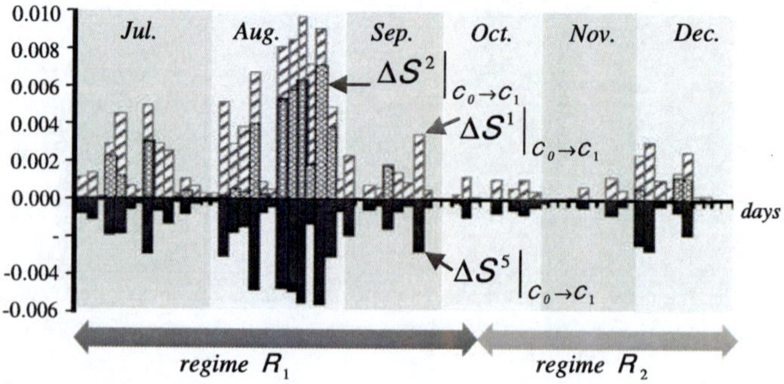

Fig. 12 Change in each area's contributions to auction surplus under $\mathbf{C}_1$

Therefore, the impacts of the change in security criterion to either $\mathbf{C}_1$ or $\mathbf{C}_2$ on the auction surplus are more pronounced for the fixed demand regime $\mathcal{R}_1$ than for the price-responsive regime $\mathcal{R}_2$, as we observe in the plots of Figs. 10, 11, 12, 13. Also, for a price-responsive demand with a uniformly low willingness to pay, the impacts may be small, and in certain cases may be negligibly so. The relaxation of the security criterion from $\mathbf{C}_0$ to $\mathbf{C}_1$ by not taking into account the double element contingencies, results in an insignificantly small relative auction surplus metric values under the regime $\mathcal{R}_2$. The tightening of the security criterion from $\mathbf{C}_0$ to $\mathbf{C}_2$ using preventive actions to replace corrective ones reduces the auction surplus. In fact, by utilizing the corrective control capabilities of the resources in the presence of price-responsive demand, the ISO-NE is able to decrease the economic impacts of the double tie line contingencies. Note that the extent of such ability depends on various factors including the topology of the system, the characteristics of the generating units and the bids/offers of the market participants.

These findings of the comparative assessment lead us to conclude that the reference criterion $\mathbf{C}_0$ is, for all intents and purposes, more appropriate for the ISO-NE DAM than either of the two security criteria considered. Through this study, we also gain important insights on the role of price-responsive demand and the selected security control action. In fact, a key finding of the ISO-NE study is that the economic efficiency of the electricity markets need not decrease when a power system is operated under a stricter criterion as long as there is price-responsive demand. The proposed approach provides good insights into the ramification of changing the security criterion on both qualitative and quantitative basis.

We next illustrate the application of the proposed approach to the study of the ISO-NE system and markets under multi-settlement environment. The objectives of our studies are to quantify the economic efficiency of the ISO-NE multi-settlement markets as a function of the security criterion in force, to investigate the impacts of the of the financial entity participation on the ISO-NE system and markets and to quantify the impacts on the ISO-NE market performance of a security criterion change from the criterion in force to a modified $(n - 1)$ security,

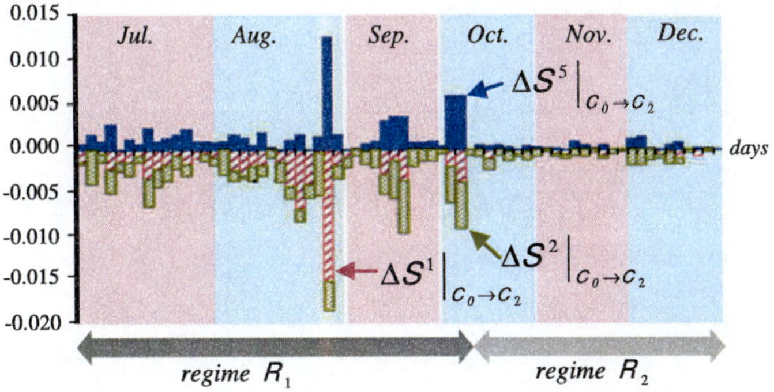

Fig. 13 Change in each area's contributions to auction surplus under $\mathbf{C}_2$

$\mathbf{C}_1$. We apply the proposed approach to quantitatively analyze the ISO-NE multi-settlement system performance. We assess the impacts of the participation of financial entities in the DAMs by performing a side-by-side comparison of the outcomes of the DAMs and the associated RTMs without and with such players. A particularly insightful aspect of the comparison is the set of values for the deviation metrics of the physical entities. In the following study, we quantify the impacts of a change in the security criterion from the current security criterion in force, $\mathbf{C}_0$, to $\mathbf{C}_1$ and compare the observed impacts under the two criteria.

We use the same 40 representative days from 2005 and 2006 to study the DAMs and their associated RTMs. For the discussion in this paper, we focus specifically on the four contiguous peak-demand hours of each selected day and analyze the values of metrics of interest for those 160 h. We start out with the evaluation of the ISO-NE multi-settlement system performance under the security criterion in force to determine the values of the metrics for the reference case for the study.

We perform market clearing for the DAMs and their associated RTMs for the selected 160 h and quantify the market performance metrics under the security criterion $\mathbf{C}_0$. We first focus on the DAM-RTM MW deviations. As the real-time demand in each RTM is considered to be fixed, the cleared demand values are not a function of the security criterion, per se, as long as the security-constrained market problem is feasible. We compare the fixed real-time demand in each of the M RTMs associated with the demand cleared in a DAM to evaluate the deviation metrics. We plot in Fig. 14 the demand values for the selected 160 h. We note that the real-time demand values exceed the DAM demand in the selected 160 h. As these hours are representative of the ISO-NE system past behavior, they correctly indicate that the RTM demands, typically, exceed the DAM demands. Therefore, there may be a need for additional physical generation in the real time over the amounts cleared in the DAMs.

We examine the physical demand and generation deviations and use the plots in Fig. 15 to gain insights into their nature. These plots indicate that both the physical

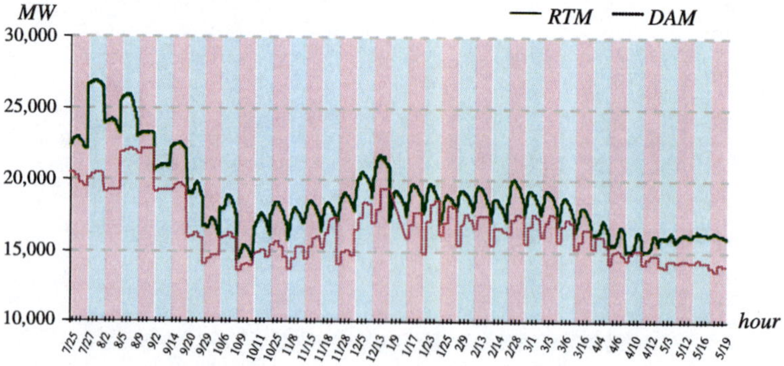

Fig. 14 Cleared demand in the DAMs and in their associated RTMs for the selected 160 hours in the study period

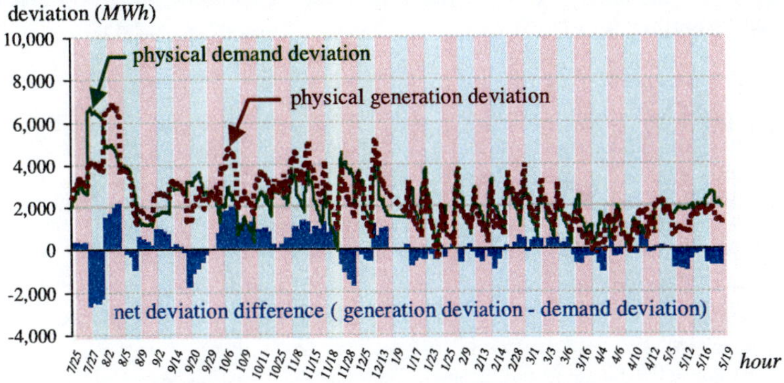

Fig. 15 The physical demand deviation, the physical generation deviation and the net deviation difference for the selected 160 hours in the study period

demand and the generation deviations are positive for the hours under consideration. Also, the positive values indicate that as much as 75% of the real-time demand and generation are cleared in the DAM. This result indicates that there is a need for additional generation in the real time. The plot of the net differences between the generation and the load deviations, Fig. 15, indicates the impacts of the net positions of the financial players in the DAMs. While there are daily variations in the financial entities' net positions, their range is up to 10% of the real-time demand, with the more pronounced impact in the higher demand days.

We now discuss the economic aspects of the secure operations of the power system. Analysis of the DAM-RTM price deviation metric indicates that, on average, the prices are higher in the DAMs than in their associated RTMs. Such results are clearly visible in the plots of the price deviation duration curves of the import areas $\mathcal{A}^1, \mathcal{A}^2$ and the export area $\mathcal{A}^5$. The area-wide price deviation measure of an area is evaluated using the load-weighted average of the prices in the area for a snapshot system. We observe that the price deviations are more pronounced for the import areas $\mathcal{A}^1$ and $\mathcal{A}^2$ than for the export area $\mathcal{A}^5$. For the study hours selected, the area $\mathcal{A}^2$ price deviations are larger than those of any other area indicating that congestion has more pronounced impacts on this area than other areas. The price deviation results also indicate that the physical sellers in the import areas capture more revenues for their real-time production in the 160 h of the study period.

Under these conditions, financial entities have more incentives to be sellers in the import area $\mathcal{A}^2$ than in any other area. If the financial sellers were to participate more intensely in the import area $\mathcal{A}^2$, then a decrease in the import area $\mathcal{A}^2$ DAM prices would result and, therefore, the price convergence would be improved. Given the nature of the physical generation and the demand deviations, we conjecture that financial entities may expect higher price deviations in the peak-demand periods and therefore they may adjust their bidding behaviors to clear more quantities in the DAMs in which they participate.

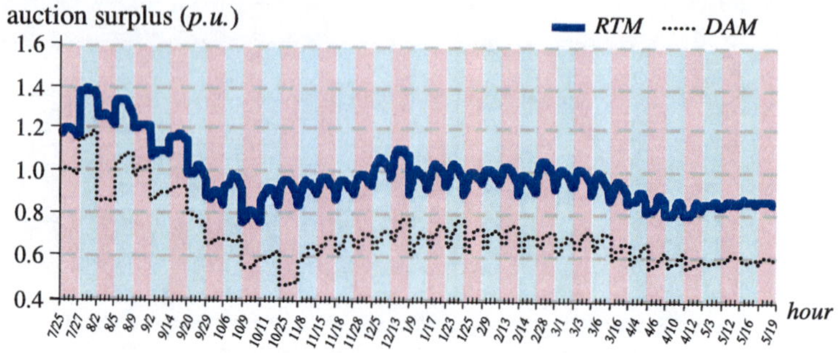

Fig. 16 The normalized auction surplus attained in the DAMs and the associated RTMs for the selected 160 hours in the study period

Next, we turn our attention to assessing the auction surplus of multi-settlement markets. We use the value of τ= 1,000 \$/MWh for the fixed demand. We choose this value on the basis of that it is the authorized bid/offer price cap in the ISO-NE markets and it is a reasonable proxy of the willingness to pay of the buyers with fixed demands. We evaluate using τ the auction surplus for the DAMs and the associated RTMs. We summarize the results in the plots given in Fig. 16 of the normalized auction surplus values for the 160 h in the study period. We normalize the auction surplus values using the average RTM auction surplus value so as to provide a meaningful comparison of the observed results. The positive load deviations and the fact that they represent fixed demands imply that the auction surplus outcomes are higher in the RTMs than in the DAMs. We next examine the individual components of the deviations of the auction surplus in the DAMs and their associated RTMs.

We investigate the deviations in the bid/offer surpluses of the physical market participants using their normalized values, with the base value being the RTM

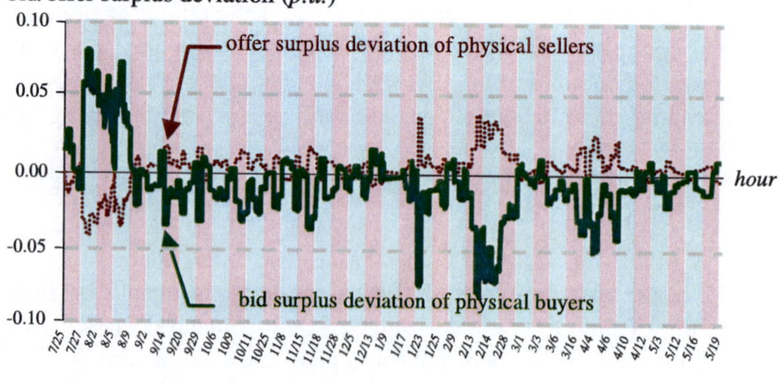

Fig. 17 Deviations in the offer (bid) surpluses of the physical sellers (buyers) for the selected 160 h in the study period

auction surplus. We provide the plots of the physical buyers and sellers in Fig. 17. We note that the physical sellers capture additional revenue for their real-time production for the majority of the hours in the simulation period. Therefore, the physical buyers pay a "premium" for that portion of their real-time demand needs that is cleared in the DAM. The plots clearly demonstrate that the sum of the bid/offer surplus deviations is not equal to zero, due to the financial entity participation, the bilateral transactions and the congestion rents. The metrics in Eqs. 18–35 serve to provide the quantification of the multi-settlement system performance for the 160 h of the study period under the reference criterion $\mathbf{C}_0$. We use these results as the reference basis for the comparative studies which we discuss next.

We examine the impacts that the financial players have on the market performance under the ISO-NE security criterion $\mathbf{C}_0$ in force. We first evaluate the impacts by considering the market operations without and with the participation of the financial entities in the DAMs. The difference between the two cases quantifies the contribution of the financial players in the multi-settlement environment. We evaluate the physical demand deviations, as well as the price deviations observed for the two cases.

Without financial entities in the DAM, lower physical demand is cleared in the DAM than in the case with the financial entity participation. Therefore, more generation is required in real time to compensate for the lower demand in order to ensure near-real-time system security. In fact, the ISO-NE study indicates that, on average, 700 MW additional output is required in real time without financial entity participation. Such an increase clearly indicates that the absence of financial entity participation makes the task to operate the near-real-time ISO-NE system securely more difficult. We find that financial entity participation leads to better forecasts of physical generation and consumption resulting in improved near-real-time system security. The plots of the cleared DAM demand without and with financial entity participation together with the real-time demand needs for the 160 h in the study period are given in Fig. 18.

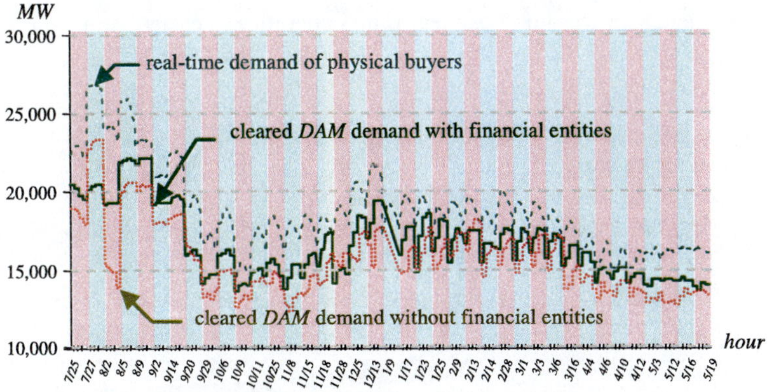

Fig. 18 Comparison of the cleared demands in the *DAMs* without and with financial entities for the selected 160 h in the study period

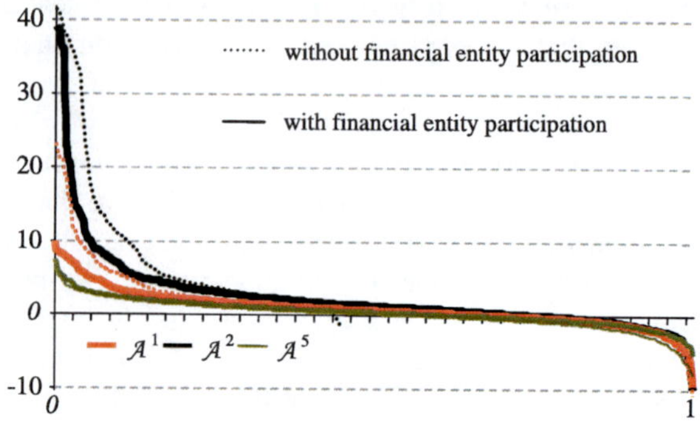

Fig. 19 Price deviation duration curves for the areas $\mathcal{A}^1, \mathcal{A}^2$ and $\mathcal{A}^5$ with and without financial entity participation

The most striking fact about financial entity participation can be discerned from examining the DAM-RTM price deviation results without and with financial entities. We superimpose in Fig. 19 the price deviation duration curves for the areas $\mathcal{A}^1, \mathcal{A}^2$ and $\mathcal{A}^5$ without the financial entity participation on those with their participation shown above in Fig. 20. The financial entity participation markedly reduces the deviation values for the areas $\mathcal{A}^1$ and $\mathcal{A}^2$. Since such a decrease corresponds to the desirable price convergence, its impact is very significant and attains the desired objective of price convergence that leads to improved market efficiency. Furthermore, our findings indicate that absent financial entity participation, the system is more congested in the DAMs, leading to the attainment of higher congestion rents than those obtained with the financial entity participation. In addition, system congestion impacts the prices of the import areas $\mathcal{A}^1$ and $\mathcal{A}^2$ more markedly than the case with financial entity participation. Therefore, financial entity participation reduces inter-area system congestion.

This side-by-side comparison results indicates very clearly the important role that financial entities play in electricity markets. Their participation decreases the magnitude of the physical demand deviations. In turn, these lower deviations make the management of near-real-time operations easier and, moreover,

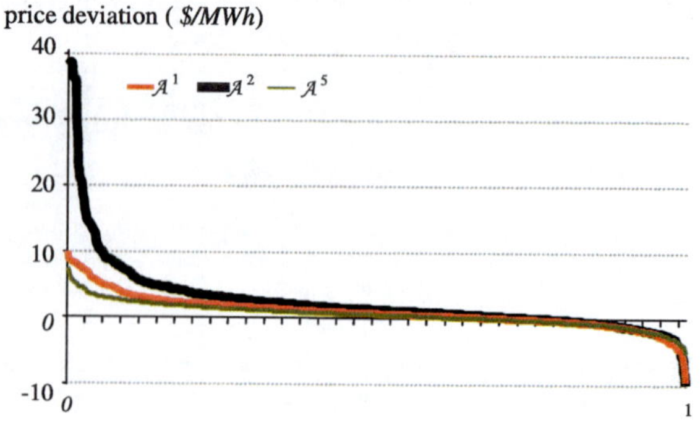

Fig. 20 Price deviation duration curves for the areas $\mathcal{A}^1, \mathcal{A}^2$ and $\mathcal{A}^5$ for the selected 160 h in the study period

improve near-real-time system security. In terms of market performance, the participation of financial entities decreases the magnitude of the price deviations.

We next study the impacts on the multi-settlement performance of a security criterion change from the reference criterion $\mathbf{C}_0$, to $\mathbf{C}_1$. For the security criterion $\mathbf{C}_1$, we perform market clearing of the DAMs and their associated RTMs for the selected hours in the study period and evaluate the market performance metrics. We compare the values of the metrics of interest with respect to those under the reference criterion $\mathbf{C}_0$.

The change from the security criterion $\mathbf{C}_0$ to $\mathbf{C}_1$ impacts the available transfer capability of the system, which, in turn, affects the ability of the import areas to bring in energy from the export area. Indeed, the examination of the ISO-NE results indicates increased import capabilities of the import areas for each hour of the given study period [35]. The increase in transfer capability has economic impacts, which we quantify from the changes in the auction surplus. We compute the hourly auction surplus values under the security criterion $\mathbf{C}_1$ for each selected hour of the study period and normalize them using the average value of the hourly auction surplus under the reference criterion $\mathbf{C}_0$. We note that the utilization of the increased import capabilities leads to increased auction surplus. We may view such an improvement as a measure of the "costs" of not violating the constraints associated with the double element contingencies in $\mathbf{J}_{\mathbf{C}_0}$. There are also a number of hours during which the change in security criterion from $\mathbf{C}_0$ to $\mathbf{C}_1$ has no impacts on the auction surplus. For such hours, the double element contingencies have zero economic impacts. We plot the changes in the normalized auction surplus values corresponding to the security criterion change in Fig. 21.

We next discuss the impacts of the change of security criterion from $\mathbf{C}_0$ to $\mathbf{C}_1$ on the market participants' bid/offer surpluses. The change of security criterion from $\mathbf{C}_0$ to $\mathbf{C}_1$ has widely varying impacts on the different market participants within the different areas. For illustration purposes, we consider five specific days to discuss the impacts. The security criterion change results in the greater utilization of the export area sellers and, therefore, in the decreased production of the

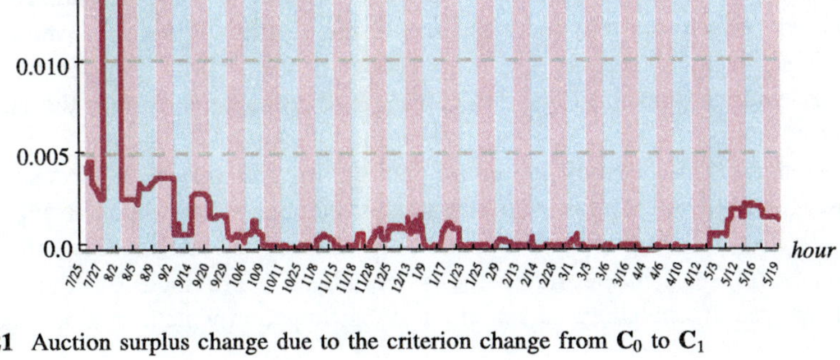

Fig. 21 Auction surplus change due to the criterion change from $\mathbf{C}_0$ to $\mathbf{C}_1$

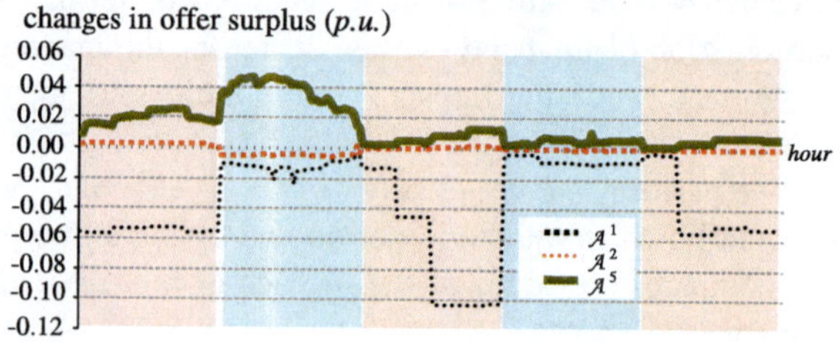

Fig. 22 Changes of the physical sellers' offer surpluses in response to the security criterion change from $\mathbf{C}_0$ to $\mathbf{C}_1$

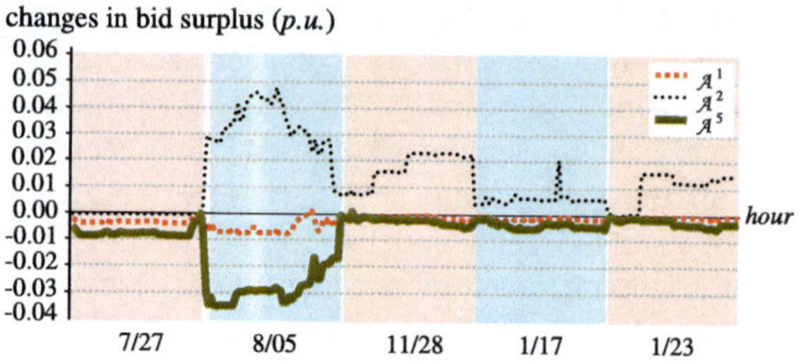

Fig. 23 Changes of the physical buyers' bid surpluses in response to the security criterion change from $\mathbf{C}_0$ to $\mathbf{C}_1$

import area physical sellers. Such a change leads to a corresponding change in the surpluses of the players in the various areas. In Fig. 22, we plot changes in the physical sellers' offer surpluses in areas $\mathcal{A}^1, \mathcal{A}^2$, and $\mathcal{A}^5$for the five selected days.

The impacts have almost the opposite effects on the surpluses of the physical buyers: while the bid surpluses of the export area physical buyers are decreasing, those in the import areas are increasing. The physical buyers within the import areas are able to meet their demand using more economic resources from the export area $\mathcal{A}^5$ to take advantage of the increased transfer capabilities, thereby decreasing their payments. In fact, the changes in the bid surpluses are particularly more pronounced for the import area $\mathcal{A}^2$ physical buyers than other areas' physical buyers as shown in Fig. 23.

The impacts of changing the security criterion on the surpluses of the financial entities are minor and of little significance compared to those impacts on the players with physical assets for a change in the security criterion. Overall, the relatively small dollar impacts due to the change of the security criterion from $\mathbf{C}_0$ to $\mathbf{C}_1$, as evident from the Figs. 21, 22, 23, furthermore justify that the current security criterion in force, $\mathbf{C}_0$, is appropriate for the ISO-NE markets [36].

Through the ISO-NE study, we gain important insights into the system security and its economics in a multi-settlement environment. The proposed approach effectively captures the impacts of the DAM clearing on the market participant bid/offer surpluses. Furthermore, the price signals, provided by the multi-settlement system, encourage financial entity participation which, in turn, leads to not only improvements in the overall market performance but also in the ability of the RTO to ensure near-real-time system security.

6 Concluding Remarks

The maintenance of secure system operations is a highly challenging task that became even more complex with the prominence of electricity markets. We use the insights we gained into the tight coupling between market and system operations under restructuring to characterize analytically the inter-relationships between the secure power system operations and the performance of the electricity markets. Such a characterization allows the development of an integrated analysis approach to quantify the economics of secure power system operations. This approach permits the quantification of the market performance as a function of security criterion and provides, for the first time, the means to provide an economic justification for a modification the security criterion. Furthermore, the approach is useful for the costs/benefits analysis of network improvements to mitigate the market performance impacts of a set of contingencies and their associated security control actions. An important application is to the assessment of the impacts of specific behavioral changes in market participants on system security. The generalization of the approach is made by its extension to quantitatively characterize the linkages between the real-time system operations and the day-ahead markets (DAMs) and their associated real-time markets (RTMs) for use in a multi-settlement environment. The extended approach provides the ability to explicitly show that the auction surplus attained in the multi-settlement system is equivalent to sum of the auction surplus attained in each RTM. Therefore, the mere presence of the DAMs results in surplus transfers among market participants. Furthermore, the extended approach provides a very useful tool to analyze the nature of the DAM-RTM price deviations and the impacts of financial entities on near-real-time system security.

We illustrate the application of the proposed approaches on the large-scale ISO-NE system in a number of studies. The results provide useful insights into the multi-faceted nature of issues that arise in the current tightly coupled market and system operations. In fact, the studies on the economics of the system security provide important insights into the role of price-responsive demand and that of specific selected security control actions measured by the economic efficiency of the electricity markets. A key result is that this efficiency need not decrease when a power system is operated under a stricter criterion, as long as there is effective price-responsive demand and appropriate utilization of the corrective control capabilities of the resources. Furthermore, the ISO-NE application study in the multi-settlement environment indicates that financial entity participation not only results in reduced DAM-RTM price deviations but also leads to DAM dispatch results that are "closer" to those of their associated RTMs. Therefore, financial player participation improves the ability of the system operator to ensure real-time system security.

References

1. FERC (2003) Remedying undue discrimination through open access transmission service and standard electricity market design. Docket No: RM01-12-000
2. The Electric Energy Market Competition Task Force (2006) Report to Congress on competition in the wholesale and retail markets for electric energy. Docket No: AD05-17-000
3. NERC, Reliability Standards. Available:http://www.nerc.com/~filez/standards/Reliability_Standards.html (Online)
4. TED Liacco (1967) The adaptive reliability control system. IEEE Trans Power App Sys vol. PAS-86, pp 517–531
5. ISO-NE, The multi-settlement system. Available:http://www.iso-ne.com/nwsiss/grid_mkts/how_mkts_wrk/multi_settle/index.html (Online)
6. Cheung KW (2004) Standard market design for ISO New England wholesale electricity market: an overview. Proc IEEE Int Conf Electric Utility Deregul Restruct Power Tech, vol. 1, pp 38–43
7. Ma X, Sun D (2005) Key elements of a successful market design. Proc IEEE/PES Trans Dist: Asia and Pacific, Dalian
8. Green R, Newbery DM (1992) Competition in the British electric spot market. J Polit Econ 100:929–953
9. Wolfram CD (1998) Strategic bidding in a multi-unit auction: an empirical analysis of bids to supply electricity in England and Wales. Rand J Econ 29:703–725
10. Mansur ET (2001) Pricing behavior in the initial summer of the restructured PJM wholesale electricity market. University of California Energy Institute, PWP-083, April 2001
11. Borenstein S, Bushnell JB, Wolak FA (2002) Measuring market inefficiencies in California's restructured wholesale electricity market. Am Econ Rev 92:1376–1405
12. Torre SDl, Conejo AJ, Contreras J (2003) Simulating oligopolistic poolbased electricity markets: a multi-period approach. IEEE Trans Power Syst 18:1547–1555
13. Baldick R, Grant R, Kahn E (2004) Theory and application of linear supply function equilibrium in electricity markets. J Regul Econ 25:143–167
14. Kaye RJ, Wu FF, Varaiya P (1995) Pricing for system security. IEEE Trans Power Syst 10:575–583
15. Wu T, Rothleder M, Alaywan Z, Papalexopoulos AD (2004) Pricing energy and ancillary services in integrated market systems by an optimal power flow. IEEE Trans Power Syst 19:339–347
16. Arroyo JM, Galiana FD (2005) Energy and reserve pricing in security and network-constrained electricity markets. IEEE Trans Power Syst 20:634–643
17. Alvarado F, Hu Y, Ray D, Stevenson R, Cashman E (1991) Engineering foundations for the determination of security costs. IEEE Trans Power Syst 6:1175–1182
18. Kirschen DS, Bell KRW, Nedic DP, Jayaweera D, Allan RN (2003) Computing the value of security. IEE Proc Gener Transm Distrib 150:673–678
19. Burns S, Gross G (1990) Value of service reliability. IEEE Trans Power Syst 5:825–834
20. Nitu P, Gross G (1993) Evaluation of reliability in power system operational planning. Proc Power Syst Comp Conf, pp 355–362
21. Kamat R, Oren SS (2002) Multi-settlement systems for electricity markets: zonal aggregation under network uncertainty and market power. Proc 35th Hawaii Int Conf Syst Sci, pp 739–748
22. Kamat R, Oren SS (2004) Two-settlement systems for electricity markets under network uncertainty and market power. J Regul Econ 25(2):5–37

23. Arciniegas I, Barrett C, Marathe A (2003) Assessing the efficiency of US electricity markets. Util Policy 11:75–86
24. Borenstein I, Bushnell J, Knittel C, Wolfram C (2004) Inefficiencies and market power in financial arbitrage: a study of California's electricity markets. Center for the Study of Energy Markets (CSEM), Working Paper 138
25. ISO-NE (2004) Impact of virtual transactions on new England's energy market, Nov. 2004
26. Saravia C (2003) Speculative trading and market performance: the effect of arbitrageurs on efficiency and market power in the New York electricity market. Center for the Study of Energy Markets (CSEM), Working Paper 121
27. Isomenger A (2006) The benefits and risks of virtual bidding in multi-settlement markets. Electr J 19(9):26–36
28. Bulow J, Levin J, Milgrom P (2009) Winning play in spectrum auction. NBER Working Paper No. 14765, Mar. 2009. Available:http://www.nber.org/papers/w14765
29. Klemperer P (2004) Auctions: theory and practice. Princeton University Press, Princeton
30. Milgrom P (2004) Putting auction theory to work. Cambridge University Press, Cambridge
31. Liu M, Gross G (2008) Congestion rents and FTR evaluations in mixed pool-bilateral systems. Int J Electr Power Energy Syst 30(8):447–454
32. ISO-NE, Value of Lost Load. Available:http://www.iso-ne.com/committees/comm_wkgrps/inactive/rsvsrmoc_wkgrp/Literature_Survey_Value_of_Lost_Load.rtf
33. Guler T (2009) Multiarea system reliability: the economic evaluation of system security criterion. Dissertation submitted in partial fulfillment of the requirements for the degree of Doctor of Philosophy in Electrical and Computer Engineering in the Graduate College of the University of Illinois at Urbana, Champaign
34. ISO-NE (2006) Operating reserves white paper, June 2006. Available:http://www.isone.com/pubs/whtpprs/operating_reserves_white_paper.pdf
35. Guler T, Gross G, Litvinov E, Coutu R (2007) Quantification of market performance as a function of system security. IEEE Trans Power Syst 22:1602–1611
36. Guler T, Gross G, Litvinov E, Coutu R (2009) On the economics of power system security in multi-settlement electricity markets. IEEE Trans Power Syst (to be published)

Probabilistic Transient Stability Assessment and On-Line Bayes Estimation

Elio Chiodo and Davide Lauria

Abstract It is a well-known fact that the increase in energy demand and the advent of the deregulated market mean that system stability limits must be considered in modern power systems reliability analysis. In this chapter, a general analytical method for the probabilistic evaluation of power system transient stability is discussed, and some of the basic contributes available in the relevant literature and previous results of the authors are reviewed. The first part of the chapter is devoted to a review of the basic methods for defining transient stability probability in terms of appropriate random variables (RVs) (e.g. system load, fault clearing time and critical clearing time) and analytical or numerical calculation. It also shows that ignoring uncertainty in the above parameters may lead to a serious underestimation of instability probability (IP). A Bayesian statistical inference approach is then proposed for probabilistic transient stability assessment; in particular, both point and interval estimation of the transient IP of a given system is discussed. The need for estimation is based on the observation that the parameters affecting transient stability probability (e.g. mean value and variances of the above RVs) are not generally known but have to be estimated. Resorting to "dynamic" Bayes estimation is based upon the availability of well-established system models for the description of load evolution in time. In the second part, the new aspect of on-line statistical estimation of transient IP is investigated in order to predict transient stability based on a typical dynamic linear model for the stochastic evolution of the

The singular and plural of names are always spelled the same; boldface characters are used for vectors; random variables (RVs) are denoted by uppercase letters.

E. Chiodo (✉) and D. Lauria
Electrical Engineering Department, University of Naples Federico II, Via Claudio 21, I-80125 Naples, Italy
e-mail: elio.chiodo@unina.it

D. Lauria
e-mail: davide.lauria@unina.it

G. Anders and A. Vaccaro (eds.), *Innovations in Power Systems Reliability*, Springer Series in Reliability Engineering, DOI: 10.1007/978-0-85729-088-5_8,

system load. Then, a new Bayesian approach is proposed in order to perform this estimation: such an approach seems to be very appropriate for on-line dynamic security assessment, which is illustrated in the last part of this article, based on recursive Bayes estimation or Kalman filtering. Reported numerical application confirms that the proposed estimation technique constitutes a very fast and efficient method for "tracking" the transient stability versus time. In particular, the high relative efficiency of this method compared with traditional maximum likelihood estimation is confirmed by means of a large series of numerical simulations performed assuming typical system parameter values. The above results could be very important in a modern liberalized market in which fast and large variations are expected to have a significant effect on transient stability probability. Finally, some results on the robustness of the estimation procedure are also briefly discussed in order to demonstrate that the methodology efficiency holds irrespective of the basic probabilistic assumptions made for the system parameter distributions.

List of main symbols and acronyms

BCI	Bayesian confidence interval
cdf	Cumulative distribution function
CSGDF	Complementary standard Gaussian distribution function
CCT	Critical clearing time (T_{cr} or T_x)
CV	Coefficient of variation
D	Set of observed data used for inference
DLM	Dynamic linear model
$E[R]$	Expectation (or "mean value") of the RV R
EV	Extreme value distribution
FCT	Fault clearing time
$F(x)$	Generic cdf
$f(x)$	Generic pdf
$g(\omega)$, $g(\omega\|D)$	Prior and posterior pdf of a generic parameter ω
$G(r, \phi)$	Gamma distribution with parameters (r, ϕ)
IID	s-Independent and identically distributed (random variables)
IP	Instability probability
LCCT	Logarithm of the CCT
LF, $L(D\|\beta)$	Likelihood function, conditional to given parameter β
L, $L(t)$	Load (at time t)
ML	Maximum likelihood
MSE	Mean square error
LN(α,β)	Log-Normal distribution with parameters α and β
$N(\mu,\sigma)$	Normal (Gaussian) distribution with mean μ and SD σ
pdf	Probability density function
RV	Random variable
s	Standard deviation of measurement errors in the DLM of the LCCT
S	Standard deviation of measurement errors in the DLM of the load

SD, σ	Standard deviation
s-independent	Statistically independent
SD[Y]	Standard deviation of the RV Y
SM	Stability margin (i.e. the quantity u defined below), for a given fault
σ^2	Denotes a variance
T_{cr} or T_x	Critical clearing time
T_{cl} or T_y	Fault clearing time
u	$(\alpha_x - \alpha_y)/(\beta_x^2 + \beta_y^2)^{1/2}$
v_x,v_y	CV values of the CCT and FCT, respectively
VST	Very short time
Var[R],$V(R)$	Variance of the RV R
w	Standard deviation of system equation error in the DLM of the LCCT
W	Standard deviation of system equation error in the DLM of the load
Λ	Peak value of the load $L(t)$, over a given time interval
WGN	White Gaussian noise
X	Logarithm of the CCT
Y	Logarithm of the FCT
α_x	$E[X]$
α_y	$E[Y]$
β_x^2	Var[X]
β_y^2	Var[Y]
ζ°	Bayes estimate of a generic parameter ζ
ζ^*	ML estimate of a generic parameter ζ
μ	Denotes a mean value (expectation)
$\hat{\mu}_k$	Bayes estimate of a "dynamic" parameter μ at time k
$\Gamma(\cdot)$	Euler–Gamma function
μ_r	Expectation of the generic RV R
$\Phi(z)$	Standard normal cdf
$\Psi(z)$	$1 - \Phi(z)$ (Complementary standard Gaussian distribution function)
$\varphi(z)$	Standard normal pdf
$R \backsim N(\alpha,\beta)$	The RV R has a Gaussian distribution $N(\alpha, \beta)$ (and similarly for the LN model, etc.)

1 Introduction

Stability assessment has long been recognized as a fundamental requirement in power system planning, design, operation and control. Transient stability can be defined as a property of an assigned power system to remain in a certain equilibrium point under normal conditions and reach a satisfactory equilibrium point after large disturbances such as faults, loss of generation, line switching, etc. [1]. Transient stability, therefore, constitutes a key aspect of modern power system reliability, and this fact is increasingly recognized in the modern power systems literature [2]. Indeed, some methods based on reliability theory are used in this chapter to perform an efficient assessment of power system stability. The traditional approach to transient stability analysis is deterministic, being based on the "worst case" approach. More specifically, transient stability quantitative assessment is generally performed on a three-phase fault on specific system buses as well as considering the load demand attaining its peak value over a prefixed time interval.

The application of probabilistic techniques for transient stability analysis was introduced in a series of articles by Billinton and Kuruganty [3–7], motivated by the random nature of:

- the system steady-state operating conditions;
- the time of fault occurrence;
- the fault type and location;
- the fault clearing phenomenon.

In fact, the steady-state operating conditions that heavily affect stability strongly depend on the load, which is a random process due to its intrinsic nature. This is especially evident in planning studies where the load level is the major source of uncertainty.

The time to clear the fault (fault clearing time, FCT), a crucial parameter in stability investigations, is also not known in advance, and so it should also be regarded as an RV. The probabilistic approach has also been explored in other significant articles such as [8], based on Monte Carlo simulations and [9], based on the "conditional probability" approach. An exhaustive account of the topic and the relevant bibliography can be found in the book by Anders [10, Chap. 12] which clearly states that: "stability analysis is basically a probabilistic rather than a deterministic problem".

The analytical computation of the probability distributions of the intermediate RVs is one of the most challenging aspects due to the complexity of the mathematical models, as also pointed out by Anders [10, p. 577].

In a few articles [11–15], some theoretical results from probability theory and statistics have been utilized in order to develop an analytical approach to the transient stability evaluation of electrical power systems by performing critical considerations on the basic probability distributions. Instability probability (IP) over a certain period of time, with regard to a given fault, is defined and calculated by means of both critical clearing time (CCT) distribution and FCT distribution. This analytical approach, overcoming the drawbacks of Monte Carlo simulations, is very useful in actual operation since it permits straightforward sensitivity analysis of IP with regard to system parameters thus highlighting those which mostly influence the system stability characteristics and providing a quantitative tool for performing proper preventive control actions. For similar reasons, the proposed analytical probabilistic approach is also a powerful tool with regard to the practical aspect of the estimation of the basic parameters relevant to the transient stability assessment like IP or other measures of "stability margin" (SM). This topic—generally neglected, or dealt with in approximate "sensitivity analyses" in the literature—was faced in [16], where an effort was made to tailor a simple, analytical, transient stability probability estimator which allows the required characteristics of both efficiency and robustness to be obtained, in the framework of classical estimation.

In this article, a new Bayesian approach is proposed in order to provide this estimation: such an approach appears to be the most suitable one for on-line transient stability assessment. Numerical application performed confirms that the estimation technique is able to adequately "track" the transient stability in time, being far more efficient than the classical maximum likelihood (ML) estimation of the IP. This could be an interesting property in a modern liberalized market in which fast and large variations are expected to have a significant effect on transient stability probability.

In the final part of the chapter, some results on the robustness of the estimation procedure are also briefly discussed in order to illustrate that the methodology efficiency holds irrespective of the basic probabilistic assumptions effected with regard to "a priori" distributions of the various system parameters.

In the four Appendices to the chapter:

1. a mathematical study of the IP versus the system parameters is illustrated in order to establish a proper "sensitivity analysis" of system stability which can be useful in the design stage.
2. some basic properties of Bayesian estimation, relevant for the problem under study, are briefly mentioned and some properties already derived by the authors in previous articles for the interval estimation of the IP are also included.

2 Probabilistic Modelling for Transient Stability Analyses

2.1 Definition and Evaluation of Transient IP

Many state variables of an electrical power system possess an intrinsically stochastic nature and, consequently, a probabilistic description of transient stability aspects is able to infer interesting deductions also in terms of control actions for improving system robustness. For instance, the steady-state conditions, fault conditions and circuit breaker clearing times are not precisely known or predictable. The various involved uncertainties should be properly taken into account using suitable probabilistic models. In a probabilistic frame, both the network configuration and faults are described as random quantities. According to this approach, all faults potentially causing instability and all the possible network states at the instant of fault (e.g. in terms of the requested loads at load buses) have to be considered, together with their probability of occurrence. Once the cost of any consequence brought about by the loss of stability is known, IP assessment allows the instability risk to be evaluated. A risk value provides a quantitative measure for undertaking adequate preventive control actions for stability improvement and avoiding the need for conservative or "worst-case" criteria like those based on the classical deterministic analyses.

Hence, in order to effectively apply a probabilistic approach, a preliminary identification of the relevant statistical parameters has to be performed. This is a crucial step since they are potentially infinite: the choice can be made according to the required degree of accuracy.

Formally, let a proper probability space (**O**,P,**S**) be defined, where **O** is the sample space of all possible outcomes, **S** a sigma Algebra of events and P an additive probability measure over **S**. For the purpose of stability investigation, the sample space **O** may be defined as the product space $\mathbf{O} = \mathbf{O_1} \times \mathbf{O_2}$, in which $\mathbf{O_1}$ is the set of all possible disturbances which can (potentially) affect system stability, $\mathbf{O_2}$ is the set of all possible "state vector" trajectories after the disturbance. This requires the definition of a proper "state vector" as a vector whose components are all the system variables whose values are the basis on which stability assessment is performed (see also following Eq. 2).

Let the random event I be the event of instability (over a given time horizon H of power system operation) and let $(C_1,\ldots, C_m)$ be a finite set of random events constituting all the credible—and mutually exclusive—disturbances ("contingencies") which can affect the system operation in H and potentially make system stability worse. Then, the IP in the interval H is provided, according to the total probability theorem, by:

$$P(I) = \sum_{j=1}^{m} P(C_j)P(I|C_j) \tag{1}$$

where $P(C_j)$ = probability of occurrence of the disturbance C_j in the time horizon under consideration; $P(I|C_j)$ = probability of instability once the disturbance C_j occurred.

The above relation may also still hold, at least as an approximation, when the disturbances C_k are not mutually exclusive random events, provided that the joint probability of two (or more) disturbances is negligible[1]: this typically happens in very short time (VST) operation which the second part of this chapter focusses on.

In the following, IP strictly denotes a term like $P(I|C)$ where C is the given fault.

As long as the fault statistics are known from available data of the system under consideration, the $P(C_j)$ terms can be considered known terms; the $P(I|C_j)$ terms

[1] Note that, without this assumption, the above equation is generally wrong, although it appears without any justification in many papers and books.

are evaluated as shown in the following so that $P(I)$ is readily obtained from the above relation.

A quite similar reasoning still applies if the fault location is modelled by an RV.

Basic RVs are load demand and the FCT. Aiming at the description of the system stability characteristics as a consequence of a given fault (for instance, a three-phase short-circuit), IP may be expressed as a function of these RVs.

For an assigned electrical power system, characterized by a state vector $\mathbf{x}_0$ at time t_0 in which the fault is supposed to occur, let us denote the stability region of the post-fault equilibrium point with S. Naturally, $\mathbf{x}_0$ is an RV (more precisely, a random vector), mainly due to the random nature of the load demand which, as previously mentioned, has a significant effect on the operation state. Let τ be the FCT: denoting by $\mathbf{x}(t, \tau; \mathbf{x}_0)$ the state vector trajectory at the time t after the fault clearing, the CCT for transient stability can be defined as follows:

$$T_{cr}(\mathbf{x}_0) = \sup\{\tau > 0 : \mathbf{x}(t, \tau, \mathbf{x}_0) \in S, \forall t > t_0 + \tau\} \tag{2}$$

This relation clearly shows the dependence of the critical time on the random initial state $\mathbf{x}_0$, thus T_{cr} is also an RV. As discussed in [11–13], the FCT should also be regarded as an RV which will be denoted by T_{cl}.

By keeping in mind that the system maintains its stability conditions if and only if the FCT is smaller than the CCT, i.e. $T_{cl} < T_{cr}$, the IP for a given fault can be expressed as:

$$q = P(T_{cl} > T_{cr}) \tag{3}$$

Formally, the model in the above relation is quite similar to the "Stress–Strength" model in reliability theory: indeed, if the failure of a certain device or system is caused by the occurrence of a "stress" T_{cl} greater than the "strength" T_{cr} of the device or system, than the above probability q represents the unreliability (failure probability) of the device or system.

Once the probability distributions of T_{cl} and T_{cr} are known, $q = P(T_{cl} > T_{cr})$ may be easily computed, as is well known in probability theory, as shown below. In fact, by describing both T_{cl} and T_{cr} as continuous non-negative RVs, with joint probability density function (pdf) $f(t_{cr}, t_{cl})$, the IP q is expressed as:

$$q = P(T_{cr} < T_{cl}) = \int_{t_{cl}=0}^{t_{cl}=\infty} \mathrm{d}t_{cl} \int_{t_{cr}=0}^{t_{cr}=t_{cl}} f(t_{cl}, t_{cr})\mathrm{d}t_{cr} \tag{4a}$$

In practice, the two RV are always considered in the literature as being statistically independent of each other since they are related to independent phenomena (as discussed in Sect. 2.3): under this assumption, let $f_{cl}(t)$ and $f_{cr}(t)$ be the marginal pdf of the RVs T_{cl} and T_{cr}, respectively, and let $F_{cl}(t)$ and $F_{cr}(t)$ be the correspondent cumulative probability distribution functions (cdf). The above expression may then be rewritten as follows:

$$q = \int_0^\infty f_{cr}(t)(1 - F_{cl}(t))\mathrm{d}t \tag{4b}$$

Alternatively, by conditioning the instability event on the values of the clearing time T_{cl}, q can also be equivalently expressed as:

$$q = \int_0^\infty f_{cl}(t)F_{cr}(t)\mathrm{d}t \tag{4c}$$

In order to evaluate the probability q, the following preliminary steps have to be taken:

- load demand and clearing time randomness have to be properly characterized in terms of distributions on the basis of realistic assumptions, by also taking the available data into account;
- the CCT distribution has to be evaluated in terms of the load distribution since there is a conceptual and analytical relationship between them (this aspect will be adequately discussed in the following sections);
- finally, the evaluation of the above integral has to be performed: often, this integration requires the use of numerical computation, but the analytical approach of this method allows it to be evaluated in a closed form.

This procedure is straightforward for the single-machine case (the so-called "one-machine infinite bus" system), as discussed in [11] since an analytical expression between T_{cr} and the load demand, based on the well-known equal area criterion, can be demonstrated. Besides, in [12], the procedure was also extended to a multi-machine system by resorting to the so-called "*Extended Equal Area Criterion*" [17]. This procedure allows difficulties arising in the evaluation of T_{cr} distribution to be overcome since T_{cr} can be analytically expressed as a function of the load demand.

2.2 Probabilistic Modelling of the CCT

In this section, the functional expression between the CCT and the load demand L at the instant of contingency is discussed. Due to the random nature of the load demand evolution and the unpredictability of the instant of fault, the load active power L has to be correctly regarded as an RV. This implies that T_{cr} is also an RV and its probability distribution may be calculated in terms of the load distribution which is generally estimated by load forecasting. In a quite general way, the load demand over time can be efficiently described through a continuous random process, $L(t)$.

With reference to a generic time instant t, the load probability cumulative distribution function is denoted by $F_L(l; t)$ and is defined over the non-negative real numbers as $F_L(l; t) = P(L(t) \leq l)$, $l \geq 0$.

On the basis of the central limit theorem, it is generally assumed that $L(t)$ can be described by a Gaussian random process [18]. The functional dependence between T_{cr} and L can be described in a compact way as $T_{cr} = g(L)$ where $g(\cdot)$ is a continuous non-negative function over the positive real axis. Moreover, it can be proved that $g(\cdot)$ is a decreasing function of the active power L. Since T_{cr} is a function of the RV L and the function $g(\cdot)$ is continuous and non-negative, the CCT is also represented by a continuous non-negative RV. Once the distribution function of L is known, in principle, the distribution function of T_{cr} can be calculated by means of well-known theorems with regard to RV transformations applied to the analytical relation. Nevertheless, since the function $g(\cdot)$ is not analytically invertible, a closed-form expression for the probability distribution function of the variable T_{cr} cannot be obtained. In these cases, the problem of the distribution evaluation is often solved using a stochastic Monte Carlo simulation.

However, in [11], an approximate method for the analytical calculation of the probability distribution function of T_{cr} has been presented. The first step for the analytical evaluation is the approximation of the true characteristic $g(L)$ with a simpler, invertible, analytical function. In particular, a log-linear model has proved to be very adequate when expressing the above characteristic for any given set of electrical parameters:

$$T_{cr} = \beta_0 \exp(-\beta_1 L) \tag{5a}$$

or:

$$\ln(T_{cr}) = a - b\,L\,(a = \ln(\beta_0)\,;\, b = \beta_1) \tag{5b}$$

The model coefficients (β_0, β_1) are positive constants (so that: a is real, b positive), depending on the electrical parameters of the system. They can be efficiently determined by performing a linear regression of the natural logarithm of T_{cr} with regard to the load; i.e. according to the least-square method, a and b are chosen as the values minimizing the sum of the square deviations:

$$S^2(a,b) = \sum_{i=1}^{n} (\ln T_{cr,i} - a - bL_i)^2 \tag{6}$$

the points ($T_{cr,i}$, L_i; $i = 1,\ldots, n$) being chosen assuming a proper step in the interval (L_1, L_n) in which they will probably occur. For instance, a $\mu \pm 4\sigma$ interval may be chosen to represent the load values generated by a Gaussian distribution with mean μ and standard deviation σ.

On the basis of the model (see Sect. 2.1), the evaluation of the cdf of T_{cr} in terms of the probability distribution function of the load L is straightforward and, for non-negative values of the CCT, it is expressed by:

$$F_{cr}(t_{cr}) = 1 - F_L\left[\frac{a - \ln(t_{cr})}{b}\right], \quad t_{cr} \geq 0 \tag{7}$$

The above cdf is, of course, equal to zero for negative values of the argument. The above relation is quite general, i.e. independent of any particular assumption made about the load distribution. Moreover, it can be seen that the cdf of the load $L(t)$—i.e. the function $F_L(\cdot)$ which appears in the right-hand side of the previous equation—is dependent, of course, on time t (even if this not explicitly expressed in the previous equation); therefore, the expression of the cdf of T_{cr} also depends on time, and it is valid for any particular time instant t in which the fault occurs. The hypothesis of a Gaussian distribution, generally adopted to describe load, implies a Log-Normal distribution for the CCT and this model will be used as illustrated in the sequel. It should be stressed, however, that this distribution varies with time, although this fact may not be apparent at first sight. Indeed, time will not always be represented explicitly in the relevant equations which are often referred to at given short time intervals in which the above RV—the load, and thus the CCT—may be considered constant, and also the corresponding distribution. However, in the successive interval, this distribution is subject to changes. This should be quite clear in the framework of dynamic estimation.

2.3 Analytical Evaluation of IP: A General Methodology

As previously stated, the time interval needed for fault clearing (comprehensive of the time for the fast reclosure of the faulted line) should also be regarded as an RV, here denoted by T_{cl}. The arc extinction phenomenon, in fact, is intrinsically not deterministic; the randomness of T_{cl} may also be due to imperfect switching which can depend on the wear conditions of the poles caused by previous faults. Moreover, the (random) environmental conditions (temperature, humidity) also influence the clearing time T_{cl}. The RV T_{cl} is assumed to be continuous, non-negative and independent from the time instant of fault occurrence since the above phenomena can be considered independent of those which cause the fault. According to the definition of the CCT, it is natural to define the probability of instability after a contingency occurring at a given time instant t as:

$$q = P(T_{cr} < T_{cl}) = P\{g[L(t)] < T_{cl}\} \tag{8}$$

In (8), the relation $T_{cr} = g[L(t)]$ between the CCT and the load at (intended as "immediately before") the instant t of the contingency is explicitly presented. In order to obtain IP over a prefixed time horizon (0, h), the statistics of the random process of faults should be taken into account. This means that (8) must be integrated with the probability distribution of the number of faults in (0, h) which is indeed a random process. A Poisson stochastic process [10, 19] may be generally assumed as valid for this purpose. However, a different and simpler approach is possible [11]. The stability event over (0, h) may be defined as the property whereby—in the whole interval–stress T_{cl} never exceeds strength T_{cr}, which depends on time t through function g. Hence, the IP can be defined as the probability that T_{cr} is exceeded by T_{cl} for at least one time instant t in (0, h): this happens if and only if, as time t varies in (0, h), T_{cl} exceeds the minimum value of T_{cr} attainable in this interval.

Mathematically speaking, the IP, q, can be expressed as follows:

$$q = P\{\inf g[L(t)] < T_{cl}; 0 < t < h\} \tag{9}$$

Let the peak load value, $\Lambda = \sup\ (L(t);\ 0 < t < h)$, over the interval $(0, h)$ be introduced: since $L(t)$ is a continuous random process, Λ is a continuous RV whose probability distribution function is denoted by $F_\Lambda(\lambda)$. As previously mentioned, the function $g[L(t)]$ is continuous and decreasing versus L. Therefore, the "minimum" CCT over $(0, h)$, again denoted by T_{cr}, can be expressed as follows:

$$T_{cr} = \inf g[L(t)] = g[\sup L(t)] = g(\Lambda) \tag{10}$$

Hence, the IP is expressed by:

$$P(T_{cr} < T_{cl}) = P\{g(\Lambda) < T_{cl}\} \tag{11}$$

Hence, by keeping in mind the expression in (7), T_{cr} can be expressed in terms of the peak load Λ, so that its probability distribution function $F_{cr}(t_{cr})$ can be written as follows:

$$F_{cr}(t_{cr}) = 1 - F_\Lambda\left[\frac{a - \ln(t_{cr})}{b}\right], \quad t_{cr} \geq 0 \tag{12}$$

where the constants a and b depend on the particular system, but are indeed constant with time (unless system topology changes; this case is excluded here but can be dealt with the same methodology, once it occurs, by simply computing the new values of a and b).

The problem can then be easily solved once the peak load and the clearing time distributions are known. The distribution function and the probability density function of the clearing time T_{cl} are denoted by $F(t)$ and $f(t)$, respectively. Assuming, as reasonable, that the variables T_{cl} and T_{cr} are statistically independent, the IP in Eq. 11 can then be calculated as in Eqs. 4b or 4c. This approach, taken from [13], expresses the IP over an arbitrarily large interval—once the pdf of peak load is known—and is useful if a planning horizon is being studied. It was presented here for the sake of completeness: the application in this chapter is in fact devoted to on-line stability assessment for VST applications, related to time intervals typically lasting 1 h or less, so that in those intervals, the load L may be considered as a constant, albeit unknown (random) value so that, in practice, it will be modelled through an RV instead of a stochastic process. However, such a distinction does not affect the methodology followed in the sequel, since—as anticipated—the Gaussian distribution which will be used here is widely employed for describing the load process since it is also common practice to describe the peak load uncertainty by means of a Gaussian RV whose expected value is the forecasted peak load. For very large time horizons, the Extreme value (EV) distribution is also a natural candidate for describing the peak load [20]: this model—as well as others—can also be handled in practice with no particular problems as shown in [13], using the same methodology illustrated here.

3 Analytical IP Evaluation for Gaussian Load and Log-Normal FCT

3.1 Analytical Expression of IP

In this section, the analytical expression of the statistical parameter q—the IP of the given system under a given fault occurrence—is discussed on the basis of reasonable assumptions for the distribution functions of the RVs L (and consequently T_{cr}) and T_{cl}. For the sake of notation simplicity, the RVs T_{cr} and T_{cl} will be, respectively, named T_x and T_y.

The load L is then assumed to be a $N(\mu_L, \sigma_L)$ RV: then, letting l be a given possible load value, L is characterized by the following pdf over $(-\infty < l < +\infty)$:

$$f_L(l) = \frac{1}{\sigma_L\sqrt{2\pi}}\exp\left[\frac{-(l-\mu_L)^2}{2\sigma_L^2}\right]$$

The interval of possible values is $(-\infty < l < +\infty)$ only theoretically, being derived from the Gaussian representation. In fact, the probability of negative values for L should, of course, be equal to zero; in practice, it is known that $P(L < l) \approx 0$ if $l < \mu - 3\sigma$.

The natural logarithm of T_x is also normally distributed on the basis of the above-discussed relationship $X = \ln(T_x) = a - bL$. Hence, the distribution of T_x can be described by a Log-Normal distribution. It can be seen that the authors have shown [11–13], by means of extended numerical simulations and adequate statistical tests, that the proposed Log-Normal model for the distribution of the critical time, when the load is a Gaussian RV, is very adequate.

The Log-Normal pdf with parameters α (scale) and β (shape) is expressed by:

$$f(t;\alpha,\beta) = \frac{1}{\sqrt{2\pi}t\beta}\,e^{\frac{-(\ln t-\alpha)^2}{2\beta^2}} \quad t \geq 0 \tag{13}$$

and the density $f(t)$ is zero for $t < 0$.

In expression (13) α and β represent the mean value and the standard deviation of the natural logarithm of the Log-Normal variable, respectively; the mean value μ and the standard deviation σ, corresponding to Eq. 13, are:

$$\mu = e^{\alpha+\frac{\beta^2}{2}}; \quad \sigma = \mu\sqrt{\exp(\beta^2) - 1} \tag{14}$$

In this section, T_x is thus assumed to follow a Log-Normal distribution, with parameters α_x and β_x.

From relationships (14), the parameters α_x and β_x can be obtained from the statistical parameters μ_L, σ_L and the regression coefficients a and b—denoting by X the natural logarithm of the CCT T_x—are expressed by:

$$\alpha_x = E[X] = a - b\mu_L; \quad \beta_x = \sqrt{\mathrm{Var}[X]} = b\sigma_L \tag{15}$$

A proper probabilistic modelling for the clearing time T_y must also be introduced.

If there is a lack of experimental data, a Gaussian distribution is often assumed. However, the Gaussian distribution does not appear to be a very adequate and flexible choice, and the Log-Normal model is used instead, as in [11–13]: the Log-Normal pdf is indeed very flexible, since it can assume a large variety of shapes with positive "skewness index" which allows for a typical long "right tail" [21] whereas the Gaussian model only allows a single shape for the distribution of the clearing time, i.e. a symmetrical (bell-shaped) distribution around the mean value which is not likely to occur in real applications. The presence of a right tail in the Log-Normal density accounts for the possibility of relatively large clearing times compared with the expected value: thus, the Log-Normal assumption corresponds to a conservative approach which is appropriate when the exact distribution is unknown. The Log-Normal assumption for T_y also permits a straightforward analytical calculation of the IP, without being restrictive, since other distributions may be adopted with the same methodology as shown in [13], requiring only elementary numerical methods.

Furthermore, if the β coefficient of the Log-Normal pdf is small enough, the pdf tends to become symmetrical and may also satisfactorily approximate a Gaussian model.

In this section, the IP computation is performed under the previously discussed hypothesis that both the clearing time T_y and the minimum CCT T_x are described by Log-Normal, independent, RVs. It is, therefore, assumed that T_y has a Log-Normal distribution with parameters α_y and β_y, with density $f_{T_y}(t_y)$ expressed by (13). As a particular case, in VST applications the FCT may be considered as a known constant (as discussed in Sect. 3). With reference to the choice of α_y and β_y, they can be related to the values of μ_{T_y} (mean value of T_y) and σ_{T_y} (standard deviation of T_y), on which some information could be known in practice.

Denoting by $v_y = \frac{\sigma_{T_y}}{\mu_{T_y}}$ the coefficient of variation (CV) of T_y, the relations specifying α_y and β_y as functions of μ_{T_y} and σ_{T_y} are the following:

$$\beta_y = \sqrt{\ln(1 + v_y^2)}, \quad \alpha_y = \ln \mu_{T_y} - \frac{\beta_y^2}{2} \tag{16}$$

Different values can be considered for the parameters μ_{T_y} and v_y, in order to establish a sensitivity analysis. The IP variability versus the mean FCT μ_{T_y} is particularly interesting since such a mean clearing time is a practical measure of the reliability level of the protection system.

The determination of the probability q, when T_x and T_y are Log-Normal and independent of each other, is now considered.

First, the following auxiliary RVs are introduced:

$$X = \ln T_x, \quad Y = \ln T_y \tag{17a}$$

$$Z = X - Y \tag{17b}$$

Under the assumed hypotheses, the probability laws of the above RV X and Y are, respectively, $N(\alpha_x, \beta_x)$ and $N(\alpha_y, \beta_y)$, where, using from now on the symbol α_x instead of α_{cr}, and generally suffixes (x, y) instead of (cr, cl):

$$\alpha_x = E[X]; \quad \alpha_y = E[Y]; \quad \beta_x = \mathrm{SD}[X]; \quad \beta_y = \mathrm{SD}[Y] \tag{18a}$$

According to the well-known properties of the Gaussian distribution, the variable Z, being the difference between two independent Gaussian RVs, is also Gaussian with mean value and standard deviation given by:

$$\mu_Z = E[Z] = E[X] - E[Y] = \alpha_x - \alpha_y \tag{18b}$$

$$\sigma_Z = \mathrm{SD}[Z] = \sqrt{\mathrm{Var}[X] + \mathrm{Var}[Y]} = \sqrt{\beta_x^2 + \beta_y^2} \tag{18c}$$

It is opportune, although obvious, to remark that in practice μ_z is always positive ($\alpha_x > \alpha_y$) since the FCT must always be small enough when compared with the CCT for the system to possess an acceptable level of stability, namely possess a very small IP value. This—being IP $= P(Z < 0)$—can occur only if $E[X]$ is larger than $E[Y]$: this intuitive fact will be confirmed by the computations in the following.

By introducing the standard Gaussian distribution function:

$$\Phi(x) = \int_{-\infty}^{x} \frac{1}{\sqrt{2\pi}} e^{-\frac{u^2}{2}} du \tag{19}$$

the IP can be easily computed as:

$$q = P(T_x < T_y) = P[\ln(T_x) < \ln(T_y)] = P(X < Y) = P(Z < 0) = \Phi\left(-\frac{\mu_Z}{\sigma_Z}\right)$$
$$= 1 - \Phi\left(\frac{\mu_Z}{\sigma_Z}\right)$$

by using the well-known property: $\Phi(-x) = 1 - \Phi(x)$, valid for each real number x.

Alternatively, the "*Complementary standard Gaussian distribution function*" (CSGCDF) can be used as we have done here:

$$\Psi(x) = 1 - \Phi(x) \tag{20}$$

Since the CSGDF $\Psi(x)$ is a strictly decreasing function of x from the value $Q(-\infty) = 1$ to the value $Q(\infty) = 0$, and $Q(0) = 0.5$ (see Appendix 1 for some curves), the IP can be expressed by the more compact expression:

$$q = \Psi(u) = \int_{u}^{\infty} \frac{1}{\sqrt{2\pi}} \exp\left(-\frac{\xi^2}{2}\right) d\xi \tag{20a}$$

where

$$u = \frac{\alpha_x - \alpha_y}{\sqrt{\beta_x^2 + \beta_y^2}} = \frac{E(X) - E(Y)}{\sqrt{V(X) + V(Y)}} \tag{20b}$$

The above quantity u, which plays a key role in the statistical assessment of the IP, can be defined as the "SM" of the system (under the given fault) since the larger the value of u, the smaller the IP. Confirmation of the fact that the relation ($\alpha_x > \alpha_y$) must always be satisfied in practice (although it is not mandatory on theoretical grounds) is that, unless this happens, the IP is greater than 0.5 (if $\alpha_x = \alpha_y$, then $u = 0 \rightarrow q = 0.5$).

3.2 A Numerical Example

As a numerical example, typical values of the mean values of FCT and CCT (which will be used in the applications in the chapter) are $\mu_x = 0.145$ s and $\mu_y = 0.10$ s, respectively.

It is, therefore, assumed that the CCT and FCT follow two independent LN distributions with mean values as above; moreover, a common CV value of 0.1 is assumed for both the CCT and the FCT (i.e. $v_x = v_y = 0.10$). The following values of (α_x, α_y, β_x, β_y) correspond to these values:

$$\alpha_x = -1.9360, \quad \beta_x = 0.0998; \quad \alpha_y = -2.3076, \quad \beta_y = 0.0998$$

having used the relations already stated:

$$\beta = \sqrt{\ln(1 + v^2)}; \quad \alpha = \ln(\mu) - \frac{\beta^2}{2}$$

It can be seen that $\alpha_x > \alpha_y$, as expected. The above values β_x and β_y in the example are equal since they depend on the CV value only.

Finally, the SM value is $u = \frac{\alpha_x - \alpha_y}{\sqrt{\beta_x^2 + \beta_y^2}} = 2.634$ and the IP is then evaluated as $q = \Psi(u) = 0.00423$.

For high values of the SM like the one above, it is worth noting that (see Appendix 1 too) the IP is very sensitive to the variations of system parameters such as the mean FCT μ_y, as can also be seen by taking the derivative of q with regard to μ_y. For instance, if the mean FCT increases from the above 0.10 to 0.11 s (a 10% increase), with the same CVs, then the IP increases to 0.0251 (a 493% increase!). The IP variation compared with both CV variations is also very high. This is just an example of some analytical remarks, briefly discussed below, which may be useful in actual practice.

3.3 Some Final Remarks on IP Sensitivity and Its Estimation

Deferring a more detailed illustration of the IP expression to Appendix 1, this section concludes by highlighting some basic facts which are easily deduced when observing the expression of q:

$$q = \Psi(u) = \Psi\left(\frac{\alpha_x - \alpha_y}{\sqrt{\beta_x^2 + \beta_y^2}}\right) \tag{21}$$

Since Ψ is a decreasing function of its argument, then, as intuitive, q decreases (and stability improves) as the SM increases, e.g. the mean CCT μ_x increases, or the mean FCT μ_y decreases; for given values of μ_x and μ_y, it can be verified, also analytically (see Appendix 1), that the IP also increases when the CV of the CCT and/or FCT increases.

In other words, IP increases as the uncertainty about the above times increases: this consideration has the practical implication that, if uncertainty in load values (which entails uncertainty in the CCT) and/or in FCT is neglected (i.e. their CV values are assumed as zero), the IP may be undesirably underestimated.

It can also be seen that $q(u)$ decreases very quickly towards 0—as exemplified in the above numerical example—especially when the SM u is large enough. This and other mathematical aspects of the relation between the IP and its parameters are discussed and also illustrated graphically, with some details in Appendix 1, in which a sensitivity analysis of the IP is also illustrated.

The great advantage of the proposed analytical approach—compared with numerical methods or Monte Carlo simulation—consists indeed in the very easy way that this approach enables us to perform this sensitivity analysis with regard to system parameters. This is clearly a very desirable property in view of an efficient system design (i.e. with regard to the protection system: taking decisions on how to improve performance of the protection system, lowering the mean value of the FCT, or improving data acquisition in order to reduce its SD or, with regard to the network topology: trying to devise the opportune actions in order to increase the mean value of the FCT and similar actions).

It can be seen that the IP value obtained by the above methodology is only a (statistical) point estimate of the "true" IP since it is obtained from estimated values of the true parameters α_y, α_x, β_y, β_x (as far as the CCT parameters are concerned, they are "forecasted" since they are obtained on the basis of a load forecast; the FCT parameters are estimated from available field or laboratory data). The topic of estimation is discussed in a Bayesian framework in the following sections where ML estimation is also mentioned.

The problem of IP sensitivity described above and estimation are closely related. The above results, and those in Appendix 1, show that, in view of the above high sensitivity of the IP to system parameters, particular attention should be paid to developing an efficient estimation of the characteristic parameters of the CCT and FCT.

4 Bayesian Statistical Inference for Transient Stability

4.1 Introduction

Bayesian inference [22–24] is becoming more and more popular as a powerful tool in all engineering applications including recent applications to power system analysis. This section and the following one, including the last ones which focus on numerical applications, are devoted to a novel methodology for the Bayesian statistical estimation, or briefly "Bayesian estimation" of the IP. In particular, here we are interested in developing a proper methodology for making inference about IP, once prior information and experimental data are available regarding the pdf of the unknown parameters of the IP, q, the transient IP.

It has been seen that the analytical expression of the IP value requires efficient statistical estimates of the true parameters α_y, α_x, β_y, β_x to be evaluated in actual practice (e.g. the CCT parameters, as pointed out before, depend on the load parameters, which are not known, but estimated as a consequence of a load forecast). The extreme IP sensitivity in the region of the values of practical interest (i.e. those yielding IP values of the order of 1e−3 or less) reinforces the need for an efficient estimation.

The aim of the inference is to establish both point and interval estimates of the unknown probability $q = Q(u)$ given that the parameters (α_x, β_x, α_y, β_y) of the two LN distributions must be estimated on the basis of the available random samples (T_{xk}: $k = 1,\ldots, n$) and (T_{yk}: $k = 1,\ldots, m$).

Bayesian inference [22–25] successfully provides a coherent and effective probabilistic framework for sequentially updating estimates of model parameters as demonstrated by the ever increasing number of publications addressed to it in both theoretical and applied fields. Bayes estimation, therefore, appears to be quite adequate for on-line sequential estimation of model parameters. For well-known reasons, moreover, it is particularly efficient (compared with traditional classic estimation, based on ML methods, briefly mentioned in the final part of this section) when rare events are of interest, as is the case here. This is so true that it is currently proposed even when there are no data (see, e.g. [26] for a recent application).

The core of the Bayesian approach is the description of all uncertainties present in the problem by means of probability, and its philosophical roots lay in the subjective meaning of probability [25]. According to such philosophy, the unknown parameters to be estimated are considered RVs, characterized by given distributions whose meaning is not a description of their "variability" (parameters are indeed considered fixed but unknown quantities) but a description of the observer's uncertainty about their true values. Let $\boldsymbol{\omega} = (\omega_1, \omega_2,\ldots, \omega_n)$ be the n-dimensional vector of the parameters to be estimated. The first step in a Bayes estimation process is to introduce—in order to express the available knowledge on the parameters before observing data—a "prior" probability distribution, characterized—in the continuous case here considered—by a joint (n-dimensional) pdf over the parameter space Ω:

$$g(\boldsymbol{\omega}) = g_{1,2,\ldots n}(\omega_1, \omega_2, \ldots, \omega_n), \quad \boldsymbol{\omega} \in \Omega \tag{22}$$

This prior pdf is often—but not always—chosen "subjectively", which does not mean "arbitrarily", but means "on the basis of the knowledge available to the analyst", also using "objective" pieces of information which in most cases could not be used in classical (frequentist) statistical estimation [22–25] which does not admit the existence of a prior pdf.

Then, the data D are observed according to a formal probability model which is assumed to represent the probabilistic mechanism for some (unknown) value of ω which has generated the observed data D. This model gives rise to the "likelihood function" (LF), $L(D|\omega)$, i.e. the conditional probability of the data, given $\boldsymbol{\omega} \in \Omega$. After observing the data D, all the new (updated) available knowledge is contained in the corresponding posterior distribution of ω. This is represented by a posterior joint probability density, $g(\omega|D)$, obtained from Bayes' theorem:

$$g(\omega|D) = \frac{L(D|\omega)g(\omega)}{\int\int\ldots\int_{\Omega} L(D|\omega)g(\omega)\mathrm{d}\omega} \tag{23}$$

where the denominator is the n-fold integral over the whole parameter space Ω. Then, if a function $\tau = \tau(\boldsymbol{\omega})$ of the parameters in $\boldsymbol{\omega}$ is the subject of estimation, according to the well-known "mean square error" (MSE) criterion, the best Bayes estimate "point" estimate—denoted by $\tau°$—is given by the posterior mean of τ, given the data D. This may be obtained by well-known rules related to the expectation of a function of RV [19] by:

$$\tau^{\circ} = E[\tau|D] = \int\int\ldots\int_{\Omega} \tau(\omega)g(\omega|D)\mathrm{d}\omega \tag{24}$$

The particular case $\tau(\omega) = \omega_\kappa$—for any given k value, $k = 1,2,\ldots, n$—yields the Bayes estimation of any single parameter ω_κ ($k = 1,2,\ldots, n$).

Alternatively, by denoting the prior pdf of τ by $h(\cdot)$, i.e. the pdf induced—by a proper manipulation of the pdf $g(\omega)$—on the space of τ values by the transformation $\tau = \tau(\omega)$, and introducing, analogously, the posterior pdf of τ, $h(\tau|D)$, the above expectation may be obtained equivalently by the following integral:

$$\tau^{\circ} = E[\tau|D] = \int_{\Xi} \tau \cdot h(\tau|D)\mathrm{d}\tau \tag{25}$$

with Ξ being the space of τ values. In practice, and also in this application, it is very difficult, if not impossible, to deduce an analytical expression for the posterior pdf of τ, and the above expectation may be more easily obtained by the integral over Ω even if, in most cases, it is evaluated numerically or by means of simulation.

Unlike classical estimation, which is inherently focussed on the point estimate, here this is only a particular piece of information: indeed, Bayesian inference aims to express all the available knowledge on the parameters not by a single value but by means of the complete posterior pdf of τ, $h(\tau|D)$, denoting by $h(\cdot)$ the pdf induced—by a proper transformation of the pdf $g(\omega)$—on the space of τ values by the transformation $\tau = \tau(\omega)$. The point estimate is only a "synthesis" of this pdf. This pdf (which would have no meaning in the classical inference since it is regarded as an unknown constant, not an RV) is the "key" information provided by Bayes estimation since it allows any probabilistic statement about the values of τ to be expressed. Typically this pdf is used to form a "Bayesian confidence interval" (BCI) or "Bayesian credible interval" of the unknown τ, defined as:

$$\mathrm{BCI}(\tau;\pi) = (\tau_1, \tau_2) \tag{26}$$

so that $P(\tau_1 < \tau < \tau_2) = \pi$, where π is a given probability. The BCI is expressed in terms of the posterior pdf of τ as follows:

$$P(\tau_1 < \tau < \tau_2|D) = \int_{\tau_1}^{\tau_2} h(\tau|D)\mathrm{d}\tau = \pi, \quad 0 < \pi < 1 \tag{27}$$

In practice, the above relation is generally not sufficient to find the BCI but further requirements, such as a search for the "*Highest Posterior Density*" regions [23], allow the determination of both the unknowns (τ_1, τ_2) above.

4.2 A General Methodology for Bayesian Inference on Transient IP

Let us transpose the above concepts of Bayesian inference to the estimation of the IP:

$$q = \Psi(u), \quad u = \frac{\alpha_x - \alpha_y}{\sqrt{\beta_x^2 + \beta_y^2}} \tag{28}$$

with u being the SM whereas $\Psi(z)$ is the CSGCDF $\Psi(z) = 1 - \Phi(z)$:

$$\Psi(z) = \int_{z}^{\infty} \frac{1}{\sqrt{2\pi}} \exp\left(-\frac{\xi^2}{2}\right) d\xi, \quad z \in (-\infty, +\infty) \tag{29}$$

For the purpose of Bayes estimation, in the most general case, all the parameters shall be considered unknowns. Therefore, in the Bayes approach, the four parameters ($\alpha_x,\alpha_y,\beta_x,\beta_y$) and also the SM u and the same IP, q, are regarded as a realization of RV which will be denoted by the following capital letters in the sequel:[2]

$$\alpha_x \to M_x; \quad \alpha_y \to M_y; \quad \beta_x^2 \to V_x; \quad \beta_y^2 \to V_y \tag{30}$$

The symbols M and V are also chosen for "mnemonic" reasons since, as mentioned below, they correspond to mean values and variances, in particular of the logarithm of the FCT (T_x) and of the logarithm of the CCT (T_y):

$$M_x = E[\ln T_x], \quad V_x = \mathrm{Var}[\ln T_x], \quad M_y = E\left[\ln T_y\right], \quad V_y = \mathrm{Var}\left[\ln T_y\right] \tag{31}$$

Bayes estimation, therefore, consists of assessing prior distributions to the above parameters (M_x,M_y,V_x,V_y), and then evaluating point and interval estimates of the unknown parameter IP, described by the RV Q, function of the RV U:

$$Q = \Psi(U), \quad U = \frac{M_x - M_y}{\sqrt{V_x + V_y}} \tag{32}$$

The above relations specify the relation between the basic RV (M_x,M_y,V_x,V_y) and the IP Q:

$$Q = Q(M_x, M_y, V_x, V_y), \text{ with} : (M_x, M_y) \in \Re^2; \quad (V_x, V_y) \in \Re^{+2} \tag{33}$$

These relations appear to be quite complicated, in particular due to the presence of the special function $\Psi(U)$, whatever the choice of the prior pdf of the basic RV, a topic which is dealt with below. The same argument also applies, a fortiori, to the posterior pdf. In practice, it is impossible to evaluate the pdf of Q—be it prior or posterior—analytically. It can be, however, handled numerically by resorting to a reasonable "Beta approximation", for example—introduced in a different study by Martz et al. [27] and also used by the authors in the above-mentioned (a.m.) article [16] in the framework of ML estimation. This approximation is illustrated in Appendix 3 and has been shown to be very adequate, although not being the only possible approximation, since every pdf over (0, 1) which can be rather smooth and flexible may be a good candidate (possible alternative choices studied by the

[2] From now on we use different symbols for the four parameters—when they are considered as RV—to avoid confusion with other symbols used in this section (see Sect. 4.4) and in Appendix 2 where the capitals (A_x, A_y, B_x, B_y) corresponding to (α_x, α_y, β_x, β_y) denote specific ML estimators (it is recalled that RV are denoted by capitals).

authors are also mentioned in Appendix 2). In any case, once the numerical pdf of Q is obtained, its usefulness in this application seems to consist, first of all, in establishing a proper "upper confidence bound" for IP, i.e. a q_{UP} value that makes the probability of the "desirable" event ($Q < q_{UP}$) high enough, say 0.95 or 0.99. Therefore, by denoting this high probability value with η, interest may be focussed on the determination of a q_{UP} value so that:

$$P(Q<q_{UP}) = \eta \tag{34}$$

With a sufficiently large probability value, we are, therefore, assured that the "true" IP, Q, is smaller than an "upper bound" q_{UP}, which is the Bayes counterpart of the confidence level of $100\eta\%$.

An important characteristic, perhaps the most important one, of the Bayes inference methods is the one, already mentioned, of allowing any probabilistic statement on the values under investigation, here the IP, to be expressed, e.g. in terms of the above BCIs.

This is the core of "Bayes inference", which is something more than pure estimation, and this is also the reason behind the heading of this section ("Bayesian inference" rather than "Bayesian estimation"). Also, in many practical cases (e.g. in order to control if some prefixed requirements or standards are met by system performances), an interval estimate may be more significant than the point estimate alone.

However, also in view of the analytical or numerical difficulties mentioned above associated with the establishment of the BCI, in actual practice, there is no doubt that the typical objective of the Bayes methods is to assess the point estimate of Q. This is the topic dealt with from now on. This point estimate of Q may be evaluated after the assessment or evaluation of the:

- prior parameters' pdf: $g_{m_x,m_y,v_x,v_y}(m_x,m_y,v_x,v_y)$, briefly denoted as $g(m_x,m_y,v_x,v_y)$;
- the LF: $L(D|(m_x,m_y,v_x,v_y)$, which is given in this case by the conditional joint pdf of the observed data (FCT values, i.e. the times T_{xk}, and CCT values, i.e. times T_{yk}, recorded in the interval of interest for the IP prediction). This joint pdf is conditional to the parameters (m_x,m_y,v_x,v_y);
- posterior parameters' pdf: $g_{M_x,M_y,V_x,V_y}(m_x,m_y,v_x,v_y|D)$, briefly denoted as $g(m_x,m_y,v_x,v_y|D)$, obtained by the prior pdf and the LF by means of the Bayes theorem as illustrated above.

Finally, the Bayes estimate, denoted by $Q°$, of the IP Q is given in principle by the four-dimensional integral:

$$Q^\circ = E[Q|\boldsymbol{D}] = \int\int\int\int_{2\Omega} Q(m_x, m_y, v_x, v_y) g(m_x, m_y, v_x, v_y|\boldsymbol{D}) \mathrm{d}m_x \mathrm{d}m_y \mathrm{d}v_x \mathrm{d}v_y \tag{35}$$

with Ω the parameter space above specified for the four parameters (m_x,m_y,v_x,v_y), $Q(m_x,m_y,v_x,v_y) = \Psi(u)$ (lowercase letters are used for the single determinations of the RV being studied), $u = \frac{m_x - m_y}{\sqrt{v_x+v_y}}$, and Ψ the above CSGDF.

As far as the choice of prior pdf both for the above parameters is concerned, it is well known that the most simple natural candidates are the so-called "conjugate prior pdf" [23, 24]: this means adopting Gaussian prior pdf for the mean values M_x and M_y and Inverted Gamma prior pdf for the variances V_x and V_y (see Appendix 2). These are indeed the prior pdf of mean and variances for both the normal and the Log-Normal sampling distributions [21–24].

The above integral may appear quite cumbersome, yet its evaluation can be made—at least in some cases—relatively simple, observing that the particular form of the random IP, $Q(U)$, can be reformulated in terms of the RV: $M = M_x - M_y$, $V = V_x + V_y$, $S = \sqrt{V}$. Hence $Q = \Psi(M/S)$.

If prior and posterior information on the four parameters (M_x, M_y, V_x, V_y) is recast into prior and posterior information on the difference $M = (M_x - M_y)$ and the sum $V = (V_x + V_y)$, the above integral, hence, reduces to a double integral (with respect to the pdf of M and S) which can be solved using methods related to Bayes estimation of Gaussian probabilities [23, 24].

The above transformation between the pdf of (M_x, M_y, V_x, V_y) and those of M and V may be effected by elementary RV transformations, taking advantage of the assumed s-independence between the CCT and the FCT which logically implies the s-independence between their mean values and variances. For instance, adopting conjugate Gaussian prior pdf both for M_x and M_y, assumed as s-independent, then the prior pdf for $M = (M_x - M_y)$ is again Gaussian with obvious values of the parameters; the same holds, as is well known, for the posterior pdf. As far as the variances are concerned, the same reasoning does not apply for the above-mentioned conjugate Inverted Gamma pdf which is typically adopted as the prior pdf. However, if one is able to express information directly in terms of the sum of variances $V = (V_x + V_y)$, by using an Inverted Gamma pdf for V, the classical results of Bayes estimation for the Gaussian model (mentioned in Appendix 2) may still be applied. In general, however, if other prior models are chosen, the above estimation must be carried out numerically. This poses no particular problem nowadays since specific codes and algorithms have been devised for such purposes [22–24].

A major simplification occurs in the particular case considered in the application of this contribution, i.e. in VST applications in which the FCT may be considered in practice as a known constant (as discussed in the following section).

4.3 A Simplified Method for Bayes Estimation of the IP in the Event of VST Stability Prediction

The general theoretical problem of the Bayes estimation for the IP, discussed above, will not be pursued here in view of the VST application of this contribution. In this case, indeed, an event of instability—in a very short interval lasting typically 1 h—is very unlikely as confirmed by the typical values of the IP illustrated in the previous section and observed in actual system practice. Observing new FCT values is a very rare event.

For practical purposes, the FCT T_y can, therefore, be considered a known constant instead of an RV. This constant value is the prior estimate of the FCT before observing data since no inference can be made out of them; i.e. T_y assumes a deterministic value, $t_y = t^*$, estimated from previous experiences.

Alternatively, the FCT may be considered as an RV, with the assumed law LN(α_y,β_y), whose parameters are characterized by their prior pdf, since it is highly improbable that new data can change our information about the FCT in a VST interval (or the data are so rarely acquired that they do not change the prior pdf much).

The two cases are equivalent, as will be shown later, so that in the sequel reference will be made to the first one (i.e. a deterministic value, t^*, of the FCT T_y is assumed).

Let t be the FCT and T_x the RV describing the CCT in the given interval under investigation. Two alternative hypotheses can be assumed for the RV describing the load: (1) the load has a constant value (i.e. it is unknown, but constant in time) due to the interval shortness; (2) the load variations with time are considered—adopting a more rigorous approach—not negligible: then, reference to the peak load is made in the interval. As previously shown, the two cases are formally equivalent.

Then, the IP value in that interval is given—by using the usual transformation from an LN cdf into a Gaussian one:

$$Q = P(T_x < t) = P[(X < \ln(t)] = \Psi\left(\frac{\alpha_x - \ln(t)}{\beta_x}\right) \tag{36a}$$

always using

$$\Psi(z) = \int_z^\infty \frac{1}{\sqrt{2\pi}} \exp\left(-\frac{\xi^2}{2}\right) d\xi$$

Obviously, the above relation could also be deduced from the general one:

$$Q = \Psi\left(\frac{\alpha_x - \alpha_y}{\sqrt{\beta_x^2 + \beta_y^2}}\right)$$

with $\alpha_y = \ln(t)$; $\beta_y = 0$ (since T_y is deterministic, as is its logarithm Y, so that its only assumed value, $\ln(t)$, coincides with its mean value whereas its SD β_y is zero).

The consequent IP expression is, therefore, equal to:

$$\Psi\left(\frac{\alpha_x - \tau}{\beta_x}\right); \quad \tau = \ln(t) \tag{36b}$$

which can be handled like that of a Gaussian cdf, as shown in the sequel, in order to perform a Bayes point estimation of the IP.

In the framework of Bayes estimation, let us assume that the mean value α_x of $X = \ln(T_x)$—being T_x the CCT—is an RV, denoted as M (with analogy with the previous section) whereas its SD $\beta_x = s$—as assumed in common practice—is known.[3]

Let the prior information about the unknown parameter M be described by a conjugate prior normal distribution with known parameters (m_0, s_0), i.e. $M \backsim N(m_0, s_0)$ so that the prior pdf of M (bear in mind that M, the mean value of the RV $X = \ln(T_x)$ can be negative[4]):

$$g(m) = \frac{1}{s_0\sqrt{2\pi}}\exp\left[-\frac{(m - m_0)^2}{2s_0^2}\right], \quad m \in \Re \tag{37a}$$

By using results in Appendix 2, the posterior pdf of M, after observing data $\mathbf{X}$, is again Gaussian:

$$g(m|\mathbf{X}) = \frac{1}{s_1\sqrt{2\pi}}\exp\left[-\frac{(m - m_1)^2}{2s_1^2}\right], \quad m \in \Re \tag{37b}$$

with posterior mean and variance given by:

$$m_1 = E[M|\mathbf{X}] = \frac{s^2 m + n s_0^2 M_n}{s^2 + n s_0^2} \tag{38a}$$

$$s_1^2 = \mathrm{Var}[M|\mathbf{X}] = \frac{s_0^2 s^2}{n s_0^2 + s^2} \tag{38b}$$

where

$$M_n = (1/N)\sum_{k=1}^{N} X_k; \quad s = \beta = \mathrm{SD}[X] \tag{38c}$$

being X_k a generic log-CCT value of the sample $\mathbf{X}$.

---

[3] The case in which the SD should be unknown poses no problems. Indeed it can be dealt with, implying only a little computational effort, by means of well-known methods like those mentioned in Appendix 2.

[4] In the numerical examples or applications of this chapter, measuring times in seconds as done here, both X and Y have negative mean values.

The Bayes point estimate of the IP is, therefore, given by:

$$\begin{aligned} Q^{\circ} &= \int_{-\infty}^{\infty} Q(m)g(m|X)\mathrm{d}m \\ &= \int_{-\infty}^{\infty} \Psi\left(\frac{m-\tau}{s}\right)\frac{1}{s_1\sqrt{2\pi}}\exp\left(-\frac{(m-m_1)^2}{2s_1^2}\right)\mathrm{d}m \end{aligned} \tag{39}$$

which, after some manipulation, after changing the variable to: $z = \frac{m-\tau}{s}$ (see [19], in chapter titled Transmission Expansion Planning: A Methodology to Include Security Criteria and Uncertainties Using Optimization Techniques), can be shown to be equal to:

$$Q^{\circ} = \Psi\left(\frac{m_1-\tau}{s_1}\right) = \int_{z}^{\infty} \frac{1}{\sqrt{2\pi}}\exp\left(-\frac{\xi^2}{2}\right)\mathrm{d}\xi, \quad z_1 = \frac{m_1-\tau}{s_1} \tag{40}$$

This estimator will be used in the final numerical application related to VST stability prediction, in which, on the basis of an adequate dynamic model of the load (and the CCT), the posterior means and variances will be updated at each time step.

In these applications, typically only one datum of the CCT is observed at each step (so in the above relations $n = 1$ will be used)—i.e. at the generic kth step—the measured or forecasted load value L_k. This "data" scarcity renders the Bayes estimation more attractive, as discussed above.

Finally, let us briefly examine the second case, mentioned above, with regard to the knowledge of the pdf of the RV T_y. Let us assume that it is an RV, and not a constant as above, letting the parameters of the RV T_y, i.e. (α_y, β_y)—denoted as (α, β) in the sequel—be distributed according to their prior pdf, which remain unchanged after every interval, since no new FCT value is obtained. Let us assume, as above, that only the mean of Y, $\alpha = \alpha_y$ is unknown with a prior conjugate Gaussian distribution $N(\mu_o, \sigma_o)$. Consequently, the pdf of Y (conditional to $\alpha_y = \alpha$) and the pdf of α are, respectively:

$$Y|\alpha \sim N(\alpha, \beta); \quad \alpha \sim N(\mu_0, \beta_0)$$

Then, using the total probability theorem for continuous RV [19] or known results from Bayesian estimation theory [22–24] (see also Appendix 2), it can be seen that the marginal pdf is still a Gaussian pdf:

$$Y \sim N(\mu_0, \beta*), \text{ with } \beta^* = \sqrt{\beta_0^2 + \beta^2}.$$

In the light of this fact, it is not difficult to show that (40) still holds with properly re-arranged values of the constants (α_x, β_x).

4.4 A Mention of the Classical Estimation of the IP

Here, only a brief account of classical (ML) estimation of Q is given in order to compare it with the one adopted here. Some details can be found in [21, 28].

As stated in Sect. 4.1, let us assume that the following data are available: $\mathbf{X} = (X_1,.., X_n)$, where $X_k = \ln(T_{xk})$, $k = 1,\ldots, n$: i.e. $\mathbf{X}$ is a random sample of n elements constituted by the natural logarithms of a CCT sample, and let $\mathbf{Y} = (Y_1,.., Y_m)$, where $Y_k = \ln(T_{yk})$, $k = 1,\ldots, m$: i.e. $\mathbf{Y}$ is a random sample of m elements constituted by the natural logarithms of the FCT sample which can be obtained from field or laboratory data on the system protection components, with regard to the assumed kind of fault.

By referring, for easier notation, to estimated quantities with capital letters, the most widely adopted estimators (A_x, B_x, A_y, B_y) of the above four parameters—for the well-known properties of the ML estimation [21, 28]—are given, for the LN variables under study, by:

$$A_x = \left(\frac{1}{n}\right)\sum_{k=1}^{n} X_k; \quad B_x = \sqrt{\left(\frac{1}{n}\right)\sum_{k=1}^{n}(X_k - A_x)^2} \tag{41}$$

$$A_y = \left(\frac{1}{m}\right)\sum_{k=1}^{m} Y_k; \quad B_y = \sqrt{\left(\frac{1}{m}\right)\sum_{k=1}^{m}(Y_k - A_y)^2} \tag{42}$$

These estimators indeed maximize, compared with any other function of the data, the LF $L[(\mathbf{X},\mathbf{Y})|(\alpha_x, \alpha_y, \beta_x, \beta_y)]$.

In practice, these estimators coincide with the sample estimators of the mean values (A_x and A_y) and standard deviations (B_x and B_y) of the Normal RV $X = \ln(T_x)$ and $Y = \ln(T_y)$, and show some desirable properties such as consistency. Moreover, the log-mean estimators A_x and A_y are also unbiased estimators of α_x and α_y, respectively.

Then, the ML estimator Q^* of q is given by:

$$Q^* = \Psi\left(\frac{A_x - A_y}{\sqrt{B_x^2 + B_y^2}}\right) \tag{43}$$

In [16], the authors analysed the classical point and interval estimates of Q based on an estimator of this type whose properties are not easy to assess.

As a final remark, it should be clear that, when prior information are available, as in most engineering applications and also in this case, the Bayes estimator definitely performs better then the ML estimators. This is especially evident in on-line estimation, as will be shown later, since very few data can be collected for inference. Typically, indeed, no data are available on FCT if the fault does not occur, and this non-occurrence is of course very likely; only one datum is available on CCT, based on the forecasted load value for the time interval under investigation.

5 Dynamic Bayesian Estimation of Mean CCT and IP for VST Applications

5.1 Introduction

In this section, devoted to VST system operation, the principle of recursive Bayesian estimation is applied for a fast and efficient on-line evaluation of the mean CCT (actually, of its natural logarithms), and thus of the IP, in a dynamic framework. This evaluation exploits:

1. the above-discussed relation between the CCT and the system load L;
2. the probabilistic knowledge of the time evolution of the load which is generally available in VST applications.

With regard to point (1), reference is made here for illustrative purposes to a single-machine system,[5] or to a system which is reducible to it. The above discussed log-linear characteristic is therefore assumed to hold—at any given instant (for a given network topology)—between the CCT T_x and the load L:

$$T_x = \beta_0 \exp(-\beta_1 L) \tag{44}$$

with model coefficients (β_0, β_1) which are positive known constants, depending on the electrical parameters of the system. As mentioned above, they can be determined by performing a linear regression of the natural logarithms of T_x with respect to the load. This is accomplished after computing the CCT values off-line, for the given network and fault, by means of an appropriate system model based on the classical Lyapunov direct methods for transient stability analysis and sensitivity. Therefore, by denoting—as before—the natural logarithm of T_x by X and the values of X and L at a given time instant t_k by (X_k, L_k), respectively, the following relation is assumed:

$$X_k = a - bL_k \tag{45}$$

with

$$X_k = \ln(T_{xk}); \quad a = \ln(\beta_0)(a \in \Re); \quad b = \beta_1(b \in \Re^+)$$

This linear relation between the logarithm of the CCT (LCCT in the following) and the system load is the basis for dynamic estimation. In particular, the proposed Bayes recursive estimation uses known results in dynamic estimation—such as the Kalman filter theory—which are well established under the hypothesis that the series $\{X_k\}$ to be estimated is a Gaussian time series.

[5] The generalization to multi-machine systems, illustrated by the authors in [12], can be accomplished without difficulties by adopting the "Extended Equal Area Criterion".

Such hypothesis is true if the load $L(t)$ is a Gaussian process as above assumed (generalizations to other kinds of load distribution are of course possible without particular problems).

With regard to point (2), namely, the load evolution in time, an adequate load evolution model must be chosen like those adopted for VST load forecasting algorithms.

In particular, we must consider the given time instants t_1, t_2,..., t_k,... of interest for VST operation (typically, the successive hours of a certain time interval in which the network topology is assumed as fixed). Then, the following "*dynamic linear model*" (DLM) [29], or "autoregressive model" is often satisfactorily adopted for the stochastic process (L_k, $k = 1,2\ldots$), which is supposed to generate the load values at times t_k according the "system equation":

$$L_{k+1} = L_k + \lambda_k \quad (k = 0, 1, 2\ldots) \tag{46a}$$

in which$\{\lambda_k\}$ is a "*White Gaussian Noise*" (WGN) sequence, i.e. a set of IID Gaussian RV with mean 0, and known SD, denoted by W. This is formally expressed as:

$$\lambda_k \sim \mathrm{WGN}(0, W) \tag{46b}$$

The sequence is "initiated" by a value L_0 (load value at time $t = 0$) which is (like all the L_k values) an RV, as appropriate in a Bayes framework, with known pdf representing our prior information. It is also assumed to be a Gaussian RV (with known mean μ_{L_0} and known SD σ_{L_0}), statistically independent of any finite set of the sequence $\{\lambda_k\}$

$$L_0 \sim N\left(\mu_{L_0}, \sigma_{L_0}\right) \tag{46c}$$

The above model tries to capture a reasonable "Markovian" dependence between the successive random values L_{k+1} and L_k in a simple way, suitable for VST applications. However, it may be extended without excessive difficulty to cover, e.g. more complex autoregressive model, such as ARIMA processes, or non-linear models [29].

Generally, the values of the load L_k are not measurable with precision but their acquisition is subject to forecasting or measurement errors (also taking into account possible time delays or even missing values in the acquisition process). The following "observations equation" is typically adopted for the estimation of the DLM:

$$Y_k = L_k + v_k \quad (k = 1, 2\ldots) \tag{47a}$$

where $\{v_k\}$ is another WGN sequence, with mean 0, and known SD sv, statistically independent of the sequence $\{\lambda_k\}$ and all the other RVs in the model:

$$v_k \sim \mathrm{WGN}(0, S) \tag{47b}$$

The above assumptions assure that both L_k and Y_k are Gaussian sequences.

Analogously to "Kalman filtering" language [29], the basic DLM equations (46a) and (47a) can be, respectively, regarded, as the "state equation" and the "measurement equation".

In order to define a similar DLM for the LCCT values, and repeating for convenience equation (45), let us define the sequences:

$$X_k = a - bL_k, \quad Z_k = a - bY_k, \quad \xi_k = -b\lambda_k, \quad \eta_k = -b\nu_k \tag{48}$$

It is easy to see that these definitions, observing that $\{\xi_k\}$ and $\{\nu_k\}$ are still WGN sequences, allow the definition of a DLM for the sequence of the LCCT values as follows:

$$X_{k+1} = X_k + \xi_k, \quad Z_k = X_k + \eta_k \tag{49a}$$

where (ξ_k, η_k) are, respectively, the system and measurement noise for the DLM of the LCCT.

The above assumptions for (X_0, ξ_k, η_k) are formally expressed as follows:

$$X_0 \sim N(\mu_0, \sigma_0), \quad \xi_k \sim \text{WGN}(0, w), \quad \eta_k \sim \text{WGN}(0, s) \tag{49b}$$

The SD w of the model and the SD s of the measures, appearing in the above relations, are clearly related to the above SD (W,S) of λ_k and ν_k by the following, obvious, linear relations[6]:

$$w = bW; \quad s = bS \tag{50}$$

Finally, the initial mean and SD of the LCCT sequence X_k, i.e. those of X_0 (first equation of 49b), denoted simply by (μ_0,σ_0) are obviously expressed in terms of the corresponding initial load L_0 parameters $(\mu_{L_0}, \sigma_{L_0})$ in (46c) as follows:

$$\mu_0 = a - b\mu_{L_0}; \quad \sigma_0 = b\sigma_{L_0} \tag{51}$$

It is apparent from the second equation (48) and the above hypotheses summarized in (49a) and (49b) that Z_k, the observed LCCT Z_k, being the sum of two Gaussian independent RV, X_k and ξ_k, is still a Gaussian RV whose marginal pdf is easily deducible (it is sufficient to compute its mean value and variance, as shown below). Moreover, if X_k should be known, the conditional distribution of Z_k—being ξ_k a Gaussian RV with zero mean—would be a Gaussian one with mean equal to X_k, and known SD w. Formally[7]:

$$(Z_k|X_k) \sim N(X_k, w)$$

Therefore, in the framework adopted here for the estimation process, X_k is the unknown (unobservable) mean value of the observable Gaussian RV Z_k, with

---

[6] Note that if $Y = a \pm bX$, where X and Y are RV and (a,b) constants, then $\text{SD}[Y] = |b|\text{SD}[X]$ (the SD is intrinsically non-negative).

[7] The notation $(R|S) \sim N(\alpha,\beta)$, being R and S two RV, denotes that the conditional distribution of R, given S, is $N(\alpha,\beta)$.

known SD w. In other words, interest here is focused on the estimation of the mean value of the LCCT, so that results mentioned above (and recalled in Appendix 2) related to estimation of the unknown mean value of a Gaussian RV may be adopted. For brevity, the term "LCCT" (both in the acronym form or in the complete one) will, however, still be used in the sequel instead of the more correct "mean LCCT".

The X_k sequence is "initiated" by a value X_0 which, based on prior information for the load L_0, is again assumed to be a Gaussian RV, as above reported, statistically independent of any finite set of the sequences $\{\xi_k\}$ and $\{\eta_k\}$.

5.2 Estimation Methodology

Once the measurements (z_1,z_2,..., z_k) have been assigned until time instant t_k, the optimal dynamical state estimate $\hat{X}_k$ of the "true" state X_k at time t_k—according to the Bayesian approach to estimation—is provided by a posteriori "MSE" minimization:

$$\mathrm{MSE} = E\left[\left(\hat{X}_k - X_k\right)^2 | z_1, z_2, \ldots, z_k\right] \tag{52}$$

This can be accomplished, as will be shown, using recursive Bayesian estimation (Appendix 2, see also [28]) which is substantially resumed by the following recursive relationship.

$$\hat{\mathrm{x}}_k = E[\mathrm{x}_k | z_1, z_2, \ldots, z_k] = {}^{-}\hat{\mathrm{x}}_k + G_k(z_k - {}^{-}\hat{\mathrm{x}}_k) \tag{53}$$

where ${}^{-}\hat{\mathrm{x}}_k$ represents the state estimate at instant t_k, before z_k knowledge, i.e. the a priori estimate at stage k, and G_k is a constant which is obtained as shown below on the basis of the above "minimum MSE" criterion. The above relation is substantially equivalent to Kalman Filter, but is obtained using the Bayes estimation process, as discussed in [29]: this method has the advantage over the classic Kalman Filter derivation of accounting for the random nature of state X and of allowing the computation of any probabilistic statement about this state. The constant G_k corresponds to the well-known "Kalman gain" [28].

The following stages to which the Bayes procedure is applied can be defined:

Stage "0", or "a priori" Stage: "Stage 0" means the initial stage before any observation is available. Therefore, in this stage the only available information is the a priori characterization for the RV X at time instant t_0:

$$X_0 \sim N(\mu_0, \sigma_0) \tag{54}$$

Thus, from a Bayesian point of view with quadratic "*Loss function*", the initial optimal estimate is $\hat{X}_0 = \mu_0$.

Stage 1: In this stage and in following stages, according to the Bayes methodology, two kinds of information are available, before and after the measurement—which here is the first observation z_1—is acquired. The first (prior) information yields the prior estimation, the latter (posterior) information yields the posterior estimation.

Before the first measurement z_1 is performed, the following a priori estimation can be given:

$$X_1 = X_0 + \xi \Rightarrow X_1 \sim N({}^{-}\mu_1, {}^{-}\sigma_1) \tag{55}$$

where the prior mean value and variance[8] are determined by:

$$\begin{aligned} {}^{-}\mu_1 &= E[X_1] = E[X_0] + E[\xi] = \hat{X}_0 = \mu_0 \\ {}^{-}\sigma_1^2 &= \mathrm{Var}[X_1] = \mathrm{Var}[X_0] + \mathrm{Var}[\xi] = \sigma_0^2 + w^2 \end{aligned} \tag{56}$$

Once the measurement z_1 is known, the aim is directed towards X_1 estimation conditional to z_1. Denoting by z_1 the observed realization of the RV Z_1. Z_1 is still a Gaussian RV, with conditional mean (given X_1) equal to X_1, and SD equal to that of η_1, i.e. s. Formally

$$Z_1 = X_1 + \eta_1 \Rightarrow (Z_1|X_1) \sim N(X_1, s) \tag{57}$$

and since $E[Z_1|X_1] = X_1$, it can be deduced that the posterior mean (i.e. the Bayes estimate) of X_1 is:

$$\hat{X}_1 = E[X_1|z_1] = \mu_0 + \frac{{}^{-}\sigma_1^2}{s^2 + {}^{-}\sigma_1^2}(z_1 - \mu_0) = \mu_0 + \frac{\sigma_1^2}{s^2}(z_1 - \mu_0) \tag{58}$$

where

$$\sigma_1^2 = \frac{{}^{-}\sigma_1^2 s^2}{{}^{-}\sigma_1^2 + s^2} = \frac{(\sigma_0^2 + w^2)s^2}{\sigma_0^2 + w^2 + s^2} \tag{59}$$

The posterior estimate is used as the prior for the next stage according to recursive Bayesian estimation, as illustrated in Appendix 2. By applying this algorithm recursively, the following result at time t_k can be obtained.

Generic Stage k: By applying recursive Bayesian estimation, we can immediately verify (e.g. by induction) that the following relation—clearly appearing as the general case of (59)—holds at time instant t_k:

$$\sigma_k^2 = \frac{(\sigma_{k-1}^2 + w^2)s^2}{\sigma_{k-1}^2 + w^2 + s^2} \tag{60}$$

Similarly, the following recursive formulation for the Bayesian estimate at time t_k can be obtained:

$$\hat{X}_k = E[X_k|z_1, z_2, \ldots, z_k] = \hat{X}_{k-1} + G_k(z_k - \hat{X}_{k-1}) \tag{61}$$

---

[8] A prior estimate of a parameter ζ is denoted here by ${}^{-}\zeta$

where

$$G_k = \frac{\sigma_k^2}{s^2} = \frac{\left(\sigma_{k-1}^2 + w^2\right)}{\sigma_{k-1}^2 + w^2 + s^2} \tag{62}$$

As is well known, this estimate exhibits the noticeable property to minimize the posterior MSE for every time instant t_k, expressed by:

$$\text{MSE} = E\left[\left(\hat{X}_k - X_k\right)^2 | z_1, z_2, \ldots, z_k\right] \tag{63}$$

Of course, the recursive procedure allows knowledge of the pdf of X_k at each stage k (needless to say, unlike the "static" Bayes estimation, the posterior pdf of X changes with time), and also allows Bayes estimation of the IP. This is directly deduced using results derived in Sect. 4.3: since (61) is the posterior mean of the LCCT and σ_k^2 the posterior variance, by re-arranging Eq. 40, the following recursive Bayesian estimate of the IP at time t_k is obtained:

$$\hat{Q}_k = \Psi\left(\frac{\hat{X}_k - \tau}{\sigma_k}\right) \tag{64}$$

Confidence intervals, particularly the previously illustrated "upper confidence bounds" may also be computed for both the LCCT and the IP. In the following numerical application, for sake of brevity the estimation procedure is illustrated only for the LCCT sequence, which is a Gaussian one, so that its results are more easily interpretable. Moreover, the confidence interval assessment is straightforward for the LCCT sequence whereas for the IP sequence it can be computed by applying the procedure illustrated in Appendix 3 at each step using the a.m. Beta approximation, since no analytical result exists. Numerical simulations results were similar as regards parameter point estimation. A numerical example of the BCI computation is still reported in Appendix 3, only for the IP, being it straightforward for the LCCT.

5.3 Concluding Remark

The proposed procedure is based on a relation (CCT–Load) which can be analysed and computed off-line—for the given network—once for all, so that on-line estimation shown here does not require the solution of the system model at each iteration. This allows the time duration of the intervals in which the stability is assessed to be shortened and is favourable to reliable and efficient security assessment. A distributed version of the proposed Kalman filtering approach can be applied in the case of large power systems [30].

6 Numerical Application of the Bayes Recursive Dynamic Model

In this section, a simple numerical application—based on typical load and CCT values and simulated patterns of the load process in time—is presented in order to illustrate the on-line estimation of the LCCT in VST operation.

The load process is assumed to follow the DLM model, with ($k = 1,2\ldots$):

$$L_{k+1} = L_k + \lambda_k \text{ and } Y_k = L_k + \nu_k \tag{65}$$

in which

- $\{\lambda_k\}$ is a "WGN" sequence, WGN(0, W);
- $\{\nu_k\}$ is another WGN sequence, WGN(0, S);

the two sequences are statistically independent of each other and the other RV in the model.

For simplicity of notation, the SDs in the above WGN sequences (λ_k, ν_k) are, respectively, denoted as (W, S) instead of (σ_λ, σ_ν) as in the previous section. The lowercase letters (w, s) will be used for the LCCT sequence.

The evolution model of the LCCT corresponding to (65) is, as already deduced:

$$X_{k+1} = X_k + \xi_k \text{ and } Z_k = X_k + \eta_k \tag{66}$$

with the already discussed basic assumptions

$$X_0 \sim N(\mu_0, \sigma_0), \quad \xi_k \sim \text{WGN}(0, w), \quad \eta_k \sim \text{WGN}(0, s) \tag{67}$$

and with the SD of the WGN sequences (ξ_k, η_k), respectively, equal to $w = bW$, $s = bS$.

For the sake of a numerical example, let us assume that the starting value of the system load, L_0, measured in p.u., is a Gaussian RV with mean $\mu_{L_0} = 0.8750$ p.u. and SD $\sigma_{L_0} = 0.0417$ p.u.

These values imply that L_0, with probability 0.9973, assumes values in the following interval (0.75–1 p.u.) of amplitude equal to $6\sigma_{L_0}$ around the mean value μ_{L_0}.

Let us also assume that, in the log-linear model $X = a - bL$, the following values of the regression coefficients have been computed: $a = 1.7242$, $b = 4.1774$.

Consequently, the mean and SD of X, i.e. the parameters of the LN pdf of the CCT, are equal to $\mu_{X_0} = -1.9310$; and $\sigma_{X_0} = 0.1741$. The mean value corresponds to a CCT of about 0.145 s, which was used for the numerical examples illustrated above.

The numerical results, obtained by means of stochastic simulation of the above sequences, will be expressed in relation to the values of the "primary" SD values (S, W) of the load model and the initial load variance, V_0, i.e. the variance of L_0, a value which is chosen by the analyst in a Bayes methodology, on the basis of her/his prior information.

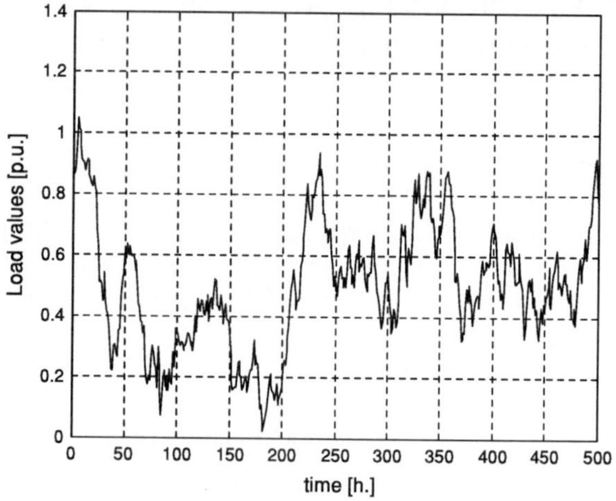

Fig. 1 Example of load pattern $\{L_k\}$, generated by a DLM (65) with parameters: $W = 0.05$, $S = 0.10$, and initial variance $V_0 = 0$

More specifically, the values (0.05 and 0.1) will be used for S and W, and these values will be swapped with each other in the course of the application to obtain at least some basic information on the sensitivity of the results in relation to variations of the model parameters. The initial variance, V_0, of the model (i.e. the variance of X_0, a value which is subjectively chosen by the analyst in a Bayes methodology) will alternatively also assume 2 values, 0 or 1, corresponding to different degrees of belief in the prior information (very strong in the first case, slight in the second one). An example of a possible sample path of the load sequence with these parameters (and $V_0 = 0$), simulated by means of the "*normrnd*" function of MATLAB®, is illustrated in Fig. 1 in a time interval covering 500 h of system operation (the corresponding series of the LCCT values will be shown in Fig. 2, darker curve).

To each load sequence, generated by a DLM corresponds, as discussed above, an LCCT sequence of X_k values, also constituting a DLM, which are estimated by the recursive approach by the values $X_k°$. In Fig. 2, for a sample path of $N = 120$ time values, the sequence of LCCT values and of its estimated values are shown, obtained with the same values of W, S, V_0 as in Fig. 1.

The efficiency of the estimation method is evaluated using extensive Monte Carlo simulations [31]. In particular, the model performance has been summarized for any given set of time instants (t_1, t_2,..., t_N) by the average squared error (ASE)[9]:

[9] The ASE index should not be confused with the MSE, which was defined in the previous section: the (theoretical) MSE evaluates the statistical mean square error between ζ_j and ζ_j^0 for any fixed time t_j with respect to the posterior conditional distribution. Instead, the ASE is an empirical measure (deduced from the sample) which takes into account the precision of estimation for *all* the RV ζ_j ($j = 1$,..., N) of the sequence.

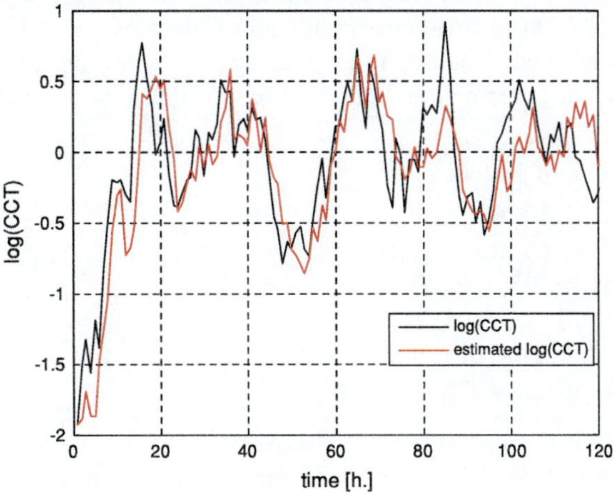

Fig. 2 Example of LCCT pattern $\{X_k\}$, and its estimated values generated by a DLM as in Sect. 1.2 in chapter titled Transmission Expansion Planning: A Methodology to Include Security Criteria and Uncertainties Using Optimization Techniques, with parameters: $W = 0.05$, $S = 0.10$, and initial variance $V_0 = 0$

$$\text{ASE} = \frac{1}{N}\sum_{j=1}^{N}\left(\zeta_j^{\circ} - \zeta_j\right)^2 \tag{68}$$

in which ζ_j is the quantity to be estimated (in this case, the LCCT X_j at any given instant t_j) and ζ_j^0 is its estimate. In practice, given the length N of a sequence (here $N = 120$ is chosen), M simulated sequences have been generated by the same algorithm and the average of the squared errors values obtained has been reported as a sample estimate of the "true" squared error. Extensive simulations were based on a number $M = 10^4$ of replications for each simulated trial; only a significant subset of the relevant results are reported in the following.

The ASE obtained using the traditional ML method has also been evaluated since the ML estimator at time k is equal—as is well known from estimation theory for the mean of a normal RV—to the sample mean of the k observed values Z_j $(j = 1,\ldots, k)$ so far. The precision (as measured by the relative bias and the maximum relative estimation error) of the dynamic Bayes estimator of the LCCT has also been verified. The basic statistics—estimated at the end of each simulation case study—which describe the efficiency of the proposed estimates, and which will be reported below—are:

- ASEB: average squared error of the bayes estimator;
- ASEL: average squared error of the ML estimator;
- ARE = ASEL/ASEB: average relative efficiency of the Bayes estimator compared with the ML estimator.

The ARE ratio, which is in a sense the dynamical counterpart of the classic "relative efficiency" of the Bayes estimator compared with the ML estimator used for a "static" parameter is indeed a synthetic measure of efficiency of the estimation method. The more the ARE value exceeds unity, the more efficient the Bayes estimate is when compared with the ML estimate.

A lot of different combinations of values of the model parameters (W, S, V_0) have been adopted to explore the estimation performances. In the following, the eight combinations indicated below will be reported for the triplet (W, S, V_0):

1. (0.05, 0.10, 0)
2. (0.05, 0.10, 1)
3. (0.10, 0.05, 0)
4. (0.10, 0.05, 1)
5. (0.025, 0.05, 0)
6. (0.025, 0.05, 1)
7. (0.05, 0.025, 0)
8. (0.05, 0.025, 1)

It is recalled that the above quantities are the SD describing uncertainty in the system load model (i.e. the equations in L_k and Y_k from which the ones for the LCCT are derived). The SD of the LCCT dynamic model, s and w, are larger than the correspondent load model parameters (W, S) values indicated here since $s = bS$ and $w = bW$, and $b = 4.1774$, as reported above.

It is seen that in the first four cases the values (0.05, 0.10) or (0.05, 0.10) are used for S and W, and every combination is obtained from the previous one by changing the value of the initial variance from $V_0 = 0$ to $V_0 = 1$, or by swapping the values of S and W. For instance, in cases 3 and 4, the values of W and S are swapped in relation to cases 1 and 2. In cases 1 and 3, $V_0 = 0$ was chosen; in cases 2 and 4, $V_0 = 1$ was chosen.

An analogous method has been used to form the combinations (5)–(8), by using the values (0.025, 0.05) for (W, S), i.e. half of the values (0.05, 0.10) used in the first four combinations. It is noticed that the combination of SD values of the first four combinations may be too high (particularly, because of the value 0.1 for S or W), especially for VST applications. Indeed some unrealistic value has been obtained in the course of the simulations for the LCCT (and, thus, for the IP). They have been reported here only to show that the estimation procedure works quite well even in these unrealistic cases, in which high SD values could imply high estimation errors.

Indeed, it is observed that—as typically occurs—the different choices do not affect the performances of the methodology.

Out of the many numerical simulations which have been performed, the most significant have been reported in the two tables of this section, Table 1 being relevant to the first four combinations, Table 2 relevant to the other four combinations.

For each case, the results of three different simulations (proofs), amongst all the ones performed, are reported. In particular: proof #1 is—for any given sample size—the one with the "worst" results (i.e. when the ARE gets the lowest observed value); proof #3 is the one with the "best" results (i.e. when the ARE gets the highest observed value); proof #2 gives the average results for the REFF, thus resulting intermediate between proof #1 and proof #3. So, a total number of 12 proofs is shown in each table. For example, in Table 1, the case 1.1

Table 1 Some results of the estimation performances with different combination of values of the (load) model parameters, with $(W, S) = (0.05, 0.10)$ or $(0.10, 0.05)$

Case	Model parameters (W, S, V_0)			ASEB	ASEL	ARE
	W	S	V_0			
1.1	0.05	0.10	0	0.0724	0.2495	3.4483
1.2	0.05	0.10	0	0.0606	0.4399	7.2570
1.3	0.05	0.10	0	0.0748	3.9853	53.269
2.1	0.05	0.10	1	0.0530	0.6227	11.761
2.2	0.05	0.10	1	0.0566	0.7962	14.065
2.3	0.05	0.10	1	0.0639	1.0124	15.840
3.1	0.10	0.05	0	0.0321	1.0379	32.344
3.2	0.10	0.05	0	0.0321	1.1004	34.317
3.3	0.10	0.05	0	0.0378	4.1745	110.44
4.1	0.10	0.05	1	0.0366	2.0879	57.070
4.2	0.10	0.05	1	0.0387	2.3414	60.542
4.3	0.10	0.05	1	0.0412	3.3678	81.825

V_0 assumes the values 0 or 1, here and in the following tables

(with ARE = 3.4485) precedes case 1.2 (with ARE = 7.2570). The three cases 1.1, 1.2 and 1.3 are all relevant to the same combination of values of (W, S and V_0) i.e. (0.05, 0.10 and 0), the first of the eight combinations above reported.

One of the results in Table 1 (the case 1.2) is also reported in Fig. 2, already mentioned.

As a general comment to the above results, it is noticed (by looking at the ASEB values) that the Bayes estimate errors are per se reasonably limited. Moreover, the relative performance in relation to the ML estimate—as measured by the ARE index—is always much greater than 1. To evaluate the precision of the estimates, other significant quantities have also been evaluated such as the average and maximum relative error of the Bayes estimates, with similar results.

It must also be remarked that, even in the "worst" cases (e.g. cases 1.1, 2.1, etc. in both tables), the ARE index is always greater than 1. Indeed, the reported results point out the efficiency of the proposed Bayesian approach, even in the case if "unrealistic" high SD as the ones in Table 1.

Finally, it has been already reported that the SD of the LCCT dynamic model, s and w, is relatively large in the application here illustrated. This could imply relatively large estimation errors whereas the reported results show that these errors are very limited, a fact which strengthens the efficiency of the estimation procedure.

For the sake of brevity, the procedure for obtaining the BCI is briefly illustrated in Appendix 3, with reference to the IP estimation.

The above good performances of the Bayes estimates with respect to the ML ones are consistent (and—to a certain extent—to be expected on theoretical grounds) with Bayesian statistical theory, as long as the Bayes estimates are evaluated assuming the "right" a priori distribution of the load, i.e. the one actually used in performing the simulation of the random samples. So, it is very

Table 2 Some results related to estimation performances with other different combination of values of the (load) model parameters with $(W, S) = (0.025, 0.05)$ or $(0.05, 0.025)$

Case	Model parameters (W, S, V_0)			ASEB	ASEL	ARE
	W	S	V_0			
1.1	0.025	0.05	0	0.0182	0.0818	4.4932
1.2	0.025	0.05	0	0.0606	0.4399	6.7520
1.3	0.025	0.05	0	0.0748	3.9853	59.619
2.1	0.025	0.05	1	0.0165	0.1736	9.280
2.2	0.025	0.05	1	0.0530	0.6227	18.167
2.3	0.025	0.05	1	0.0566	0.7962	24.506
3.1	0.05	0.025	0	0.0079	1.0096	127.52
3.2	0.05	0.025	0	0.0122	26.1034	2138.5
3.3	0.05	0.025	0	0.0134	43.40	3236.0
4.1	0.05	0.025	1	0.0090	0.1438	15.936
4.2	0.05	0.025	1	0.0102	0.4296	42.226
4.3	0.05	0.025	1	0.0089	0.9090	102.27

opportune to assess, as mentioned in the introduction, the robustness of the proposed methodology when the "a priori" hypotheses about such distribution are not valid. This is the object of the following section.

7 Some Numerical Robustness Analyses

Finally, also a simple "robustness" analysis of the proposed methodology has been performed: with respect to the initial prior distribution of the load (see Sect. 7.1), and with respect to the system load random errors prior distribution (see Sect. 7.2). For the purpose, many simulations (more than those shown here) have been carried out, assuming different (Extreme value, Log-Normal, Uniform and others) prior pdf for the initial load value L_0 or the system equation errors, instead of the Gaussian one assumed for the calculations.

7.1 A Numerical Robustness Analysis with Respect to the Initial Load Value Prior Distribution

First, some results relevant to a robustness analysis with respect to the initial load value prior distribution are reported. Since such kind of robustness is generally well established and accepted, for brevity only four cases are presented for each table, corresponding to the mean values of ASEB, ASEL, and ARE. As in the previous section, the first table is relevant to the SD values $(W, S) = (0.05, 0.10)$ or $(0.10, 0.05)$; the second table is relevant to the SD values $(W, S) = (0.025, 0.05)$ or $(0.05, 0.025)$.

Table 3 Mean values of ASEB and ASEL, and the correspondent ARE values, related to a robustness analysis with respect to a Uniform prior pdf for the initial load value. Table relevant to the SD values (W,S)=(0.05,0.10) or (0.10,0.05).

Case	Model parameters (W, S, V_0)			ASEB	ASEL	ARE
	W	S	V_0			
1	0.05	0.10	0	0.0689	1.1970	17.380
2	0.05	0.10	1	0.0683	1.0618	15.551
3	0.10	0.05	0	0.0327	2.9910	91.572
4	0.10	0.05	1	0.0382	4.2411	111.01

In tables 3 and 4, some significant results relevant to the Uniform pdf as a prior pdf for L_0 are reported, with satisfying results which were indeed expected since the well-known robustness properties of the Kalman filter. In previous section, it was assume that the starting value of the system load, L_0, measured in p.u., is a Gaussian RV with mean $\mu_{L_0} = 0.8750$ p.u. and SD $\sigma_{L_0} = 0.0417$ p.u. In the present case, the values of X_0 were generated, in each simulation trial, according to a Uniform prior pdf on an interval (a_L, b_L), with the same mean and SD, so that: $a_L = 0.8028$, $b_L = 0.9472$.

Tables 3 and 4 report some results (the mean values of ASEB and ASEL obtained in all the simulations, and the correspondent ARE value) related to a robustness analysis of the proposed methodology with respect to a Uniform prior pdf for the initial load value L_0 instead of the Gaussian one assumed for the calculations.

The calculations were performed as if the Gaussian model, which is a basic assumption of the procedure, was the "true" model generating system errors, whilst in fact the Uniform model was the true one.

The results of this robustness analysis—and more other simulation results with different prior pdf, not shown here—still confirm the adequacy of the estimation procedure.

7.2 A Numerical Robustness Analysis with Respect to System Equation Random Errors Distribution

In addition to the previous ones, a similar robustness analysis has been performed also with respect to the random errors pdf, assumed this time to be a Uniform or an EV distribution instead of a Gaussian one, with the same mean and variance (it is reminded indeed that such parameters are assumed known). Being the pdf referred to errors distribution, they all have zero mean.

By the term "*Extreme Value*" model, it is meant the "*Largest Extreme Value*" model one characterized by the following cdf:

$$F(t;\chi,\delta) = \exp\{-\exp[-(t-\chi)/\delta]\}(-\infty < t < +\infty) \tag{69}$$

with parameters: χ real, δ positive.

Table 4 Mean values of ASEB and ASEL, and the correspondent ARE values, related to a robustness analysis with respect to a Uniform prior pdf for the initial load value: (0.025, 0.05) or (0.05, 0.025)

Case	Model parameters (W, S, V_0)			ASEB	ASEL	ARE
	W	S	V_0			
1	0.025	0.05	0	0.0301	0.2962	**9.8333**
2	0.025	0.05	1	0.0158	0.9587	**60.677**
3	0.05	0.025	0	0.0082	0.1743	**21.179**
4	0.05	0.025	1	0.0068	0.8481	**124.72**

As already mentioned at the end of Sect. 2, the EV distribution is another natural candidate for the probabilistic description of the load, if interest is focussed on the peak load value [20]. It is indeed the most suitable for large time horizons, but it has already been shown (see Sect. 2.3) that a rigorous approach should be referred to peak load values, independently from the time horizons width. However, in this case, for the purpose of a robustness analysis, the EV model has been reported principally as an interesting alternative model since its pdf has a very different shape from the Gaussian one.

Also these results, shown in Tables 5 through 8—where they are reported in the same format of those in Sect. 6—confirm the estimation robustness. Figure 3 is referred to a typical case of those in Table 5.

All the above results show that the performances of the Bayes estimates are always scarcely sensitive to the assumed prior distribution or even to the model

Table 5 Some results related to a robustness analysis of the proposed methodology with respect to a Uniform error pdf for the system model of load sequence L_k instead of the Gaussian one assumed for the calculations

Uniform load case	Model parameters (W, S, V_0)			ASEB	ASEL	ARE
	W	S	V_0			
1.1	***0.05***	***0.10***	**0**	0.0697	0.3421	**4.9078**
1.2	*0.05*	*0.10*	0	0.0620	0.4757	**7.6703**
1.3	*0.05*	*0.10*	0	0.0671	0.9426	**14.038**
2.1	***0.05***	***0.10***	**1**	0.0653	0.1392	**2.1324**
2.2	*0.05*	*0.10*	1	0.0849	0.5900	**6.9462**
2.3	*0.05*	*0.10*	1	0.0664	0.7982	**12.015**
3.1	***0.10***	***0.05***	**0**	0.0386	1.2199	**31.592**
3.2	*0.10*	*0.05*	0	0.0344	3.0792	**89.486**
3.3	*0.10*	*0.05*	0	0.0406	7.2856	**179.46**
4.1	***0.10***	***0.05***	**1**	0.0312	0.6694	**21.475**
4.2	*0.10*	*0.05*	1	0.0386	1.2199	**31.592**
4.3	*0.10*	*0.05*	1	0.0406	7.2856	**179.46**

The different combination of values of the model parameters (W, S, V_0) are as in the tables of previous section

Table 6 Some results related to a robustness analysis of the proposed methodology with respect to a Uniform error pdf for the system model

Uniform load case	Model parameters (W, S, V_0)			ASEB	ASEL	ARE
	W	S	V_0			
1.1	***0.025***	***0.05***	***0***	0.0149	0.0627	**4.2113**
1.2	*0.025*	*0.05*	*0*	0.0175	0.1189	**6.7963**
1.3	*0.025*	*0.05*	*0*	0.0177	0.1952	**11.017**
2.1	***0.025***	***0.05***	***1***	0.0197	0.1218	**6.1739**
2.2	*0.025*	*0.05*	*1*	0.0202	0.2244	**11.128**
2.3	*0.025*	*0.05*	*1*	0.0137	0.5487	**40.095**
3.1	***0.05***	***0.025***	***0***	0.0089	0.4189	**47.058**
3.2	*0.05*	*0.025*	*0*	0.0091	0.8031	**88.279**
3.3	*0.05*	*0.025*	*0*	0.0114	1.1545	**100.10**
4.1	***0.05***	***0.025***	***1***	0.0109	0.5393	**49.414**
4.2	*0.05*	*0.025*	*1*	0.0106	2.4593	**232.22**
4.3	*0.05*	*0.025*	*1*	0.0094	3.4521	**366.19**

The table is similar to Table 5, with different values of the model parameters (W, S, V_0)

Table 7 Some results related to a robustness analysis of the proposed methodology with respect to an EV error pdf for the system model of load sequence L_k instead of the Gaussian one assumed for the calculations

EV load case	Model parameters (W, S, V_0)			ASEB	ASEL	ARE
	W	S	V_0			
1.1	***0.05***	***0.10***	**0**	0.0655	0.5195	**7.9359**
1.2	*0.05*	*0.10*	0	0.0963	1.0422	**10.822**
1.3	*0.05*	*0.10*	0	0.0810	2.7890	**34.220**
2.1	***0.05***	***0.10***	**1**	0.0735	0.2974	**4.0455**
2.2	*0.05*	*0.10*	1	0.0852	0.4999	**5.8661**
2.3	*0.05*	*0.10*	1	0.0594	0.3604	**6.0696**
3.1	***0.10***	***0.05***	**0**	0.0291	0.6494	**22.287**
3.2	*0.10*	*0.05*	0	0.0333	1.9626	**58.859**
3.3	*0.10*	*0.05*	0	0.0380	9.1490	**241.01**
4.1	***0.10***	***0.05***	**1**	0.0350	0.9561	**27.318**
4.2	*0.10*	*0.05*	**1**	0.0260	1.3422	**51.698**
4.3	*0.10*	*0.05*	**1**	0.0388	5.0203	**129.38**

The different combination of values of the model parameters (W, S, V_0) are as in the tables of previous section

distribution. In practice, the ASEB values are not significantly changed with respect to the previous cases, and it must be remarked that, even in the "worst" cases, the "efficiency" index, ARE, is always greater than 1. Similar results were obtained also for different model parameters values—i.e. different values of (W, S, V_0)—and for different random errors pdf. So, the Bayes estimator appears to be robust with respect to "wrong" prior assumptions.

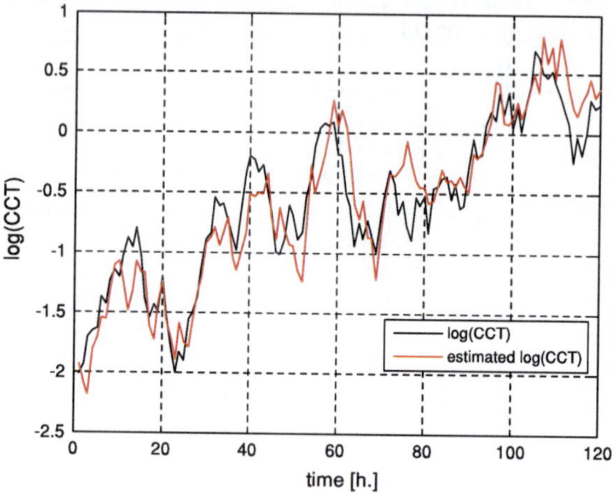

Fig. 3 Curve of LCCT and its estimates, relevant to a robustness analysis with respect to a Uniform error pdf for the system model of load sequence L_k, with parameters: $W = 0.05$, $S = 0.1$, $V_0 = 0$ (1st row of Table 5)

Table 8 Some results related to a robustness analysis of the proposed methodology with respect to an EV error pdf for the system

EV load case	Model parameters (W, S, V_0)			ASEB	ASEL	ARE
	W	*S*	V_0			
1.1	***0.025***	***0.05***	***0***	0.0199	0.1612	**8.1102**
1.2	*0.025*	*0.05*	*0*	0.0179	0.3679	**20.533**
1.3	*0.025*	*0.05*	*0*	0.0127	0.5091	**40.193**
2.1	***0.025***	***0.05***	***1***	0.0159	0.0781	**4.9187**
2.2	*0.025*	*0.05*	*1*	0.0198	0.1898	**9.6012**
2.3	*0.025*	*0.05*	*1*	0.0263	0.6546	**24.870**
3.1	***0.05***	***0.025***	***0***	0.0090	0.3848	**42.742**
3.2	*0.05*	*0.025*	*0*	0.0097	2.2534	**231.80**
3.3	*0.05*	*0.025*	*0*	0.0100	2.458	**245.70**
4.1	***0.05***	***0.025***	***1***	0.0068	0.2267	**33.245**
4.2	*0.05*	*0.025*	*1*	0.0116	0.9562	**82.116**
4.3	*0.05*	*0.025*	*1*	0.0110	2.6351	**238.72**

The table is similar to Table 5, with different values of the model parameters (W, S, V_0)

7.3 A Final Comment

The authors believe that, although deduced under simple hypotheses and/or system models, the above results—along with those already established in the field of power system or component reliability studies [32, 33]—could encourage new advanced applications of Bayesian inference in Power System analysis. Its use is indeed not yet widespread in stability or security studies, although a Bayesian classifier has been recently proposed for power system probabilistic security assessment [34]. Further developments of Bayes applications in the field of Stability surely require advanced computational tools, which are nowadays increasing in number and efficiency, as recently illustrated in [35].

Acknowledgements The authors wish to thank sincerely prof. Francesco Gagliardi, of the University of Naples Federico II, Italy, for encouraging them in undertaking the researches which constituted the foundations of this study.

8 Appendix 1: An Analytical Study of the IP

Under the assumed hypothesis of LN pdf for both the CCT and the FCT, it was shown that the IP, $q = P(\text{CCT} < \text{FCT}) = P(T_x < T_y)$, can be expressed by:

$$q = \Psi(u) = \int_u^\infty \frac{1}{\sqrt{2\pi}} \exp\left(-\frac{\xi^2}{2}\right) \mathrm{d}\xi \tag{70}$$

where u is the "SM":

$$u = \frac{\alpha_x - \alpha_y}{\sqrt{\beta_x^2 + \beta_y^2}} = \frac{E(X) - E(Y)}{\sqrt{V(X) + V(Y)}} \tag{71}$$

being $X = \ln(T_x)$; $Y = \ln(T_y)$, and denoting by $V(X)$ and $V(Y)$ their variances.

The function $q = q(u)$ is shown in Fig. 4. Since $q(u)$ decreases very quickly towards 0, especially when u is large enough, two different curves are shown: one

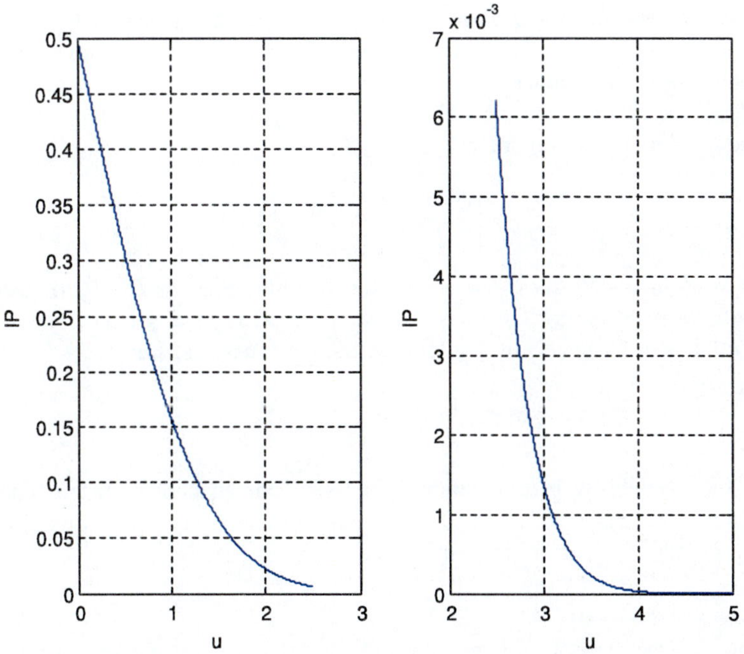

Fig. 4 A curve of the IP as a function $q = q(u)$, u being the SM

(left curve) is relevant to the interval ($0 < u < 2.5$), the other (right curve) relevant to the interval ($2.5 < u < 5$): the latter is the one which often occurs in practice since, in this interval, the IP typically assumes realistic small values, less than 6e−3. To appreciate the quickness with which $q(u)$ decreases, the following values are given as examples:

- $q(2.0) = 2.28\text{e}{-2}$;
- $q(2.5) = 6.20\text{e}{-3}$;
- $q(3.0) = 1.30\text{e}{-3}$;
- $q(5.0) = 2.85\text{e}{-7}$

In order to appreciate the variation of the IP as a function of the SM u, the following well-known asymptotic approximation of the Ψ function which may be found in many books (e.g. [36]) is given:

$$\Psi(u) \approx \frac{1}{u\sqrt{2\pi}} \exp\left(-\frac{u^2}{2}\right) = \frac{\phi(u)}{u}, \quad \text{for } u \text{ large enough} \tag{72}$$

with $\phi(u)$ being the standard Gaussian pdf. In practice, the above approximation is satisfactory for $u \geq 3$ (e.g. it yields 0.0015 for $u = 3$, with a relative error of less than 0.8%). From the above relation, it is readily shown (as discussed in the following) that the relative variation of q with u is in practice linear in u for typical values of u: therefore, the larger—as desired—the SM is, the more abrupt the variation (decrease) in the IP value. Indeed, a curve of the relative variation in the IP versus the argument u is shown in Fig. 5. It is apparent that such a function, which is of course negative (and decreasing), is approximately linear with u, especially for high u values.

Indeed, the relative variation of the function $Q(u)$ may be analysed using the derivative of its logarithm since:

$$\frac{\mathrm{d}Q}{Q} = \left[\frac{Q'(u)}{Q}\right]\mathrm{d}u = D[\ln Q(u)]\mathrm{d}u. \tag{73}$$

For what has been discussed above, the function: $K(u) = D[\ln(Qu)]$ tends to approach the value $(-u)$ if $u \to \infty$. However, $K(u)$ is readily expressed for any finite value of u by means of available statistical functions, since:

$$K(u) = D[\ln \Psi(u)] = -\frac{\phi(u)}{\Psi(u)} \tag{74}$$

So, the availability of the standard Gaussian pdf and cdf (e.g. the functions "*normpdf*" and "*normcdf*" in MATLAB) provides an easy computations of $K(u)$[10] and the possibility of drawing graphs such as the one in Fig. 5.

[10] In practice, the function $K(u)$ coincides with the "Hazard Rate function" of a standard Gaussian RV, as defined in Reliability applications (see, e.g. [37], where also the linearity of $h(t)$ is discussed).

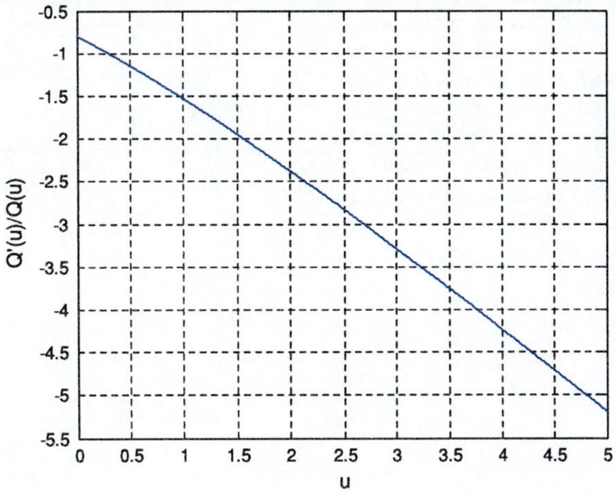

Fig. 5 A curve of the relative variation of the IP as a function of the SM

Moreover, by virtue of the above asymptotic approximation, we get the above-mentioned linear approximation of $K(u)$, which is also confirmed by Fig. 5:

$$K(u) \to -u, \quad \text{as } u \to +\infty \tag{75}$$

The above relationship between the IP and the SM u can be readily expressed (see also [13]) as a function of the basic statistical parameters of the CCT—or the ones of the load on which the CCT depends—and the clearing time.

This allows a rapid sensitivity analysis of the IP for these parameters. For instance, using the already relations between the LN parameters and the mean value and the CV of the LN distributions given above, the dependence of q on μ and v (the CV value) of both the FCT and CCT is straightforward. The following curves are obtained assuming, for illustrative purposes only, a common value v of the CV, i.e. from the expression:

$$q = \Psi(u); \quad u = \frac{\ln \mu_x - \ln \mu_y}{\sqrt{2\ln(1+v^2)}} \tag{76}$$

The curves in Fig. 6 describe the variation of q (in %) as a function of the mean FCT, for a fixed value of the mean CCT, chosen equal to 0.1 s. as in the numerical examples of the chapter and with 2 different values of the (common) value of the CV, i.e. $v = 0.10$ and $v = 0.12$.

These curves illustrate the high or extreme variability of the IP versus the mean FCT and also the CV. This last aspect is confirmed by the curve depicted in Fig. 7 which expresses the IP—on a logarithm scale—versus the CV, assuming mean CCT value $= 0.1$ s and mean FCT value $= 0.145$ s. All the above aspects are very important in view of the estimation process, and this is why they have been illustrated in detail.

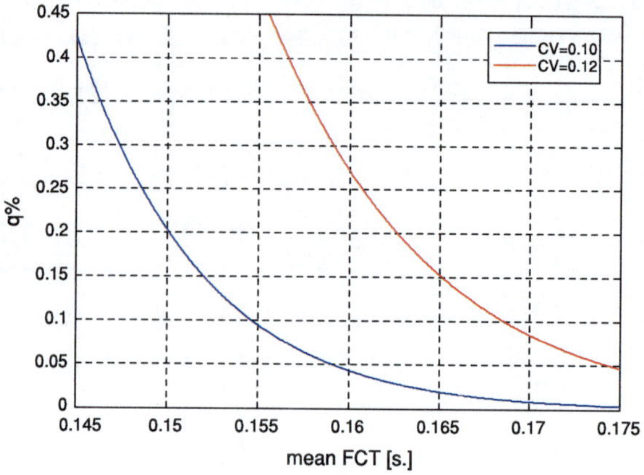

Fig. 6 Curves of the IP (in %) versus mean FCT (in s) with mean CCT = 0.1 s. Each curve refers to a given (common) value of the CV for FCT and CCT, namely CV = 0.10 (below) and 0.12 (above)

9 Appendix 2: Bayes Point Estimation for the Gaussian Model

9.1 Known Variance

For the purposes of making inference in the application in this chapter, some known results [22–24] on Bayes point estimation for the Gaussian model have been applied. Indeed, the Log-Normal model assumed for both FCT and CCT can be easily converted in the Gaussian one by means of a logarithmic transformation. Some results which are specific to the Log-Normal model are given, for example, in [21].

Let us assume that $\mathbf{X} = (X_1,\ldots, X_n)$ is a random sample of n elements generated by a Gaussian model with the same mean μ, and SD σ; let μ be an unknown to be estimated whereas σ is known. Therefore, for each $k = 1,\ldots, n$, the conditional pdf of X_k, for a given value of μ is a $N(\mu, \sigma)$ pdf. Formally $X_k|\mu \sim N(\mu,\sigma)$.

Let the prior information about the unknown parameter μ be described by a prior Normal distribution with known parameters (μ_0, σ_0), i.e. $\mu \sim N(\mu_0, \sigma_0)$,[11] so that the prior pdf is:

$$g(\mu) = \frac{1}{\sigma_0\sqrt{2\pi}}\exp\left[-\frac{(\mu-\mu_0)^2}{2\sigma_0^2}\right], \quad \mu \in \Re \tag{77}$$

The prior parameters (μ_0, σ_0) are also denoted "hyper-parameters" and are assumed to be known. Before observing data, the "best" estimator of μ cannot be that its prior mean: $E[\mu] = \mu_0$, with prior variance:

$$\mathrm{Var}[\mu] = \sigma_0^2 + \frac{\sigma^2}{n} \tag{78}$$

[11] The suffix "0" is typically used to denote prior parameters, e.g. (μ_0, σ_0) in this case.

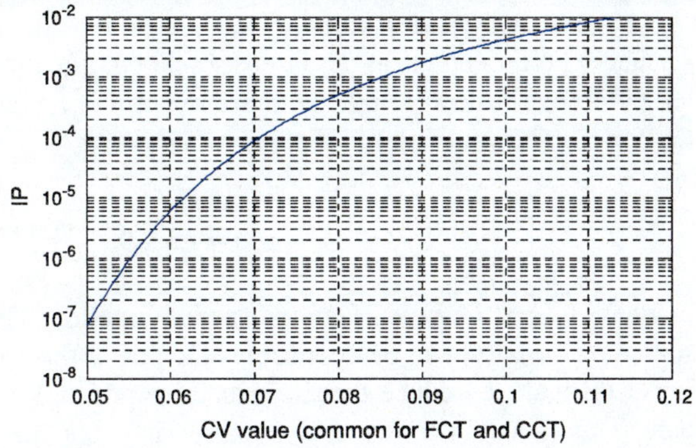

Fig. 7 Curves of the IP (in %)—on a logarithm scale—versus the (supposed common) value of the CV of FCT and CCT, assuming a mean CCT value of 0.1 s and a mean FCT value of 0.145 s

The LF of the observed sample $\mathbf{X} = (X_1, \ldots, X_n)$, conditional to μ, is expressed by:

$$L(\underline{X}|\mu) = \left(2\pi\sigma^2\right)^{\frac{-n}{2}} e^{-\sum (x_i-\mu)^2/2\sigma^2} \tag{79}$$

Then, multiplying the above two functions for applying the Bayes theorem, after some algebra, the following well-known result is obtained for $\mu°$, the Bayes estimator of μ, i.e. the posterior mean:

$$\mu° = E[\mu|X] = \frac{\sigma^2\mu_0 + n\sigma_0^2 M_n}{\sigma^2 + n\mu_0^2} = \frac{\frac{\sigma^2}{n}\mu_0 + \sigma_0^2 M_n}{\frac{\sigma^2}{n} + \sigma_0^2} \tag{80}$$

where M_n is the sample mean (which is equal in this case to the classical ML estimator of μ):

$$M_n = (1/N)\sum_{k=1}^{N} X_k \tag{81}$$

In the final equation expressing $\mu°$, the prior variance (σ_0^2) and the one of Mn (σ^2/n) are clearly indicated. In this form, the above relationship shows the known property that the Bayesian estimator of μ can be, in a suggestive way, expressed as the weighted mean (a linear convex combination, in fact) of the prior estimator and the sample mean. The posterior variance is given by:

$$\mathrm{Var}[\mu|X] = \frac{\sigma_0^2\sigma^2}{n\sigma_0^2 + \sigma^2} = \frac{\frac{\sigma^2}{n}\sigma_0^2}{\frac{\sigma^2}{n} + \sigma_0^2} \tag{82}$$

9.2 Unknown Variance

Although unknown variance is not considered in the application of this chapter, it seems opportune to mention it, even if very briefly.

Let us first consider known mean, $\mu = m$. The conjugate prior pdf for the variance, here denoted by V, is the so-called "*Inverted Gamma*" model, characterized by the following pdf, with argument v (a realization, of course positive, of the RV V) and (positive) parameters r and ϕ:

$$g(v; r, \phi) = \frac{\phi^r}{v^{(r+1)}\Gamma(r)} \exp(-\phi/v), \quad v > 0 \tag{83}$$

in which $\Gamma(\cdot)$ is the Euler–Gamma special function. The "Inverted Gamma" pdf is so denoted since it can be deduced as the pdf of the reciprocal of a Gamma RV. It is not difficult to deduce that the posterior pdf of V is again an Inverted Gamma pdf. This result derives from expressing the LF of the observed sample $\mathbf{X} = (X_1,\ldots, X_n)$ as above—Eq. 79—but conditioning to $V = v$:

$$L(\underline{X}|v) = (2\pi v)^{\frac{-n}{2}} e^{-\sum (x_i - m)^2/2v} \tag{84}$$

(here, the mean m is assumed to be a known constant whereas the variance v is the argument under investigation). By multiplying the above prior pdf and LF, it is apparent that the posterior pdf of V is again an Inverted Gamma pdf and the updated values of r and ϕ are obvious.

Then, let us also consider the general case of unknown mean μ, i.e. both mean μ and variance V are unknown. Here, the most adopted prior model for μ is again described—conditionally to the variance V—by a Gaussian prior pdf. This prior model, multiplied by the above Inverted Gamma pdf for V, constitutes the so-called "Normal Inverted Gamma" prior pdf. This is indeed the conjugate prior model, as the joint posterior pdf of μ and V is again a Normal Inverted Gamma pdf [23, 24]. A similar model can also be developed in the dynamic framework [29].

10 Appendix 3: A BCI for the IP Using the Beta Distribution

In order to establish a BCI, a numerical procedure derived from a similar one, proposed in [27] and already proved satisfactory by the authors in [16], is illustrated. In [16], it was used in a different context (the one of classical statistic estimation) whereas here it is revised in the Bayesian framework. As discussed above, in the Bayesian approach the IP Q is an RV in (0, 1), depending on the four random parameters $(\alpha_x, \alpha_y, \beta_x, \beta_y)$. The need for a numerical procedure is based on the fact that an analytical expression of such a pdf is impossible to find. A reasonable choice for its characterization is the approximation of its true pdf with a suitable distribution such as the Beta which is very flexible for describing RV in (0, 1) and is capable of producing a large variety of shapes. The Beta is in fact the most commonly used distribution for describing random probabilities because it is also a conjugate pdf under a Binomial sampling [22–24, 36]. The analytical expression of the Beta pdf, as a function of the values q assumed by the RV Q in (0, 1), is [19, 36, 37]:

$$f(q;\omega,\xi) = \begin{cases} \Gamma(\omega+\xi)\frac{q^{\omega-1}(1-q)^{\xi-1}}{\Gamma(\omega)\Gamma(\xi)} & (0<q<1) \\ 0 & \text{elsewhere} \end{cases} \tag{85}$$

where $\Gamma(\cdot)$ is the already introduced Gamma special function, and ω and ξ are positive shape parameters. Mean value and variance of the Beta distribution are given by:

$$\mu_{\mathrm{B}} = \frac{\omega}{\omega+\xi}; \quad \sigma_{\mathrm{B}}^2 = \mu_{\mathrm{B}}^2\left\{\frac{\xi}{\omega(\omega+\xi+1)}\right\} \tag{86}$$

In order to choose the approximating Beta pdf for Q, an adequate choice of the two parameters (ω, ξ) must be made, for instance—as proposed in [27]—by equating the above Beta statistical parameters (μ_{B}, σ_{B}^2) to opportune (as explained in the following) values of the mean value M and variance V, thus obtaining the following equations which give the Beta parameters as functions of the mean M and variance:

$$\omega = M\frac{(M-M^2-V)}{V}; \quad \xi = \frac{\omega(1-M)}{M} \tag{87}$$

The above mean value M and variance V of the RV Q cannot be of course obtained from its (unknown) distribution, but an excellent approximation for them is obtained: the resulting approximate values are, respectively, denoted as (M', V'). They are obtained, still following [27], by considering an expansion of $Q = \Psi(U)$ expressed as a function, say $G = G(\alpha_x,\alpha_y,\beta_x,\beta_y)$, of the four variables $(\alpha_x,\alpha_y,\beta_x,\beta_y)$ in a Taylor series about the point $\Pi_0 = (A_x,A_y,B_x,B_y)$, being (A_x,A_y,B_x,B_y) the a.m. ML estimators (see Sect. 4.4) of the random parameters $(\alpha_x,\alpha_y,\beta_x,\beta_y)$. In particular, expanding Q in a Taylor series about Π_0 up to second-order terms, the following values (M', V') are obtained by the well-known "Delta method" or the "statistical differentials" method [19, 36]:

$$M' = \Phi(U) - 0.5\Phi(U)\left[\frac{A_y - A_x}{B^3}\right]\chi \tag{88}$$

$$V' = \Phi(U)^2\left[\frac{B_x^2}{nB^2} + \frac{B_y^2}{mB^2} + 0.5(A_y - A_x)^2\frac{B_x^4}{n_1B^6} + 0.5(A_y - A_x)^2\frac{B_y^4}{m_1B^6}\right] \tag{89}$$

being: $m_1 = m - 1$, $n_1 = n - 1$;

$$U = \frac{A_x - A_y}{\sqrt{B_x^2 + B_y^2}}. \tag{90}$$

$$B = \sqrt{\left(B_x^2 + B_y^2\right)}; \tag{91}$$

$$\chi = \left[\frac{B_x^2}{n} + \frac{B_y^2}{m} + \frac{1}{2}(A_y - A_x)^2 \frac{B_x^4}{n_1 B^4} - 1.5\frac{B_x^4}{n_1 B^2} + \frac{1}{2}(A_y - A_x)^2 \frac{B_y^4}{m_1 B^4} + 1.5\frac{B_y^4}{m_1 B^2}\right]$$

$\Phi(x)$ and $\phi(x)$, respectively, the already introduced standard Normal cdf and pdf.

Indeed, let us denote by F_B $(q;\ \omega,\ \xi)$ the generic Beta cdf of the RV Q, evaluated in q, with parameters $(\omega,\ \xi)$, i.e.

$$F_B(q;\omega,\xi) = P(Q<q) \tag{92}$$

This Beta distribution is used for inference on the BCI. For example, the estimated η-quantile of the above IP is given—still denoting by τ' the value of a true parameter τ estimated by this procedure—by:

$$Q' = F_B^{-1}(\eta\cdot;\omega',\xi') \tag{93}$$

i.e. by the inverse function of the above Beta cdf $F_B(x;\ \omega',\ \xi')$ evaluated in η, namely the solution, q^*, of: $\eta = F_B(q^*;\ \omega',\ \xi')$. So, any "upper confidence bound" for the IP mentioned in Sect. 4 (see Eq. 34) can be computed easily, since the Beta quantiles are largely available in most software packages (e.g. using the function "*Betainv*" of MATLAB). The above equation is equivalent indeed to the following:

$$P(Q<Q'\eta) = \eta \tag{94}$$

And thus $Q'_\eta = F_B^{-1}(\eta;\ \omega',\ \xi')$ coincides with the upper confidence bound of probability (degree of belief) η. Of course, this procedure also allows easy estimation of the whole distribution of Q and establishes any desired confidence interval for the IP, also a bilateral one.

A simple practical numerical example is given to evaluate the BCI. In this example the FCT T_y is therefore assumed to be deterministic, with value $t_y = 0.1$ s, and the LCCT mean $\alpha_x = E[\ln(T_y)]$ is assumed to follow a prior Gaussian distribution with a mean value equal to 0.145 s and an SD equal to 1% of the mean value (i.e. a CV value equal to 0.01). The values (0.10 s, 0.145 s) of the CCT and of the mean FCT, respectively, are typical values, equal to those used in the computations already performed in the VST application of the present chapter. A CV value equal to 0.01 for the mean FCT may be also a reasonable value for describing uncertainty in such kind of on-line applications.[12] The following values of parameters $(\alpha_x,\alpha_y,\beta_x,\beta_y)$ correspond to the above CCT and FCT values:

[12] It should be remarked, however, that—in the Bayesian setting here adopted—the choice of prior parameters only reflects the information of the analyst, or her/his degree of uncertainty. So, this choice—at least from a "philosophical" point of view [25]—does not need to be "reasonable", neither it must be necessarily accepted by others. An effort has been made nonetheless, here as in he whole chapter, to choose "realistic" values from a practical engineering point of view.

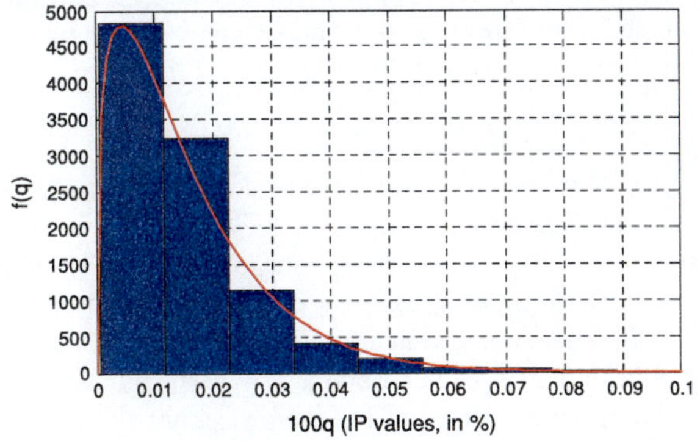

Fig. 8 Frequency histograms of the sampled IP values (in %) and the corresponding hypothetical frequency distribution obtained by the theoretical approximating Beta pdf

$\alpha_y = \ln(0.1) = -2.3026$, $\beta_y = 0$, $\beta_x = 0.0998$ whereas α_x is an RV with the above pdf.[13]

For illustrative purposes, a simulated sample of $N = 10^4$ values of $M = \alpha_x$ was generated and the corresponding empirical pdf of the IP has been evaluated and compared with the approximated theoretical Beta pdf obtained as mentioned above. The goodness of fit of this Beta pdf to the random sample has been validated through the Kolmogorov–Smirnov test of hypothesis [28]. For graphical evidence, this is also confirmed by histograms such as the one in Fig. 8 in which the frequency histograms of the sampled IP values (measured in per cent) and the corresponding hypothetical frequency distribution obtained by the a.m. Beta pdf are superimposed. Also the "Q–Q plots" [28] confirmed this adequacy, as also shown in [16].

To be more specific, the values M' and V' of the above mean and variance approximations resulted equal, for such an example, to: $M' = 0.0155\%$; $V' = (0.0155)^2$. A value of 0.0155 is therefore obtained for the SD $S' = \sqrt{V}$, with a corresponding CV equal to 0.8326, much higher than the CV = 0.1 of the basic RV M, thus confirming the already discussed high variability of the IP. This is confirmed by the values assumed by the 5th and 95th percentiles of the IP sample, i.e. 0.0032 and 0.0399, respectively, with an increase of 1147% from the former to the latter.

The Beta pdf corresponding to values of $M = M'$ and $V = V'$, which is shown in Fig. 8, has the following values of the two parameters (ω, ξ), obtained from (M_0, V') as described above: $\omega = \omega' = 1.4047$, $\xi = \xi' = 89.2197$. This computation closes the procedure of finding a BCI.

For instance, the 0.95 upper confidence bound of the above IP is given by:

$$Q_{0.95} = F_{\mathrm{B}}^{-1}(0.95; \omega', \xi') = 0.041\% \tag{95}$$

[13] In this example, in which the mean FCT is assumed to be the only RV f the problem, the ICB could be easily computed by means of the Gaussian cdf, by using known results some known results on Bayesian inference [23, 24, 32]. However, the presented example is kept simple on purpose, since it only serves to illustrate a methodology, which we have proven to be valid also in the general case (in which no analytical solution exists) as far as we know.

which is very close to the sample value above reported (0.0399), with a relative difference of less than 3%. So, under this approximation:

$$P(Q<0.041\%) = 0.95 \tag{96}$$

In other words, we can be confident, with a subjective probability equal to 0.95, that the IP is less than 0.041%. As apparent, and as already discussed in Sect. 4, this information has a greater meaning than a simple point estimate, and is consistent with the establishment of possible standards.

Of course, approximately the same values of the statistical parameters of the IP and of its BCI could be obtained by performing a Monte Carlo simulation instead of computing a Beta cdf but the procedure should be repeated—in the dynamical framework here focussed on—at each time step, which is at least tedious if not time-consuming. A much more important advantage of the Beta approximation over Monte Carlo simulation is that the former allows an analytical sensitivity analysis which is cumbersome when done by simulation.

Of course, other pdf approximations can be devised for the above purposes. In our studies, an LN approximation also seemed to be adequate for describing IP randomness. However, the Beta pdf has the advantage of being theoretically limited to the interval (0, 1). Another possible adequate model in this interval is the "Negative Log-Gamma" Distribution; it was introduced in [37] and already discussed and satisfactorily adopted by the authors in studies on uncertainty characterization in reliability analyses [38], and is worth being studied also in the present context.

11 Appendix 4: Recursive Application of Bayes Estimation

Recursive Bayes estimation is based on repeated application of the Bayes theorem which shows the coherence of the updating process and its adequacy for a "dynamical" estimation, i.e. an estimation procedure involving stochastic processes. Let us perform a statistical inference for an unknown parameter θ, characterized by a prior pdf $g(\theta)$. Once a set of data D is observed, let the posterior pdf be $g(\theta|D)$:

$$g(\theta|D) = g(\theta)L(D|\theta)/P(D) \tag{97}$$

In view of dynamical applications, we can imagine acquiring data D in a two-stage process so that D consists of two sets of data—denoted by D_1 and D_2—observed in succession. Then, by repeatedly using the Bayes theorem and the "chain rule" for joint probabilities, the above posterior pdf for θ may be alternatively obtained by expanding the previous equation in the following "two stage" process:

$$\begin{aligned} g(\theta|D_1 \cap D_2) &= g(\theta)L(D_1 \cap D_2|\theta)/P(D_1 \cap D_2) = g(\theta|D_1)L(D_2|\theta \cap D_1)/P(D_2|D_1) \\ &= g_1(\theta)L_1(D_2|\theta)/P_1(D_2) \end{aligned} \tag{98}$$

having denoted with the suffix "1" every probability (or pdf) conditional to data D_1, i.e.

$$g_1(\theta) = g(\theta|D_1); \quad L_1(D_2|\theta) = L(D_2|\theta \cap D_1); \quad P_1(D_2) = P(D_2|D_1) \tag{99}$$

As can be seen, the posterior pdf $g(\theta|D_1 \cap D_2)$ can be obtained by applying the Bayes theorem "starting" with a prior pdf $g(\theta|D_1)$, which is the posterior pdf after observation of D_1, then applying the same conditioning to the LF $L(D_2|\theta)$ and the probability of data D_2. The updating process may be indefinitely continued in this way through successive stages, transforming every posterior information gained at the end of stage k into prior information for the next stage:

$$g(\theta|D1 \cap D2 \cdots \cap D_{k+1}) = g(\theta|D_1 \cap D_2 \cdots \cap D_k)L(D_{k+1}|\theta \cap D_1 \cdots \cap D_k)/C \tag{100}$$

where C is the "constant" (with respect to θ):

$$C = L(D_{k+1}|D_1 \cdots \cap D_k) \tag{101}$$

References

1. Kundur P (1993) Power system stability and control. Electric Power System Research, Power System Engineering Series, McGraw-Hill
2. Amjady N (2004) A framework of reliability assessment with consideration effect of transient and voltage stabilities. IEEE Trans Power Syst 19(2):1005–1014
3. Billinton R, Kuruganty PRS (1979) An approximate method for probabilistic assessment of transient stability. IEEE Trans Reliab 28(3):255–258
4. Billinton R, Kuruganty PRS (1980) A probabilistic index for transient stability. IEEE Trans PAS 99(1):195–206
5. Billinton R, Kuruganty PRS (1981) Probabilistic assessment of transient stability. IEEE Trans PAS 100(5):2163–2170
6. Billinton R, Kuruganty PRS (1981) Probabilistic assessment of transient stability in a practical multimachine system. IEEE Trans PAS 100(5):3634–3641
7. Billinton R, Kuruganty PRS (1981) Protection system modelling in a probabilistic assessment of transient stability. IEEE Trans PAS 100(7):3664–3671
8. Anderson PM, Bose A (1983) A probabilistic approach to power system stability analysis. IEEE Trans PAS 102(4):2430–2439
9. Hsu YY, Chang CL (1988) Probabilistic transient stability studies using the conditional probability approach. IEEE Trans PAS 3(4):1565–1572
10. Anders GJ (1990) Probability concepts in electric power systems. Wiley, New York
11. Chiodo E, Gagliardi F, Lauria D (1994) A probabilistic approach to transient stability evaluation. IEE Proc Generat Transm Distrib 141(5):537–544
12. Chiodo E, Lauria D (1994) Transient stability evaluation of multimachine power systems: a probabilistic approach based upon the extended equal area criterion. IEE Proc Generat Transm Distrib 141(6):545–553
13. Allella F, Chiodo E, Lauria D (2003) Analytical evaluation and robustness analysis of power system transient stability probability. Electr Eng Res Rep NR16:1–13
14. Chiodo E, Gagliardi F, La Scala M, Lauria D (1999) Probabilistic on-line transient stability analysis. IEE Proc Generat Transm Distrib 146(2):176–180

15. Ayasun S, Liang Y, Nwankpa CO (2006) A sensitivity approach for computation of the probability density function of critical clearing time and probability of stability in power system transient stability analysis. Appl Math Comput 176:563–576
16. Allella F, Chiodo E, Lauria D (2003) Transient stability probability assessment and statistical estimation. Electric Power Syst Res 67(1):21–33
17. Pavella M, Murthy PG (1994) Transient stability of power systems. Theory and practice. Wiley, New York
18. Breipohl AM, Lee FN (1991) A stochastic load model for use in operating reserve evaluation. In: Proceedings of the 3rd international conference on probabilistic methods applied to electric power systems, London, 3–5 July 1991, IEE Publishing, London
19. Papoulis A (2002) Probability, random variables, stochastic processes. McGraw Hill, New York
20. Belzer DB, Kellogg MA (1993) Incorporating sources of uncertainty in forecasting peak power loads. A Monte Carlo analysis using the extreme value distribution (with discussion). IEEE Trans Power Syst 8(2):730–737
21. Crow EL, Shimizu K (1988) Lognormal distributions. Marcel Dekker, New York
22. Robert CP (2001) The Bayesian choice. Springer Verlag, Berlin
23. Press SJ (2002) Subjective and objective Bayesian statistics: principles, models, and applications, 2nd edn. Wiley, New York
24. O'Hagan A (1994) Kendall's advanced theory of statistics: vol 2b, Bayesian inference. E. Arnold Editor, London
25. De Finetti B, Galavotti MC, Hosni H, Mura A (eds) (2008) Philosophical lectures on probability. Springer Verlag, Berlin
26. Jin M (2009) Estimation of reliability based on zero-failure data. In: Proceedings of the 8th international conference on reliability, maintainability and safety, ICRMS 2009, 20–24 July 2009, pp 308–309
27. Martz HF, Hamm LL, Reed WH, Pan PY (1993) Combining mechanistic best-estimate analysis and level 1 probabilistic risk assessment. Reliab Eng Syst Saf 39:89–108
28. Rohatgi VK, Saleh AK (2000) An Introduction to Probability and Statistics, 2nd edn. Wiley, New York
29. West M, Harrison J (1999) Bayesian forecasting and dynamic models. Springer Verlag, Berlin
30. Khan UA, Moura JMF (2008) Distributing the Kalman filter for large-scale systems. Part I. IEEE Trans Signal Process 56(10):4919–4935
31. Robert CP, Casella G (2004) Monte Carlo statistical methods. Springer Verlag, Berlin
32. Erto P, Giorgio M (2002) Generalised practical Bayes estimators for the reliability and the shape parameter of the Weibull distribution. In: Proceedings of PMAPS 2002: probabilistic methods applied to power systems, Napoli, Italy
33. Chiodo E, Mazzanti G (2006) Bayesian reliability estimation based on a Weibull stress-strength model for aged power system components subjected to voltage surges. IEEE Trans Dielectric Electric Insulat 13(1):146–159
34. Kim H, Singh C (2005) Power system probabilistic security assessment using Bayes classifier. Electric Power Syst Res 74:157–165
35. Robert CP (2007) Bayesian core: a practical approach to computational bayesian statistics. Springer Verlag, Berlin
36. Johnson NL, Kotz S, Balakrishnan N (1995) Continuous univariate distributions, 2nd edn. Wiley, New York
37. Martz HF, Waller RA (1991) Bayesian reliability analysis. Krieger Publishing, Malabar, FL
38. Allella F, Chiodo E, Lauria D, Pagano M (2001) Negative log-gamma distribution for data uncertainty modelling in reliability analysis of complex systems: methodology and robustness. Int J Qual Reliab Manag 18(3):307–323

Updating System Representation by Trajectory Acquisition in a Dynamic Security Framework

Sergio Bruno and Massimo La Scala

1 Introduction

After 20 years since the first re-regulating attempts, the restructuring process of electric industry can be considered completed in most countries in the world. These years brought profound modifications in the way generating plants are managed, energy resources are exploited, and the three segments, generation, transmission and distribution, are coordinated.

In addition, the way the transmission network is operated during daily normal conditions has changed. The re-regulation of the energy sector introduced new economy-driven constraints, diminishing the control that SOs have on generation and blurring the overall vision of the system and its main characteristics.

In this fast-evolving scenario, a long sequence of large, nation-wide or even larger, blackouts forced many researchers active in the field of power system dynamic and security to answer to the question if restructuring were responsible for such events [1] and, in general, if competition and security were mutually exclusive.

Some of the reports made by the Institutions that were called to investigate on the 2003 blackouts events [2, 3] showed how national and transnational grids have been managed with lacks of data and in the presence of a large number of new uncertainties. Reliable real-time data, oriented to the monitoring of system dynamics, were not available and the operators had not enough time to take decisive and appropriate remedial actions. Appropriate automated and coordinated

S. Bruno (✉) and M. La Scala
Dipartimento di Elettrotecnica ed Elettronica (DEE), Politecnico di Bari,
via Re David 200, Bari 70125, Italy
e-mail: bruno@deemail.poliba.it

M. La Scala
e-mail: lascala@poliba.it

G. Anders and A. Vaccaro (eds.), *Innovations in Power Systems Reliability*,
Springer Series in Reliability Engineering, DOI: 10.1007/978-0-85729-088-5_9,

controls were scarce or failed the chance to take immediate remedial actions against events which trigger the fast response of the system.

Similar conclusions were also address by the IEEE PES Power System Dynamic Performance Committee [4]. What it has become clear after the blackouts of 2003 is that, despite the revolution driven by electric industry restructuring and energy market re-regulation, the general approach to power system security has not changed [4]. Regardless all technological improvements, very few new countermeasures and approaches to system security have been implemented to improve immunity from blackouts [5].

Solutions to the enhancement of power systems security include adaptive relays [5], real-time measurements and control systems [wide-area measurement systems (WAMS) and wide-area measurement and control systems (WAMC)] [5–7], FACTS (flexible AC transmission system) and HVDC (high-voltage direct current) technology [5], methodologies for automation and control [8], communication systems for real-time data exchanges [9], online dynamic security assessment (DSA) [10].

In [11], the authors have shown the potentials of exploiting Phasor measurement units (PMUs) and WAMS in monitoring and controlling power system dynamic security. In particular, the authors have shown how the emerging technology on communication systems and fast computing allows the implementation of an online environment where control center operators have the capability to monitor in real-time the power system dynamic behavior, recognize threats to its integrity, evaluate and implement suitable control actions.

A suitable approach for evaluating and implementing such control actions is given by online DSA. DSA can be defined as the evaluation of the ability of the system to withstand specified contingencies, surviving to the subsequent transients and reaching an acceptable steady state operating point [12]. The DSA analysis entails the evaluation of the ability to keep system trajectories in an acceptable state space domain and gives indication about remedial actions when necessary [13]. DSA differs from static security assessment (SSA) because the latter involves just the evaluation of a secure steady state operating point (preventive control) or a post-contingency secure steady state operating point (corrective control), usually referred as secure equilibrium point, with no regard to transient phenomena.

DSA requires the prediction (simulation) of system transient trajectories that must be contained in a domain that satisfies stability and a set of practical requirements. Relying on simulations (time–domain, energy functions, hybrid methods, etc.), the dynamic behavior of the system is predicted, power systems are planned and operated, limitations are established and the need for stability countermeasures is assessed.

Clearly, the accuracy of static and dynamic parameters (in other words the database of parameters necessary for modeling power system components) can be crucial in the assessment of system transient behavior. A major criticism about the use of simulation tools for dynamic assessment and control is based on the observation that initial databases linked to the dynamic parameters are not enough reliable, whereas there is enough confidence about steady state and topology data since they are continuously updated by state estimators.

The same objection can be brought up for all those methods that make use of system representations for assessing control actions in real-time or in extended real-time. Estimation and periodic verification of the synchronous machine parameters and control parameters are necessary for guaranteeing reliable simulations and the results in preventive and corrective dynamic control, response-based control schemes [14, 15] and WAMC.

The problem of assessing dynamic parameters is particularly relevant for synchronous machines and controls, whose characteristics and main dynamic parameters are in general not known or inaccurate, as some may drift over time or with operating condition [16]. In addition, some of the data that are adopted in dynamic simulations simply do not exist and must be estimated. Very often, to simplify and speed up the analysis, dynamic simulations are carried out adopting external equivalents that represent large portions of the network external to the area of interest. Dynamic parameters adopted for external equivalents are seldom updated and, in most of cases, are based on offline studies.

System wide measurements of power system disturbances are frequently used in event reconstruction to gain a better understanding of system behavior [17, 18]. In undertaking such studies, measurements are compared with the behavior predicted by a model. Differences are used to tune the model, i.e., adjust parameters to obtain the best match between the model and the measurements.

The main difficulty in treating power system estimation is represented by model non-linearities and discontinuities. Parameter estimation techniques are well established for linear models [19, 20], whereas parameter estimation for large nonlinear systems is a relatively open field.

There are many nonlinear components in a typical generator unit. The capability of the transfer function identification method to estimate such nonlinear generator parameters (windup/non-windup limiters and exciter saturation, etc.) is not clear. Furthermore, the derivation of the actual parameters, such as exciter gains and time constants from the transfer functions is cumbersome and needs symbolic manipulation of dynamic models. In [21], an estimation strategy based on stochastic approximation methods is proposed. This method copes with noise and nonlinear features of exciter and voltage regulator models. In [22], the authors adopt a Gauss–Newton method to compute a set of model response and trajectory sensitivities for identifying parameters that can be reliably estimated from available measurements.

In this chapter, the authors propose an approach based on a nonlinear optimization for estimating dynamic parameters. It is also assumed that WAMS is adopted to collect measurements able to update dynamic parameters every time a disturbance occurs. In the proposed procedure, time–domain simulation trajectories are compared with actual recorded data to update the parameters. The adopted nonlinear programming optimization methodology permits to minimize the quadratic function given by the difference between measured and simulated trajectories following the weighted least square (WLS) approach.

The prominent feature of the proposed method is a wide flexibility. Due to its formulation, the methodology can be implemented even in the case of missing or bad data, and can make use of any measured trajectory. The methodology can also be implemented for estimating almost any parameter that influence power system dynamic in the transient stability time framework.

2 Methodology and Architecture Structure

The proposed approach has a simple design structure. Whenever a significant perturbation is experienced, the measurement system (likely a WAMS) has to record selected trajectories of main system variables subsequent to the disturbing event. These perturbed trajectories are then sent to the control center and stored. An ordinary state estimator and a topology processor, due to signals and measurements acquired in steady state conditions, estimate system state and topology before the disturbance. Knowing the initial state and the nature of the disturbance, the control center simulates the same dynamics that has just been experienced and makes a comparison between measured and simulated data.

If this comparison reveals a significant dissimilarity, the proposed optimization algorithm can be run to minimize the difference between simulations and reality, and to find the best system representation that, at that moment, describes the system.

Please note that the time requirements for implementing this procedure are not very strict. Even though modern fast communication and computing systems guarantee technological feasibility for developing this procedure in a real-time framework, there is probably no need for it. In ordinary power system operation, the update of dynamic parameters is operated very seldom. Therefore, updating dynamic parameters on daily or hourly base would be sufficient to bring significant improvements in power system dynamic security control.

Probably, the ideal solution is to apply the procedure in the extended real-time framework. This could mean that at any programmed operation on the system (such as disconnection or closure of transmission lines, injection of significant amount of reactive power, synchronization with a group of generators), the control center operator can run the proposed procedure for updating the power system dynamic model. The main advantage is that the procedure would be carried out in correspondence with sufficiently large electromechanical oscillations, when the dominant modes are not masked by system noise. Another advantage is that, since the time schedule of a programmed operation is known, the procedure can be carried out manually and not automatically, with a great simplification in the required monitoring architecture. Of course, large oscillations due to faults can provide more in deep information, especially on wide-area phenomena (for example inter-area oscillations). In this case, an automatic detection and triggering of the estimation procedure is needed.

The procedure for estimating dynamic parameters can be schematized as in Fig. 1. The output z^{meas} represents system trajectories experienced by the power

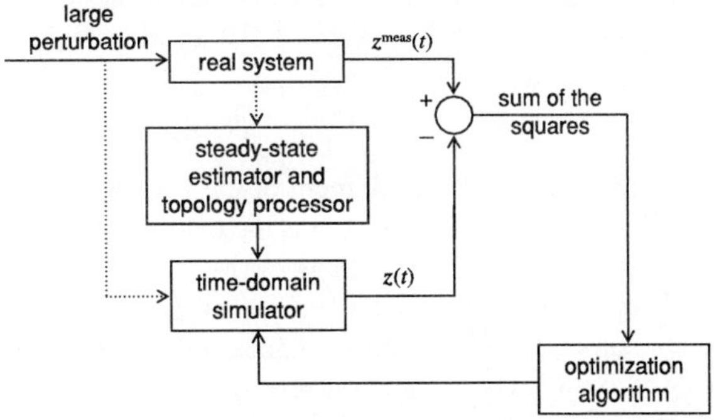

Fig. 1 Architecture of the proposed methodology

system after a large disturbance, whereas z represents the system response to the same disturbance as evaluated with a time–domain simulator. The two trajectories z and z^{meas} are the input for a recursive optimization algorithm that updates system dynamic parameters by minimizing the difference between them.

Simulated trajectories z and measured variables z^{meas} must be compared at the same time instant t to correctly detect mismatches and update the model. Consequently, it is unavoidably that synchrophasor technology has to be adopted [i.e. Phasor measurement units (PMUs)] to refer each sampled measurement at a specific time instant with sufficient accuracy. The same technology ensures that it is possible to sample the trajectory with a frequency sufficient to capture the information associated with basic electromechanical oscillation modes. Finally, WAMS technology provides a system-through vision on large area able to capture not only the local response of the system, but also the wide area one (such as inter-area oscillations).

2.1 Mathematical Formulation

The proposed approach is based on the formulation and the solution of an optimization problem that aims at obtaining the best match between measured and simulated trajectories. This optimization problem is a non-linear problem in the discrete time–domain solved by a recursive algorithm.

The mathematical formulation is based on the common power system dynamic representation. Power system electromechanical behavior on transient time-scale can be modeled through a set of nonlinear differential and algebraic equations (DAEs), represented in Eq. 1. The equations take into account the dynamic behavior of all main system components, generators, exciters, network and loads.

$$\begin{aligned} \boldsymbol{x}(t) &= f(\boldsymbol{x}(t), \boldsymbol{V}(t), \boldsymbol{p}) \\ g(\boldsymbol{x}(t), \boldsymbol{V}(t), \boldsymbol{p}) &= 0 \end{aligned} \tag{1}$$

In (1), $\boldsymbol{x}$ is the vector of state variables, whereas $\boldsymbol{V}$ is the $2n$-dimensional vector of nodal voltages, being n the total number of nodes.

$$\boldsymbol{x} = \begin{bmatrix} \boldsymbol{x}_1^{\mathrm{T}} & \boldsymbol{x}_2^{\mathrm{T}} & \cdots & \boldsymbol{x}_m^{\mathrm{T}} \end{bmatrix}^{\mathrm{T}} \tag{2}$$

$$\boldsymbol{V} = \begin{bmatrix} \boldsymbol{V}_1^{\mathrm{T}} & \boldsymbol{V}_2^{\mathrm{T}} & \cdots & \boldsymbol{V}_n^{\mathrm{T}} \end{bmatrix}^{\mathrm{T}} \tag{3}$$

In (2), the generic i^{th} element of the $\boldsymbol{x}$ vector ($\boldsymbol{x}_i$) represents the state vector for the ith synchronous machine, m is the total number of generators. The element $\boldsymbol{V}_j$ in (3) represents the nodal voltage at the jth bus.

The vector $\boldsymbol{p}$ represents the set of parameters that must be updated in the optimization algorithm. For example, the element of vector $\boldsymbol{p}$ could be represented by the basic dynamic parameters of a synchronous machine (inertia, damping, reactances, time constants, etc.). In this case $\boldsymbol{p}$ is formulated as:

$$\boldsymbol{p} = \begin{bmatrix} \boldsymbol{p}_1^{\mathrm{T}} & \boldsymbol{p}_2^{\mathrm{T}} & \cdots & \boldsymbol{p}_m^{\mathrm{T}} \end{bmatrix}^{\mathrm{T}}. \tag{4}$$

In general, the method is flexible enough to be applied to other parameters, such as load modeling constants, control device reference signals, etc.

The proposed method is based on the discretization of the DAEs set (1). The DAEs set can be discretized through any implicit rule (such as the trapezoidal rule) and written in implicit form:

$$\hat{\boldsymbol{H}}(\hat{\boldsymbol{y}}, \boldsymbol{p}) = 0 \tag{5}$$

where

$$\hat{\boldsymbol{y}} = \begin{bmatrix} \boldsymbol{y}_0^{\mathrm{T}} & \boldsymbol{y}_1^{\mathrm{T}} & \cdots & \boldsymbol{y}_i^{\mathrm{T}} & \cdots & \boldsymbol{y}_{n_T}^{\mathrm{T}} \end{bmatrix}^{\mathrm{T}} \tag{6}$$

$$\hat{\boldsymbol{H}} = \begin{bmatrix} \boldsymbol{H}_0^{\mathrm{T}} & \boldsymbol{H}_1^{\mathrm{T}} & \cdots & \boldsymbol{H}_i^{\mathrm{T}} & \cdots & \boldsymbol{H}_{n_T}^{\mathrm{T}} \end{bmatrix}^{\mathrm{T}}. \tag{7}$$

with

$$\boldsymbol{H}_i(\boldsymbol{y}_i, \boldsymbol{p}) = \boldsymbol{0} \quad i = 0, 1, 2, \ldots, n_T \tag{8}$$

$$\boldsymbol{y}_i = \begin{bmatrix} \boldsymbol{x}_i^{\mathrm{T}} & \boldsymbol{V}_i^{\mathrm{T}} \end{bmatrix}^{\mathrm{T}} \tag{9}$$

and n_{T} representing the total number of the time steps relative to the integration interval [0, T].

The optimization problem is aimed at minimizing the difference between measured and simulated trajectories by varying the dynamic parameter vector $\boldsymbol{p}$. The optimal solution is obtained minimizing an objective function that represents the mismatch between the two trajectories. In this paper, the WLS criterion has been adopted, leading to the definition of the objective function $J(z)$

$$J(z) = \sum_{i=1}^{n_T} \sum_{j=1}^{n_S} \frac{\left(z_{i,j}^{\text{meas}} - z_{i,j} \right)^2}{\sigma_{i,j}^2}. \tag{10}$$

In (10), $z_{i,j}^{\text{meas}}$ represents the jth measurement at the ith time step, $z_{i,j}$ is the jth simulated data at the ith time step, $\sigma_{i,j}^2$ is the variance at the ith time step for the jth

measurement, whereas n_S is the number of measured quantities at each time step. The total number of independent measurements is given by the product $n_S n_T$.

In the proposed formulation, trajectory mismatches are weighted by their variance. Implicitly, it has been assumed that measurements are affected by Gaussian white noise. For the sake of simplification, it was also assumed that measurements are not correlated between themselves (i.e. are statistical independent) and with respect to different time instants. Nevertheless, the approach can be easily extended to more complex formulations. Although the formulation of more complex covariance matrices is feasible, it implies the necessity of acquiring a huge amount of data to assess correlation factors. Under the current assumptions, the formulation of the proposed optimization problem assumes the characteristic of a maximum likelihood estimation.

The proposed optimization problem minimizes the objective function $J(z)$ in the presence of equality and inequality constraints. The equality constraints are given by the discretization of the differential–algebraic set of equations at each time step, as already formulated in Eq. 5. Time-varying inequality constraints were introduced to define a feasibility domain on parameters. For this reason, the parameter space Ω_P can be defined as the permissible range of all parameters to be estimated. This feature improves the convergence behavior of the overall algorithm avoiding trials too far from the final solution during iteration.

Under these assumptions, the optimization problem can be summarized as follows:

$$\min_{p \in \Omega_P} J(z) \tag{11}$$

subjected to

$$\hat{\boldsymbol{H}}(\hat{\boldsymbol{y}}, \boldsymbol{p}) = \boldsymbol{0} \tag{12}$$

$$z = \mathrm{z}(\hat{\boldsymbol{y}}) \tag{13}$$

$$k(\hat{\boldsymbol{y}}, \boldsymbol{p}) \leq 0 \tag{14}$$

The variable $\hat{\boldsymbol{y}}$, whose formulation is given in Eq. 6 represents the composition of all vectors of state variables and voltages evaluated at each time step. Therefore, $\hat{\boldsymbol{y}}$ represents the discretization of the whole system trajectory during the transient.

Equation (13) takes into consideration the dependence of simulated data vector z on the simulated trajectory $\hat{\boldsymbol{y}}$. Since $\hat{\boldsymbol{y}}$ is function of the parameter vector $\boldsymbol{p}$, z is an implicit function of $\boldsymbol{p}$. This can be easily shown considering that if the same simulation is carried out with different values in $\boldsymbol{p}$, the simulated trajectories are different and hence is the data vector z.

Equation (14) takes into consideration the presence of inequality constraints that can be referred to minimum and maximum technical requirements (for example the inertia of a synchronous machine cannot be negative or the power flow on a line should not exceed the threshold that triggers protection relays).

All inequality constraints and objective function can be taken into consideration in the formulation of the optimization problem by introducing a penalty function $C(\hat{\boldsymbol{y}},\boldsymbol{p})$. This method, known as "penalty factor method", allows to treat the whole problem as a minimization in the presence of sole equality constraints.

With the introduction of the penalty function, the optimization problem can be written in implicit form as

$$\min_{p \in \Omega_P} C(\hat{\boldsymbol{y}},\boldsymbol{p}) \tag{15}$$

subjected to

$$\hat{\boldsymbol{H}}(\hat{\boldsymbol{y}},\boldsymbol{p}) = \boldsymbol{0} \tag{16}$$

The optimization problem, as represented by Eqs. (15) and (16) is a non-linear optimization problem in the discrete domain that can be solved with the use of Lagrangian multipliers.

If λ is the Lagrangian multiplier vector, the Lagrangian function can be written as:

$$L = C(\hat{\boldsymbol{y}},\boldsymbol{p}) + \boldsymbol{\lambda}^{\mathrm{T}} \cdot \hat{\boldsymbol{H}}(\hat{\boldsymbol{y}},\boldsymbol{p}) \tag{17}$$

From (17), the set of necessary first order conditions follows:

$$\frac{\partial L}{\partial \hat{\boldsymbol{y}}} = \frac{\partial C(\hat{\boldsymbol{y}},\boldsymbol{p})}{\partial \hat{\boldsymbol{y}}} + \boldsymbol{\lambda}^{\mathrm{T}} \cdot \frac{\partial \hat{\boldsymbol{H}}(\hat{\boldsymbol{y}},\boldsymbol{p})}{\partial \hat{\boldsymbol{y}}} = 0 \tag{18}$$

$$\frac{\partial L}{\partial \boldsymbol{p}} = \frac{\partial C(\hat{\boldsymbol{y}},\boldsymbol{p})}{\partial \boldsymbol{p}} + \boldsymbol{\lambda}^{\mathrm{T}} \cdot \frac{\partial \hat{\boldsymbol{H}}(\hat{\boldsymbol{y}},\boldsymbol{p})}{\partial \boldsymbol{p}} = 0 \tag{19}$$

$$\frac{\partial L}{\partial \boldsymbol{\lambda}} = \hat{\boldsymbol{H}}(\hat{\boldsymbol{y}},\boldsymbol{p}) = 0 \tag{20}$$

The set of Eqs. (18–20) can be solved by adopting the gradient-based method through an iterative algorithm described in the next section.

2.2 *The Solving Algorithm*

The algorithm starts when a new significant complete set of measurements, suitably normalized and synchronized, is sent from the WAMS to the control center. In the proposed approach, the control center utilizes the dynamic parameters database, measurements, signals, and the topology of the network, to simulate the same trajectory that has just been acquired. This trajectory depends on the initial value given to the dynamic parameters, represented in Fig. 2 as $\boldsymbol{p}_0$. This value is the value stored in the database, and is referred to historical data or to previous runs of this same algorithm.

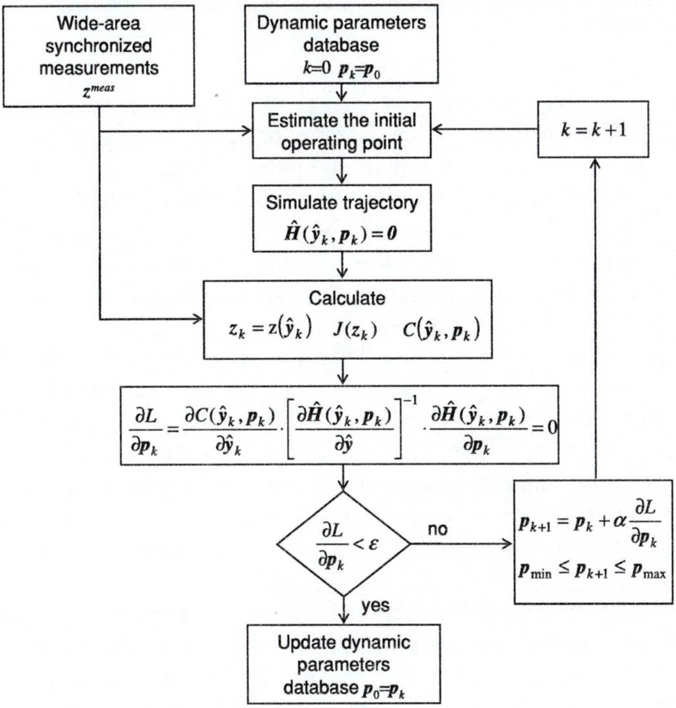

Fig. 2 Flow chart of the optimization algorithm

After this first simulation, the algorithm compares measurements with simulated trajectories and evaluates objective and penalty functions. The algorithm solves the linear set of equations (18–20) and evaluates the gradient $\partial L/\partial p_k$.

If this gradient is greater than a certain tolerance ε, the algorithm updates the $\boldsymbol{p}_k$ vector, goes back and simulates a new trajectory. The gradient $\partial L/\partial p_k$ can also be accelerated through a coefficient α which can be estimated basing on heuristics or on second-order conditions.

The algorithm is iterative and stops only when the check on tolerance is positive. At the end of the algorithm, the updated value of the $\boldsymbol{p}$ vector is stored in the database and will be used for future previsions of power system dynamics.

This approach has been widely adopted by the authors for the solution of security constrained dynamic optimization problems, exhibiting some interesting implementation properties [10, 13]. In fact, as it has been shown, the structure of the algorithm is such that ordinary time–domain simulators can be easily adapted with a plug-in module, consisting basically in the evaluation of $\partial L/\partial p_k$. This interesting feature is not common to other more sophisticated algorithms based on the second-order derivatives.

Finally, the gradient exhibits good convergence properties for this problem as it will be shown in test results.

3 Test Results

The proposed approach has been tested on a representation of the Italian national grid. The model is characterized by a degree of detail adequate for showing the feasibility of the approach in full-scale power systems. The model includes detailed information on the external systems and is characterized by about 1,333 nodes, 1,762 lines, 273 generators and 769 transformers. Non-linear loads were assumed for both active and reactive power. The parameter estimation is carried out using synchronous machines represented with a fourth order detailed model and second- and third-order excitation systems.

The set of system measurements (z^{meas}) was evaluated with a dynamic power system simulator considering an operating condition (base case) obtained from the state estimator of the Italian control center. The set of dynamic parameters adopted for the base case represents the "real" value of $\boldsymbol{p}$ that for the sake of clarity will be noted as $\boldsymbol{p}_{real}$. Tests with real measurements could be carried out only if a WAMS system were actually operated on the Italian national grid. A WAMS is necessary because it can provide a complete set of measurements, suitably synchronized and normalized.

The next step in testing the proposed approach was to build a new set of parameters. This second set represents, in the optimization algorithm, the initial value $\boldsymbol{p}_0$ (the "wrong" set of parameters stored in the dynamics database).

Knowing the initial value $\boldsymbol{p}_0$, it is possible to carry out a new simulation and calculate the first set of simulated signals z_0. With all, these inputs, it is possible to start the optimization algorithm and check if, at the final k^{th} interation, the assessed value $\boldsymbol{p}_{ass}$ is close to $\boldsymbol{p}_{real}$.

3.1 Test A

The first case was developed to show how this procedure allows in estimating model parameters (for example the model of an equivalent generator) or parameters related to power system elements that are located outside the national power system. In particular, the test was aimed at estimating the main parameters of a French generator located in Albertville, just across the Italian border.

To estimate these parameters, the objective function was built considering the mismatches between measured and simulated line flows (active and reactive) on 25 transmission lines located at the Italian border and in the mainland of the peninsula.

The algorithm was tested considering the power system behavior after a three-phase fault located in the Turin area, cleared after 0.10-s by tripping the faulted transmission line (Castelnuovo-Trino). During simulations, a feasibility domain was assumed for the parameters that had to be estimated. This domain, represented in the mathematical formulation with Eq. (14), is a very important feature of this

Table 1 Test A: Estimation of dynamic parameters for the generator in Albertville

	H (MWs)	D (p.u.)	X_d (p.u.)	Xq (p.u.)	X'_d (p.u.)	X'_q (p.u.)	T'_{d0} (s)	T'_{q0} (s)
$\boldsymbol{p}_{\text{real}}$	1,800	0	333	333	1667	1,167	500	21
$\boldsymbol{p}_0$	3,000	0	100	100	1,000	1,000	100	10
$\boldsymbol{p}_{\text{ass}}$	1,795	0	330	329	1,650	1,125	503	21

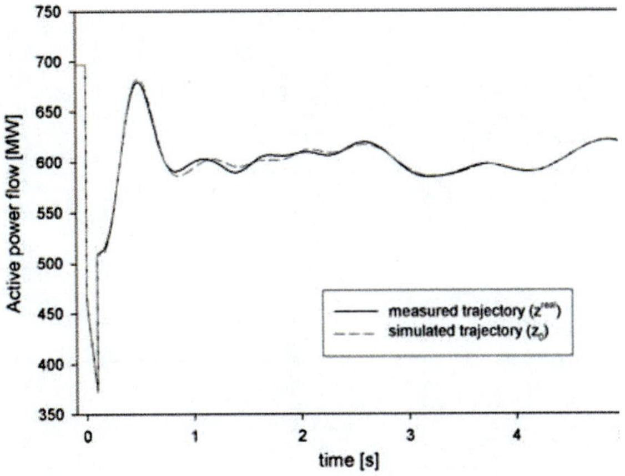

Fig. 3 Test A: measured trajectory versus simulated trajectory for the Albertville-Rondissone 400 kV transmission line (active power flow)

procedure because it allows in eliminating gross bad data and speed up the convergence of the overall algorithm.

In the followings, it is shown how the algorithm was adopted to assess all parameters of a generator located in France, right across the Italian border. Table 1 shows the value adopted for the base case ($\boldsymbol{p}_{\text{real}}$) and for the first simulation ($\boldsymbol{p}_0$), together with the assessed value $\boldsymbol{p}_{\text{ass}}$.

Note that this test can be considered a case with multiple bad data, since most of the parameters which had to be assessed were affected by very large systematic error which exceeded the usual statistics linked to the hypothesis of small drifts and random Gaussian fluctuations.

In Figs. 3 and 4, active and reactive power flow trajectories are referred to the 400 kV Albertville–Rondissone interconnecting line. In these figures, the simulated trajectories (dashed lines) refer to the first iteration of the algorithm whereas, for the sake of clarity, the optimized trajectories are not shown because they overlap the measured ones. In Fig. 4, the scale adopted is not able to show the reactive power flow trajectory during the fault. The sudden growth of reactive power exchange is due to the necessity to sustain the voltage on the nearby faulted bus.

In Table 2, the convergence behavior of the algorithm is shown. The algorithm had a satisfactory performance in terms of convergence behavior (number of iterations) and computational time. The algorithm converged in around 2 min (computational time for each iteration, considering a 5-s time window and a 0.02-s integration step, is about 20-s) when run on a Workstation XP-1000 characterized

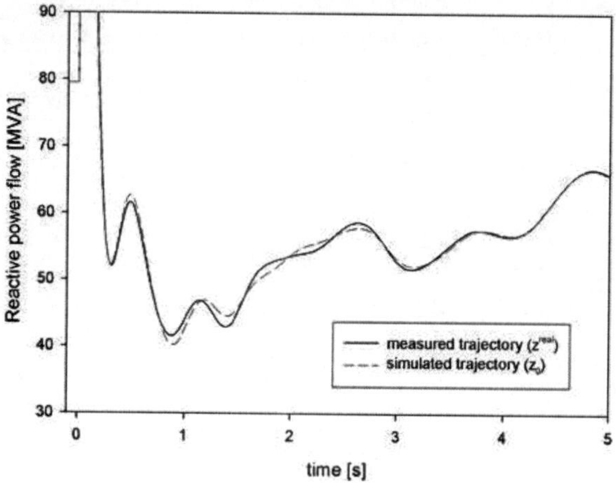

Fig. 4 Test A: measured trajectory versus simulated trajectory for the Albertville-Rondissone 400 kV transmission line (reactive power flow)

Table 2 Test A: convergence behavior

Iteration k	Objective function (p.u.)
1	208.337
2	12.441
3	0.028
4	0.049
5	0.009
6	0.001

by a CPU type 21264 ALPHA 667 MHz, 4 MB L2 cache, 512 MB RAM and 18.2 GB 10000 RPM Ultra Wide disk.

3.2 Test B

The second test has been aimed at estimating constants for voltage- and frequency-dependent load models. It has been assumed that the voltage and frequency dependence of loads, with respect to reactive and active power, can be described with the equations

$$P = P_0 \left(\frac{V}{V_0} \right)^{p_v} \left(\frac{f}{f_0} \right)^{p_\omega} \tag{21}$$

$$Q = Q_0 \left(\frac{V}{V_0} \right)^{q_v} \left(\frac{f}{f_0} \right)^{q_\omega} \tag{22}$$

where p_v, p_ω, q_v and q_ω usually depend on the nature of loads.

The simulation was carried out considering the transient subsequent to the tripping of one of the two 400-kV circuits of the transmission line Latina-Garigliano in Central Italy. The steady state condition refers to a system configuration during a Wednesday (peak-load day) at the first daily peak. The transmission line, at the moment of the tripping, was carrying about 410 MW.

The simulation was carried out considering the hypothesis that loads were wrongly modeled with a constant impedance model ($p_{v0} = 2$ and $q_{v0} = 2$), instead of the real values $p_{vreal} = 1$ and $q_{vreal} = 3$. Frequency dependence constants (p_ω and q_ω) were not considered because transients under study are too short for allowing a good assessment of frequency dependent variables.

Differently from Test A, where trajectories were assumed not affected by noise to evaluate the intrinsic accuracy of the algorithm, in this case it has been assumed, more realistically, that measurements are affected by random white noise. Considering that the accuracy reached on variables measured by a PMU can be estimated in about 0.1% [23], measurements z^{meas}, that should be coming from a WAMS, have been considered conservatively affected by white noise having variance 0.03%, with respect to the measured value. This hypothesis is also consistent with the results shown in [11].

The objective function to be minimized was built considering post-fault measured and simulated values of electric power output at 30 selected generating units widespread all over the grid.

Figure 5 shows, for a single trajectory, the difference between the real and the simulated system response. Figure 6 shows this same difference at the end of the optimization algorithm. The trajectories represented in Figs. 5 and 6 are related to a generator in Torvaldaliga (in the Rome area) close to the tripped transmission line.

Table 3 shows the convergence behavior of the algorithm. As remarked for Test A, the algorithm showed a good performance and converged in about 140-s. It can also be observed that, even the in presence of large deviations of parameters and simulated white noise, the algorithm is able to assess a correct value for the parameter under investigation.

Also, Test B can be considered a case characterized by multiple bad data since gross errors were hypothesized on the coefficients to be evaluated on load nodes.

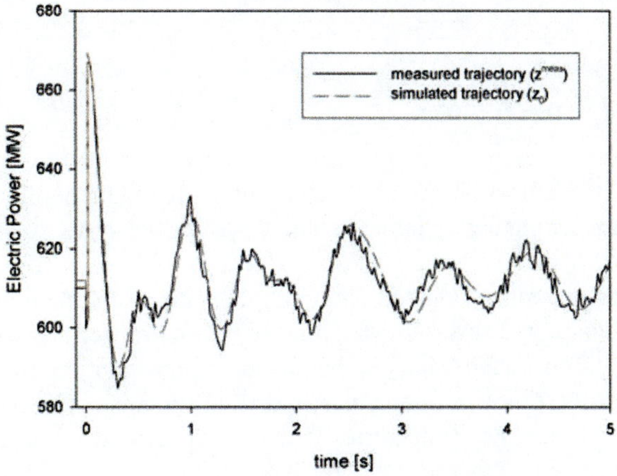

Fig. 5 Test B: measured trajectory versus simulated trajectory for the Torvaldaliga generator

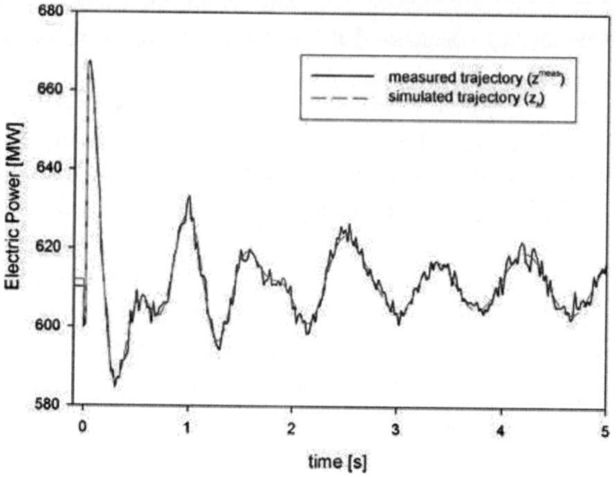

Fig. 6 Test B: measured trajectory versus optimized trajectory for the Torvaldaliga generator

Table 3 Test B: convergence behavior

Iteration k	Objective function (p.u.)	Estimated value p_{vk} (p.u.)	Estimated value q_{vk} (p.u.)
1	1,678.122	2.131	2.914
2	412.114	1.559	3.500
3	9.296	1.157	3.321
4	0.453	1.130	3.163
5	0.147	1.072	3.085
6	0.011	1.036	3.029
7	0.003	1.009	2.988

Note that this assumption is not too far from the reality since load modeling is characterized by the most uncertain parameters in dynamic simulations. Furthermore, load representation changes continuously, instant by instant, due to the nature of electric load. Assessing these parameters accurately is unavoidable for DSA calculations since they affect significantly transient simulations.

4 Conclusions

The authors proposed a methodology, based on nonlinear programming technique, for estimating dynamic parameters of power systems. Identification and updating of such parameters are crucial aspects of dynamic security assessment and control architecture. Updated data on both static and dynamic parameters guarantee more reliability in assessing preventive and control actions based on transient simulations.

The methodology is based on the formulation and the solution of an optimization problem aimed at minimizing the mismatch between online measurements and simulated trajectories. The control variable of such problem is given by the set of parameters that must be estimated.

The methodology was tested on a real representation of the Italian power system and proved to be effective and flexible enough to treat different problems such as the identification of generators external to the national power system or voltage-dependent load constants. The methodology is also flexible enough to be implemented with measurements that are commonly available at power system control centers (electric power output at generators, active and reactive power flow, bus voltage, etc.).

Even though the method does not require the implementation on an online time framework, the presence of a WAMS is unavoidable since it can provide the complete sets of synchronized and normalized measurements that are necessary to execute the proposed procedure. Therefore, the proposed architecture could be suitably implemented in a control center that makes use of a WAMS system.

References

1. Bialek JW (2004) Recent blackouts in US and Continental Europe: is liberalisation to blame? Cambridge Working Papers in Economics, No. CWPE 407. Department of Applied Economics, University of Cambridge, Cambridge
2. Union for the Coordination of Transmission of Electricity—UCTE (2004) Final Report of the Investigation Committee on the 28 September 2003 Blackout in Italy, April 2004
3. US Canada Power System Outage Task Force (2004) Final Report on the 14 August 2003 Blackout in the United States and Canada: Causes and Recommendations, 5 April 2004
4. Andersson G et al (2005) Causes of the 2003 major grid blackouts in North America and Europe, and recommended means to improve system dynamic performance. IEEE Trans Power Syst 20(4):1922–1928
5. Horowitz SH, Phadke AG (2003) Boosting immunity to blackouts. IEEE Power Energy Mag 1:47–53
6. Novosel D, Begovic MM, Madani V (2004) Shedding light on blackouts. IEEE Power Energy Mag 2:32–43
7. Taylor CW, Erickson DC, Martin KE et al (2005) WACS—wide-area stability and voltage control system: R&D and online demonstration. Proc IEEE 93(5):892–906
8. Ilić MD, Allen EH, Chapman JW et al (2005) Preventing future blackouts by means of enhanced electric power systems control: from complexity to order. Proc IEEE 93(11): 1920–1941
9. De La Ree J, Liu YL, Mili L et al (2005) Catastrophic failures in power systems: causes, analyses, and countermeasures. Proc IEEE 93(5):956–964
10. Bruno S, De Benedictis M, Delfanti M, La Scala M (2005) Preventing blackouts through reactive rescheduling under dynamical and protection system constraints. In: Proceedings of IEEE PowerTech 2005 conference, St. Petersburg, Russia, 27–30 June 2005
11. Bruno S, De Benedictis M, La Scala M (2009) Advanced monitoring and control approaches for enhancing power system security. In: Messina AR (ed) Inter-area oscillations in power systems. Springer, New York
12. David AK, Lin XJ (2002) Dynamic security enhancement in power-market systems. IEEE Trans Power Syst 17(2):431–438
13. La Scala M, Trovato M, Antonelli C (1998) On-line dynamic preventive control: an algorithm for transient security dispatch. IEEE Trans Power Syst 13(2):601–610
14. Bronzini M, Bruno S, De Benedictis M, La Scala M, Bose A (2004) Load shedding scheme for response-based control of transient stability. In: Proceedings of the IREP symposium

2004, bulk power system dynamics and control—VI, Cortina d'Ampezzo, Italy, 22–27 August 2004
15. Bruno S, De Benedictis M, La Scala M, Bose A (2004) A dynamic optimization approach for wide-area control of transient phenomena, CIGRÉ Session 2004, Rep. C2-208, Paris, France, 29 August–3 September 2004
16. Shen M, Venkatasubramanian V, Abi-Samra N et al (2000) A new framework for estimation of generator dynamic parameters. IEEE Trans Power Syst 15(2):756–763
17. Kosterev DN, Taylor CW, Mittelstandt WA (1999) Model validation for the August 10, 1996 WSCC system outage. IEEE Trans Power Syst 14(3):967–974
18. Craven RH, George T, Price GB (1994) Validation of dynamic modeling methods against power system response to small and large disturbances. In: Proceedings of CIGRÈ general session, Paris
19. Ljung L (1987) System identification. Theory for the user. Prentice-Hall, Englewood Cliffs
20. Smith JR, Hauer JF, Trudnowski DJ (1993) Transfer function identification in power system applications. IEEE Trans Power Syst 8(3):1282–1290
21. Wang JC, Chiang HD, Huang CT et al (1995) Identification of excitation system models based on-line digital measurements. IEEE Trans Power Syst 10(3):1286–1293
22. Hiskens IA (2001) Nonlinear dynamic model evaluation from disturbance measurements. IEEE Trans Power Syst 16(4):702–710
23. Depablos J, Centeno V, Phadke AG et al (2004) Comparative testing of synchronized phasor measurement units. IEEE PESGM, Denver, 6–10 June 2004

Reliable Algorithms for Power Flow Analysis in the Presence of Data Uncertainties

A. Dimitrovski, K. Tomsovic and A. Vaccaro

1 Introduction

A robust and reliable power flow analysis represents an essential requirement for many power systems applications as far as network optimization, voltage control, state estimation, and service restoration are concerned.

The most common power flow approach, referred to here as a deterministic power flow (PLF), requires precise or 'crisp' values chosen by the analyst for each input variable. The solution provides precise network voltages and flows through each line. The specified values rest upon assumptions about the operating condition derived from historical measurements or predictions about future conditions and thus, cannot be considered accurate. Even in the case where the inputs are based on measurements, inaccuracies arise from time-skew problems, three-phase

A. Dimitrovski (✉)
Oak Ridge National Laboratory, Power and Energy Systems Group, ETSD,
One Bethel Valley Road, P.O. Box 2008, Oak Ridge, TN 37831-6070, USA
e-mail: dimitrovskia@ornl.gov

K. Tomsovic
Department of Electrical Engineering and Computer Science, University of Tennessee,
414 Ferris Hall, Knoxville, TN 37996, USA
e-mail: tomsovic@tennessee.edu

A. Vaccaro
Department of Engineering, University of Sannio, Piazza Roma 21,
82100 Benevento, Italy
e-mail: vaccaro@unisannio.it

G. Anders and A. Vaccaro (eds.), *Innovations in Power Systems Reliability*,
Springer Series in Reliability Engineering, DOI: 10.1007/978-0-85729-088-5_10,

unbalance, static modeling approximations of dynamic components (e.g., transformer tap changers), variations in line parameters, and so on. The advent of deregulation and competitive power markets will only exacerbate this problem as well-known generation patterns change, loading becomes less predictable and the transmission paths grow more diverse.

Conventional methodologies proposed in literature address tolerance analysis of power flow solution by means of detailed probabilistic methods, accounting for the variability and stochastic nature of the input data, and sampling based approaches.

In particular uncertainty propagation using sampling based methods, such as the Monte Carlo, requires several model runs that sample various combinations of input values. Since the number of model runs can sometimes be very large, the required computer resources can sometimes be prohibitively expensive resulting in substantial computational demands.

As far as probabilistic methods are concerned, they represent a useful tool, especially for planning studies, but, as evidenced by the many discussions reported in literature, they could reveal some shortcomings principally arising from:

- the non-normal distribution and the statistical dependence of the input data
- the difficulty arising in accurately identifying probability distributions for some input data, such as the power generated by wind or photovoltaic generators.

All these could result in time consuming computations with several limitations in practical applications especially in power flow analysis of complex power networks.

In order to try and overcome some of these limitations, obtaining thereby comprehensive power flow solution tolerance analysis at adequate computational costs, self validated computation could play a crucial role.

Armed with such a vision, this chapter will analyze two advanced techniques for power flow analysis in the presence of data uncertainty namely the boundary power flow and the affine arithmetic power flow.

2 Problem Formulation

Power flow analysis deals with the calculation of the voltage angle and magnitude for each network bus under steady states given a set of parameters such as load demand and real power generation. Once this information is known, the network operating condition (i.e., real and reactive power flow on each branch, generator reactive power output, etc.) can be analytically determined.

The input (output) variables of the power flow problem are typically:

- the real and reactive power (voltage magnitude and phase) at each load bus (a.k.a. PQ buses);

- the real power generated and the voltage magnitude (reactive power generated and voltage phase) at each generation bus (a.k.a. PV buses);
- the voltage magnitude and phase (the real and reactive power generated) at the slack bus (a.k.a. Vδ bus);

The equations adopted to solve the power flow problem are the real power balance equations at the generation/load buses and the reactive power balance at the load buses. These equations, under some hypothesis (i.e. balanced system operation), can be written as:

$$\begin{cases} P_i = V_i \sum_{j=1}^{N} V_j Y_{ij} \cos(\delta_i - \delta_j - \gamma_{ij}) = P_i^{SP} & i \in n_P \\ Q_j = V_j \sum_{k=1}^{N} V_k Y_{jk} \sin(\delta_j - \delta_k - \gamma_{jk}) = Q_j^{SP} & j \in n_Q \end{cases} \quad (1)$$

where N is the total bus number; n_Q is the list of the buses in which the reactive power is specified; n_P is the list of the buses in which the active power is specified; P_i and Q_j are the real and reactive power injections calculated at ith and jth bus; P_i^{SP} and Q_j^{SP} are the real and reactive power injections specified at ith and jth bus; $\bar{V}_i = V_i \angle \delta_i$ is the ith bus voltage (in polar coordinates); $\bar{Y}_{ij} = Y_{ij} \angle \gamma_{ji}$ is the $[i, j]$th element of the bus admittance matrix.

Due to the non-linear nature of the power flow equations, numerical methods are employed to obtain a solution that is within an acceptable tolerance. This solution is known as "unconstrained power flow solution" since it has been obtained without taking into account the limits on the output variables (i.e., max/min values of the reactive power at generation buses, max/min voltages module at load buses, etc.).

Therefore if the obtained "unconstrained power flow solution" is not feasible with a correct power system operation, a new feasible solution satisfying the limits on the output variables (namely, a "constrained power flow solution") should be identified.

In this connection, the feasibility of the reactive power limits at the generation buses is one of the most difficult and most important issues to address [1].

To solve this problem many optimization based solution methods are proposed in the literature. The proposed methods are typically based on one of the following solution strategies:

1. *Bus-type switching* In any iteration of the numerical solution algorithm, if the reactive generation limits are violated, then the generation bus is switched to a PQ-type bus with the reactive generation set at the limiting value. In any consequent iteration if the reactive generation limits are satisfied at such a bus, then that bus is reverted back to PV-type with the original bus voltage specification.
2. *Adjusting the specified voltage at generation buses* In any iteration of the numerical solution algorithm, if the reactive generation limits are violated, then the specified voltage at the violated generation bus is adjusted in such a way that it remains as PV bus and meets the reactive power constraints.

3 Source of Uncertainty in Power Flow Analysis

Uncertainties in power flow analysis stem from several sources both internal and external to the power system [2].

Many uncertainties are induced by the complex dynamics of the active and reactive load power profiles that can vary in a fast and disordered way due to the:

- Overall economic activities and population in the analyzed area (long-term effect);
- Weather conditions (short-term effect);
- Price of electricity in relation to prices of other goods and competing energy sources (medium-term effect);
- Technological improvements of the energy end use (long-term effects).

To forecast even one of these variables over relatively medium/long periods involves large uncertainties in load profiles prediction.

A further source of uncertainty derives by the increasing number of smaller geographically dispersed generators connected to the power system. This growth is motivated by the fact that dispersed generation systems are considered today one of the most important developing areas of electric power systems in liberalized energy market, able to meet the increase of power demand and ever more pressing social and environmental constraints. The significant growth of the number of dispersed generators connected to electrical networks could considerably raise the number of power transactions and increase the complexity of controlling, protecting and maintaining existing power systems.

Besides, if the distributed generators are based on intermittent/non-programmable energy sources (i.e., wind, solar), they generate electrical power profiles that vary over time with the natural fluctuations of the sources. In details, solar insulation is subject to random coverage of clouds, which makes short-term variations of solar energy difficult to forecast. Wind speed variations may follow a generally well-known daily or seasonal pattern, but specific short-term, minute-to-minute and hourly changes are hard to predict [3]. Besides these temporal variations, wind and solar resources vary spatially, and the output from the same intermittent technologies could vary from site to site.

The difficulties arising in prediction and modeling of the electricity market operators' behavior, governed mainly by unpredictable economic dynamics, represent another relevant uncertainty source in power flow analysis.

Further uncertainties are induced by the models errors. In particular, model errors result from approximations of the equivalent model of the transmission lines and transformers, the approximations of the values of the resistance, reactances and shunt capacitances, etc. These approximations are likely sources of inaccuracies for the final network model (i.e., Y-admittance matrix) [4].

Since these uncertainties could affect the deterministic power flow solution to a considerable extent, reliable solution algorithms, incorporating the effect of data uncertainties into the power flow analysis, are therefore required [5].

4 Boundary Power Flow

4.1 BPF Formulation

The two sets of power flow equations can also be expressed in the following vector form:

$$\mathbf{Y} = \mathbf{g}(\mathbf{X}) \tag{2.1}$$

$$\mathbf{Z} = \mathbf{h}(\mathbf{X}) \tag{2.2}$$

where $\mathbf{X}$ is the vector of unknown state variables (voltage magnitudes and angles at PQ buses; and voltage angle and reactive power outputs at PV buses), $\mathbf{Y}$ is the vector of predefined input variables (real and reactive injected nodal powers at PQ buses; and voltage magnitudes and real power outputs at PV buses), $\mathbf{Z}$ is the vector of unknown output variables (real and reactive power flows in the network elements), and $\mathbf{g}$, $\mathbf{h}$ are the power flow functions.

When the uncertainty of input variables is of some "unknown-but-bounded" type, i.e., of non-statistical nature, interval numbers can be used for its modeling. This, in turn, will render all the other variables in the power flow equations as interval values as well. Furthermore, the degree of uncertainty that these intervals cover can also be varied from, say, "very likely" to "highly unlikely." As a result, the interval numbers can be nested and lumped into fuzzy numbers.

Fuzzy numbers are defined by membership functions, also known as "possibility distributions." Usually, for the sake of simplicity, trapezoidal membership functions like the one shown in Fig. 1 are assumed.

Each of the nested intervals that make the fuzzy number has an α-degree of possibility, $0 \le \alpha \le 1$. From this viewpoint, interval numbers and interval mathematics are a special case of fuzzy numbers and fuzzy mathematics. This fact is utilized when performing numerical computations with fuzzy numbers. The fuzzy numbers are broken down into several intervals over which computations are carried out and the resultant fuzzy numbers are obtained by lumping together the resultant intervals. Interval computations, in turn, consist of two or more ordinary, single point ('crisp') computations. The objective of these computations is to find

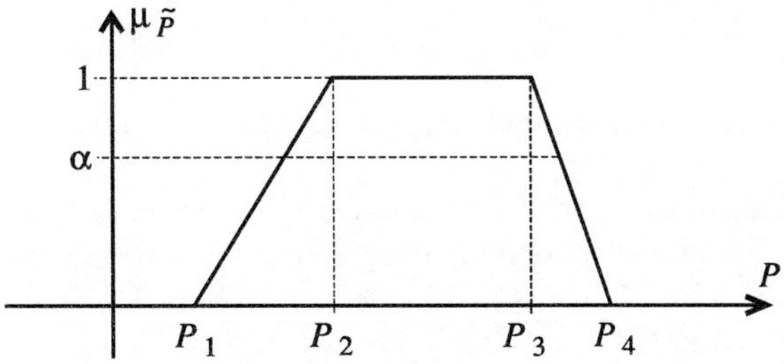

Fig. 1 Trapezoidal membership function of a fuzzy load $\tilde{P}$ expressing the possibility that load may occur between P_1 and P_4 ($\alpha = 0$), but more typically between P_2 and P_3 ($\alpha = 1$)

the boundaries of the resultant interval. Thus, one solution of an interval based power flow leads to the boundary power flow.

4.2 BPF Solution Methodology

4.2.1 Motivation

As is well known, the major power flow problem is the solution of the system of Eq. 2.1 because **X** cannot be explicitly expressed in terms of **Y** and so is instead found by an iterative process. Given a solution for **X**, the solution of 2.2 can then be found analytically.

In a deterministic power flow at iteration i, from the previous or initial trial solution, $\mathbf{X}^i$, the error is calculated as:

$$\Delta\mathbf{Y}^i = \mathbf{g}(\mathbf{X}^i) - \mathbf{Y}. \tag{3}$$

If a Newton–Raphson (N–R) iteration procedure is used, Eq. 2.1 is linearized around $\mathbf{X}^i$ and an update for the new solution is found by driving the error in Eq. 3 to zero:

$$\Delta\mathbf{X}^i = \mathbf{K}^i \cdot \Delta\mathbf{Y}^i, \tag{4}$$

where $\mathbf{K}^i = \mathbf{J}_\mathbf{g}^{-1}(\mathbf{X}^i)$ is the inverse Jacobian of $\mathbf{g}$ evaluated at $\mathbf{X}^i$. The iteration process then continues with the new point:

$$\mathbf{X}^{i+1} = \mathbf{X}^i + \Delta\mathbf{X}^i = \mathbf{X}^i - \mathbf{K}^i\Delta\mathbf{Y}^i, \tag{5}$$

and the process repeats until the convergence criterion is met or the number of iterations exceeds some predefined value.

To observe how **Z** varies with iterations, we may also linearize Eq. 2.2 and substitute from Eq. 3 to find:

$$\mathbf{Z}^{i+1} = \mathbf{Z}^i + \mathbf{S}^i\mathbf{K}^i\Delta\mathbf{Y}^i = \mathbf{Z}^i + \mathbf{L}^i\Delta\mathbf{Y}^i \tag{6}$$

where $\mathbf{S}^i = \mathbf{J}_\mathbf{h}^{-1}(\mathbf{X}^i)$ is the Jacobian of $\mathbf{h}$ evaluated at $\mathbf{X}^i$. The matrix $\mathbf{L}^i$ is a sensitivity coefficient matrix that will not be used to update **Z** at each iteration, but rather to determine the range for the outputs.

4.2.2 Linear Boundary Power Flow

Equations 5 and 6 can be used to approximately find the intervals (ranges) of values for state and output variables, given the intervals (ranges) of values of input variables. This was first implemented in the context of a probabilistic power flow [6]. The intervals of values of input variables were derived from their probability distributions. The results were then used to determine multiple points of linearization for the power flow equations in order to improve the accuracy of the probabilistic power flow solutions for the tail regions of probability distributions. This algorithm is presented briefly in the following.

Starting from a crisp point for input variables $\mathbf{Y}_0$ (the point of expected values) first find the deterministic solution for the state variables $\mathbf{X}_0$ that satisfies Eq. 2.1:

$$\mathbf{Y}_0 = \mathbf{g}(\mathbf{X}_0) \tag{7}$$

Linearizing Eq. 2.1 around the point ($\mathbf{X}_0$, $\mathbf{Y}_0$) yields:

$$\mathbf{X} = \mathbf{X}_0 + \mathbf{K} \cdot (\mathbf{Y} - \mathbf{Y}_0), \tag{8}$$

where now $\mathbf{K}$ is evaluated at $\mathbf{X}_0$. Each state variable X_i of the vector $\mathbf{X}$ is given by:

$$X_i = X_{0i} + \sum_{j=1}^{m} K_{ij} \cdot (Y_j - Y_{0j}), \tag{9}$$

where m is the dimension of $\mathbf{Y}$ (and $\mathbf{X}$) and K_{ij} are elements of the sensitivity coefficient matrix $\mathbf{K}$.

The range of values for each input variable Y_j in Eq. 9 is defined and it can be represented by an interval [$Y_j^{\min}$, $Y_j^{\max}$]. Now suppose that the minimum value of X_i associated with this linearization is desired. The minimum value of X_i can be obtained based on the sign of K_{ij}. If K_{ij} is positive, clearly, X_i will be minimum when Y_j is minimum. Likewise, if K_{ij} is negative, X_i will be minimum when Y_j is maximum. A similar reasoning applies if the maximum value of X_i is desired.

So, for a given X_i and point of linearization $\mathbf{X}_0$, there exists a certain set of boundary values for $\mathbf{Y}$ which gives the minimum (maximum) value of X_i. Let us denote this particular $\mathbf{Y}$ with $\mathbf{Yb'}_i$. By using Eq. 8, for this $\mathbf{Yb'}_i$ we can calculate the new values of $\mathbf{X}$, $\mathbf{Xb'}_i$:

$$\mathbf{Xb}'_i = \mathbf{X}_0 + \mathbf{K} \cdot (\mathbf{Yb}'_i - \mathbf{Y}_0). \tag{10}$$

The new point ($\mathbf{Xb'}_i$, $\mathbf{Yb'}_i$), however, does not satisfy Eq. 2.1. Therefore, the corresponding new value $\mathbf{Xb'}_i$ must be evaluated using Eq. 2.1:

$$\mathbf{Yb}'_i = \mathbf{g}(\mathbf{Xb}'_i). \tag{11}$$

This process can be repeated using the new point ($\mathbf{Xb'}_i$, $\mathbf{Yb'}_i$) as the second point of linearization with an updated value $\mathbf{Xb''}_i$ evaluated.

In the case of the output variables, $\mathbf{Z}$, a similar reasoning can be applied, provided a linear relationship between $\mathbf{Z}$ and $\mathbf{Y}$ has been established. Then, linearizing both Eqs. 2.1 and 2.2 around the points ($\mathbf{X}_0$, $\mathbf{Y}_0$) and ($\mathbf{X}_0$, $\mathbf{Z}_0$) gives:

$$(\mathbf{Z} - \mathbf{Z}_0) = \mathbf{S} \cdot (\mathbf{X} - \mathbf{X}_0) = \mathbf{S} \cdot \mathbf{K} \cdot (\mathbf{Y} - \mathbf{Y}_0) \tag{12}$$

and, finally,

$$\mathbf{Z} = \mathbf{Z}_0 + \mathbf{L} \cdot (\mathbf{Y} - \mathbf{Y}_0), \tag{13}$$

where $\mathbf{S}$ is the Jacobian of $\mathbf{h}$ at $\mathbf{X}_0$ and $\mathbf{L} = \mathbf{S} \cdot \mathbf{K}$.

Equation 13 has the same form as Eq. 8 and a procedure for finding the minimum (maximum) value of some Z_i, similar to that for X_i described above, can be followed.

In some cases, some or all of the coefficients K_{ij} (L_{ij}, in the case of output variables) change their signs from iteration to iteration. This phenomenon reflects the high degree of non-linearity associated with certain variables, especially in the case of voltage magnitudes and reactive power flows. It also presents convergence difficulties as values of $\mathbf{Y}$ oscillate from one boundary value to the other. An approach to overcome this problem, proposed in [6], is to set those input variables for which the sign of the coefficient oscillates to a fixed midpoint.

4.2.3 Non-linear Boundary Power Flow

It can be seen that finding the boundary values in a power flow problem is a process of locating the constrained extrema of implicitly defined vector functions of vector arguments. In our notation, we want to find the extreme values for the elements of $\mathbf{X}$ and $\mathbf{Z}$ implicitly expressed in Eqs. 2.1 and 2.2, in terms of the elements of $\mathbf{Y}$ which, in turn, are constrained.

Although the elements of $\mathbf{X}$ and $\mathbf{Z}$ cannot be explicitly expressed in terms of the elements of $\mathbf{Y}$, their partial derivatives are available; namely, the elements K_{ij} of the sensitivity coefficient matrix $\mathbf{K}$ in Eq. 9 are actually the partial derivatives of X_i with respect to Y_j. Similarly, the elements L_{ij} of the sensitivity coefficient matrix $\mathbf{L}$ in Eq. 13 are the partial derivatives of Z_i with respect to Y_j.

Similar to derivative based optimization procedures, by iteratively following the direction of the gradient, extreme points (possibly local) of the state or output variable can be found [7]. Here, as in the first approach presented above, only the signs of the partial derivatives that comprise the gradient are used. Experience has shown that the values of these partials are not useful for efficiently determining the step length. Further, procedure is needed to maintain feasibility of the solution, i.e., ensure the input variables are within constrains for all iterations.

Suppose that the *minimum* value of X_i is sought. If K_{ij} is positive (negative), then decrease (increase) the value of Y_j. After repeating for all Y_j, using the same notation as before, we obtain a new vector of input variables $\mathbf{Y}$, $\mathbf{Yb}'_i$, from which a new vector of state variables $\mathbf{X}$ from Eq. 2.1 can be found, $\mathbf{Xb}'_i$. From this new power flow solution point ($\mathbf{Xb}'_i$, $\mathbf{Yb}'_i$), the above steps are repeated until one of the following is true for *each* input variable:

- the partial derivative is positive and the associated variable is at a minimum;
- the partial derivative is negative and the associated variable is at a maximum;
- the partial derivative is zero.

If the final condition does not hold for any of the variables, then the solution is a vertex of the X_i's domain and clearly a point of constrained minimum. Because of the non-linearity of the function, this point may not be the only minimum, i.e., there may be other vertices that are also points of local constrained minima. Still, our experience has shown that the physical nature of the power flow problem dictates either a unique solution or a solution, which is dominated by a few input variables in a unique manner.

When one or more of the partial derivates are zero, the solution point lies somewhere on the boundary surface. Such a point is either a local constrained extremum (either minimum or maximum) or a saddle point. Though it seems highly unlikely that proceeding in a downhill direction one will end up trapped in a local maximum or a saddle point, theoretically such a possibility exists. Thus, additional conditions are imposed and an approach than can be characterized as local search with memory is used. Previous values of X_i shall be recorded and compared with the newly obtained one. If X_i fails to decrease, then different length steps are to be employed.

Finally, in the special case when all the partial derivatives are zero, a solution cannot be obtained due to the singularity of the Jacobian. Such a point typically indicates infeasibility of the power flow and a loading limit for the system considered. Singularity of the Jacobian may occur even if not all of the partial derivatives are zero. Such point typically indicates a point at or beyond some system flow limit, as in the nose of the P–V curve. In such cases, the ranges of values of the input variables are too great and one must repeat the calculations with reduced variations for some or all of the variables.

Based on the above discussion, a simple procedure to find the minimum value of the state variable X_i is as follows. Each input variable Y_j from $\mathbf{Y}$ is increased or decreased according to the sign of its partial derivative to the extent possible before the partial derivative changes its sign. At this point, the procedure should attempt to drive the derivative toward zero. An algorithm to achieve this is presented as pseudo-code in Fig. 2.

The algorithm drives the partial derivative toward zero by embracing the input variable within an interval, which is obtained by halving the interval from the previous step. The initial interval is the predefined range of values for the input variable, $[Y_j^{min}, Y_j^{max}]$. If at some step a variable shows the tendency to fall outside one of the boundaries of the current interval, the boundary is reset to the initial value (either Y_j^{min} or Y_j^{max}). Furthermore, once a variable is found to lie on one of the initial boundaries, it will keep its value as long the associated partial derivative does not change sign. In the case, when the maximum of X_i is sought, a simple sign change is needed to proceed. Therefore, in the algorithm presented in Fig. 2, the parameters 'up' and 'down' will change to 1 and -1, respectively. When an extreme value of the output variable Z_i is sought, the procedure is identical with X_i replaced by Z_i and K_{ij} by L_{ij}.

Remarks It should be clear that the procedure presented here, like that of [6] presented in the previous section, must be repeated for each state and output

```
up := –1;      down := 1; % directions for minimum values
for j := 1, m               % m – number of input variables
    Yl(j) := Ymin(j); Yh(i) := Ymax(j);          % set initial interval boundaries
    olddirection(j) := 0;   % reset directions
end for
repeat for Y, X, K          % new solution of the power flow
    for j := 1, m
            direction(j) := sign(K(i,j));          % new direction from inverse Jacobian
            if direction(j) = up
                    Yl(j) := Y(j);                 % set lower bound from the current value
                    if olddirection(j) = up
                            Y(j) := Yh(j)% avoid poor convergence, try boundary
                            Yh(j) := Ymax(j);      % reset upper bound
                    else
                            Y(j) := (Yl(j)+Yh(j)) / 2;% new value at mid interval
                    end if
            else if direction(j) = down
                    Yh(j) := Y(j);                 % set upper bound from the current value
                    if olddirection(j) = down
                            Y(j) := Yl(j)% avoid poor convergence, try boundary
                            Yl(j) := Ymin(j);% reset lower bound
                    else
                            Y(j) := (Yl(j)+Yh(j)) / 2;% new value at mid interval
                    end if
            end if
    end for
until abs(ΔX(i)) < tolerance
```

Fig. 2 Psuedo-code of algorithm that minimizes the state variable X_i by driving the input variables Y_j, $j = 1,\ldots,m$, towards their boundary values or values where partial derivatives are zero

variable considered. Therefore, finding boundary values involves several power flow solutions for each variable and is computationally intensive. This is the cost of a more accurate solution than that from a linearized fuzzy/interval power flow.

After finding the new point of **Y,** the new solution of the power flow is found straight from Eq. 2.1, instead of using Eq. 10 and then Eq. 11 for correction. Using Eq. 10 and 11 may save a few power flow iterations, but one still needs to calculate the inverse Jacobian to obtain the new **K**. On the other hand, when close to the boundary solution, Eq. 11 may result in some of the variables from **Y** falling outside their predefined ranges. Those variables have to be corrected and a new solution from Eq. 2.1 is needed anyway.

Generally, the above algorithm works best if the first few iterations are simplified by letting the Y_j's obtain only the boundary values from their ranges, i.e., not narrowing the initial intervals to the midpoints. In this way, the process settles down before starting to chase values that diminish the partial derivatives. The proposed algorithm may occasionally fail to find the right solution if the function exhibits extreme changes during the course of solution. Still, this can be recognized by keeping track of intermediate solutions and checking the values of partial derivatives. In such cases, a warning should accompany the obtained solution.

4.2.4 Slack Bus Treatment

An important issue that can have significant influence on the results is the treatment of the slack bus [8]. As is well known, the concept of slack bus is a mathematical necessity that has no physical relationship to any generator bus. Exception arises when a small system is linked to a much bigger system via a single tie line (single bus). In this case, one can represent the large system with an equivalent generator, which can hold the voltage constant and generate as much power as needed, i.e., the slack bus characteristics. Similarly in a distribution network fed by a substation, the transmission network acts as a slack bus with respect to the distribution network.

The slack bus allows the solution of the non-linear set of Eq. 2.1 to be feasible. Since the power losses in the network are not known in advance, its role is to pick up the 'slack' and balance the active and reactive power in the system. This usually does not represent a problem in a well defined deterministic power flow problem. However, in the case with uncertain nodal powers, the slack bus also must absorb all the resulting uncertainties from the solution. As a result, it has the widest nodal power possibility (probability) distributions in the system. This will frequently result in operating points well beyond its generating margins. This also defeats the purpose behind the study of uncertainties, which is to investigate the impact on practical operating scenarios. In the following, two ways of satisfying the constraints imposed on the slack bus are explained.

Slack Bus–PV Bus Conversion

This method is analog to that of PV bus to PQ bus conversion for PV buses with reactive power limits. During the course of solution of a power flow, when a PV bus's produced (or consumed) reactive power extends beyond its limits, it is fixed at the violated limit and its voltage magnitude is relaxed. Thus, the PV bus has been converted to a PQ bus, bus with specified active and reactive power. Later, during the solution, if the bus voltage shows tendency to return and the reactive power again falls within the limits, the bus will be converted back from PQ to PV.

Following the same approach as in PV bus to PQ bus conversion above, if the slack bus real power generation (or, theoretically, consumption) extends beyond its predefined limits, it is fixed at the violated limit. Some other PV bus's active power generation (or consumption) then must be relaxed in order to be able to solve the power flow problem. The PV bus to choose seems to be a matter of preference, but it is logical to pick the one that has the highest margin from the current production (consumption) to either its lower or upper limit, depending on which limit was violated at the slack bus.

With the choice of a PV bus to relax, it is now possible to redefine the power flow problem in Eq. 2.1 by swapping only the equation for the real power at the chosen PV bus with the equation for the slack bus real power, without changing the unknown state variables. In other words, the slack bus becomes a PVδ bus and the PV bus becomes just a V bus. We still have a system of n equations with n unknowns, only the known and unknown variables have changed and Jacobian loses some symmetry. In this case, the system of equations corresponding to Eq. 4 will have the following form:

$$\begin{bmatrix} \Delta P_{\text{Slack}} \\ \Delta \mathbf{P}_{\text{PV}-1} \\ \Delta \mathbf{P}_{\text{PQ}} \\ \Delta \mathbf{Q}_{\text{PQ}} \end{bmatrix} = - \begin{bmatrix} \frac{\partial P_{\text{Slack}}}{\partial \boldsymbol{\delta}^{\text{T}}_{\text{PV+PQ}}} & \frac{\partial P_{\text{Slack}}}{\partial \mathbf{V}^{\text{T}}_{\text{PQ}}} \\ \frac{\partial \mathbf{P}_{\text{PV}-1}}{\partial \boldsymbol{\delta}_{\text{PV+PQT}}} & \frac{\partial \mathbf{P}_{\text{PV}-1}}{\partial \mathbf{V}^{\text{T}}_{\text{PQ}}} \\ \frac{\partial \mathbf{P}_{\text{PQ}}}{\partial \boldsymbol{\delta}^{\text{T}}_{\text{PV+PQ}}} & \frac{\partial \mathbf{P}_{\text{PQ}}}{\partial \mathbf{V}^{T}_{\text{PQ}}} \\ \frac{\partial \mathbf{Q}_{\text{PQ}}}{\partial \boldsymbol{\delta}^{\text{T}}_{\text{PV+PQ}}} & \frac{\partial \mathbf{Q}_{\text{PQ}}}{\partial \mathbf{V}^{\text{T}}_{\text{PQ}}} \end{bmatrix} \cdot \begin{bmatrix} \Delta \boldsymbol{\delta}_{\text{PV+PQ}} \\ \Delta \mathbf{V}_{\text{PQ}} \end{bmatrix} \tag{14}$$

where PV is the set of all PV buses; PV $-$ 1 is the set of all PV buses without the one with relaxed real power; PQ is the set of all PQ buses, PV + PQ is the set of all PV and PQ buses; $\mathbf{P}$, $\mathbf{Q}$ are the vectors of known input variables (real and reactive nodal power vector functions), elements of $\mathbf{Y}$ in Eq. 1; $\mathbf{V}$, $\boldsymbol{\delta}$ are the vectors of unknown state variables (voltage magnitudes and angles), elements of $\mathbf{X}$ in Eq. 1, $\partial/\partial(\cdot)^{\text{T}}$ denotes Jacobian of the corresponding vector function.

The problem formulation as in Eq. 14 keeps the reference angle at the slack bus (usually 0°). Another approach will be to relax the voltage angle of the slack bus and declare the voltage angle of the PV bus with relaxed real power as the reference (i.e., known). This can simply be done by replacing it in Eq. 14 with the now unknown angle at the slack and retaining its current value. This will result in a complete slack to PV bus and PV to slack bus conversion. In this case the system of equations has the usual symmetry, with the slack bus completely swapped.

In the second approach, the original slack will change its voltage angle from the initial value during the course of solution. However, since angles are relative to each other, we can force it back to the initial value if we desired, by subtracting that difference from each voltage angle obtained from the solution. In this way, we will obtain exactly the same solution as with the previous formulation.

Distributed Slack Bus

Instead of assigning the excess load (or, generation) to only one PV bus as in the previous method, we can also choose a number of PV buses that will share it in a predetermined manner. Two methods of sharing are: (1) proportional to the current injections, and (2) proportional to the margin between the current injections and the lower or upper limits, accordingly. Of course, there are many other combinations that may be used if deemed appropriate for some particular application. In any case, there is no bus-type conversion with this method. If the slack bus production (consumption) extends beyond its limits, it is relieved by redistributing the excess load (generation) to the other PV buses. The slack bus remains the same during the power flow solution process.

It should be noted that in order to maintain the feasibility of the problem, the available generation should always match the load requirement. Cases when this is not always true are not considered here. For example, a case with excess generation (if each generator has some minimum limit and their sum is bigger than the total load) requires a different unit commitment. A case with too little generation requires a procedure for load shedding and/or some kind of adequacy assessment.

4.3 Application of the Boundary Power Flow

4.3.1 System Adequacy Indices

The results obtained from the boundary power flow allow us to integrally assess the performance of the system for the assumed uncertainties in input variables [9]. Since the boundary power flow is a static tool, we are confined to the steady state operation of the system. This, in turn, means that we consider system adequacy. The two measures directly available from the results are bus voltages and branch currents. Given the predefined operational constraints for these variables, we may express the *adequacy* or, conversely, *inadequacy* of the system components and the system as a whole to accommodate the assumed uncertainty. Other measures derived from bus voltages and branch flows can also be used following the same principles, for example, the margin to voltage collapse in connection with the continuation power flow [10].

4.3.2 Voltage Inadequacy

Given the range of values for the voltage at a particular bus and, for example, an under-voltage operational constraint, the voltage inadequacy will be equal to the part of the voltage range which is below that constraint. If we have assumed different levels of uncertainty, the voltage will be given with a possibility distribution, i.e., a fuzzy number. This will turn the inadequacy into a fuzzy number also. In this case, instead of the hard under-voltage constraint, it is possible to use a soft one if deemed more appropriate. This is illustrated in Fig. 3.

Using fuzzy arithmetic, the voltage inadequacy index of bus i can be expressed as:

$$\tilde{V}_{\mathrm{IN}i} = \max\{\tilde{V}_{\min i} - \tilde{V}_i, 0\}, \quad \text{for each } \alpha \in [0, 1] \tag{15}$$

where $\tilde{V}_{\min i}$ is the minimum acceptable voltage constraint, which can also be a fuzzy number (soft constraint), and $\tilde{V}_i$ is the fuzzy voltage magnitude at bus i.

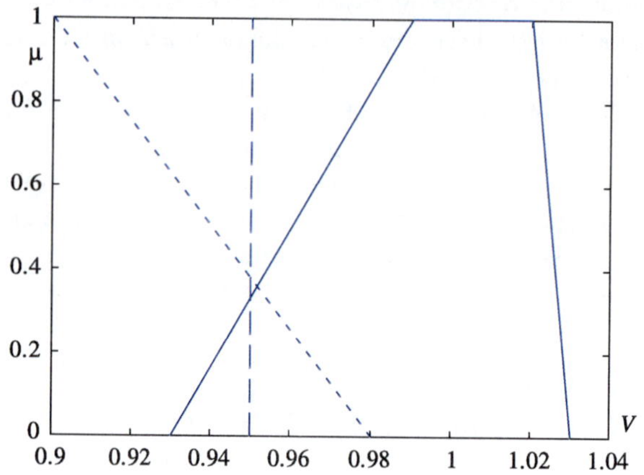

Fig. 3 Possibility distribution of a bus voltage expressed by a trapezoidal membership function (*solid line*); hard constraint for low voltage (*dashed line*); soft constraint for low voltage (*dotted line*)

Having defined the fuzzy 'low voltage' criterion, the degree of 'low voltage' can be found by applying the fuzzy *and* operator on both membership functions. This operator, while falling within the family of *t-norms*, can be defined in various ways. Here, the *min* operator appears to be adequate. The result is the intersection of the fuzzy voltage and fuzzy constraint sets.

The system voltage inadequacy index can be defined as the fuzzy sum of voltage inadequacies for all buses in the system:

$$\tilde{V}_{\mathrm{INsys}} = \sum_i \tilde{V}_{\mathrm{IN}i} \tag{16}$$

The above system index depends on the size of the system. For two systems with similar voltage conditions but different sizes (number of buses) it will be bigger for the bigger system. If this is not desirable the fuzzy sum in Eq. 16 can be replaced with a *t-conorm* operator, for example, the *max* operator.

The over-voltage inadequacy index is defined analogously to the under-voltage inadequacy index.

4.3.3 Current Inadequacy

Similarly to voltage inadequacy, current inadequacy indices for each branch in the system can be defined, given the possibility distribution for the current and the maximum loading criterion for the branch:

$$\tilde{I}_{\mathrm{IN}i} = \max\{\tilde{I}_i - \tilde{I}_{\max i}, 0\}, \quad \text{for each } \alpha \in [0, 1] \tag{17}$$

where $\tilde{I}_{\max i}$ is the maximum current constraint (in general, fuzzy number), and $\tilde{I}_i$ is the fuzzy current magnitude in branch i.

The system current inadequacy index is:

$$\tilde{I}_{\mathrm{INsys}} = \sum_i \tilde{I}_{\mathrm{IN}i} \tag{18}$$

Again, the above index can be made independent of the system size if the fuzzy sum is replaced with a *t- conorm* operator. These indices can be applied both for expansion planning and operation planning purposes. Of course, the time frames used and uncertainties considered in these two applications are quite different, as are the decisions to be made. In expansion planning context, decisions concern building and reinforcing facilities. In operations planning, decisions are about modifying operating conditions. The need for including uncertainty in the latter has only recently been recognized and may not be fully acknowledged. Nevertheless, it is expected to start gaining wider acceptance. Recently, it has been proved that in a steady state security[1] assessment the use of deterministic procedures results in significant inconsistency in terms of the risk involved [11].

4.3.4 Risk Indices

Risk can be defined as the hazard to which we are exposed because of uncertainty [12]. It is associated with some set of decisions and it has the following two dimensions:

- The likelihood of making a regrettable decision;
- The margin by which the decision is regrettable.

The decision in power system expansion planning is the particular system configuration. In that context there are two distinct measures for the two risk dimensions. *Robustness* is the likelihood of making a regrettable decision, and *exposure* is the amount by which the decision is regrettable. In operations planning context risk has been defined with a single measure which simply combines the two measures from above. If probabilities are used, that measure corresponds to the expected value obtained by multiplying the likelihoods with corresponding regrets and summing them up [11].

4.4 Simulation Study

Let us apply the boundary power flow to the IEEE 118-bus test system shown in Fig. 4. The system data and the base case descriptions can be found elsewhere (for example, http://www.ee.washington.edu/research/pstca/). Table 1 presents some of the results for the boundary values of voltage magnitudes when all specified nodal powers in the network vary in the range [90–110%] of the base case values. Shown are columns with minimal, base case, and maximal voltages. At some

[1] The term 'steady state security' is widely adopted in the industry although it actually denotes *adequacy*. *Security* deals with dynamic conditions and *adequacy* with static conditions. Both are different aspects of the overall system *reliability*.

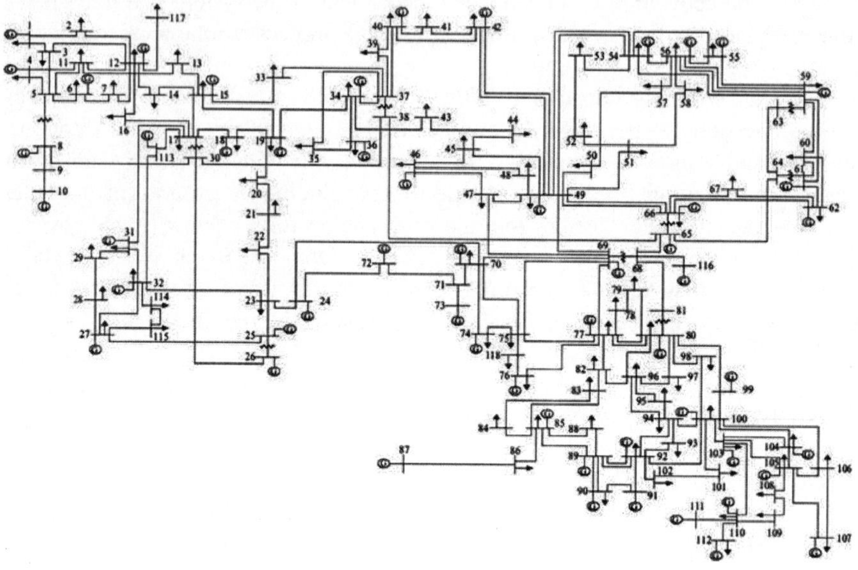

Fig. 4 IEEE 118-bus test system

Table 1 Boundary values [p.u.] for some voltages in the IEEE 118-bus system

Bus voltage	Input nodal power range of variation: [90–110%]		
	V_{min}	$V_{100\%}$	V_{max}
V_1	0.9550	0.9550	0.9550
V_2	0.9707	0.9714	0.9720
V_3	0.9670	0.9677	0.9683
V_4	0.9980	0.9980	0.9980
...	...	...	...
V_{58}	0.9562	0.9590	0.9615
V_{59}	0.9741	0.9850	0.9850
V_{60}	0.9928	0.9932	0.9934
V_{61}	0.9950	0.9950	0.9950
...	...	...	...
V_{115}	0.9594	0.9603	0.9609
V_{116}	1.0050	1.0050	1.0050
V_{117}	0.9719	0.9738	0.9758
V_{118}	0.9457	0.9494	0.9517

buses, the voltages are constant because they are of PV-type. However, that is not always the case as some of the PV buses cannot hold the voltage at the specified value due to reactive power limitations. For example, in this particular case, bus 36 is of PV-type and its minimum voltage is 0.9309 p.u. although the specified value is 0.98 p.u. This bus has the smallest voltage among all the buses in the system.

Table 2 presents some of the results for the boundary values of real and reactive power flows in the system for the same nodal power variation as before.

Again, shown are columns with minimal, base case, and maximal power flows. Note that the extreme values for real and reactive power shown do not necessarily occur simultaneously. In other words, conditions when, for example, the maximum real power in branch 63–64 occurs are not necessarily the same when the maximum reactive power in the same branch occurs. Furthermore, note that the maximum and minimum depends on the sign. For example, the minimum real and reactive power in branch 38–65 is actually the maximum power in the opposite direction and vice versa. When there is a change in sign between the maximum and minimum values, as in branch 76–118 for example, then the respective values show the maximum values in either direction. In such branches, powers reverse their flows depending on the conditions in the network defined by the input power variation. Power flows are continuous functions so their minimum values will be, obviously, zero.

Let us now assume that the nodal powers are trapezoidal fuzzy numbers with (90, 95, 105, and 110%) of the base case values. That is, the powers will fluctuate within ±10% of the base case, but most likely within ±5%. We will calculate the system voltage inadequacy for the load buses (PQ-type buses) according to Eq. 16, assuming the hard under-voltage constraint shown in Fig. 3 with dashed line. The results are shown in Fig. 5 for two different types of fuzzy aggregation, fuzzy summation, and max operator. The solid line is the result of using fuzzy summation as given in Eq. 16 while the dotted line is the result of applying the max operator. From here we can see that there exists some non-zero possibility that the

Table 2 Boundary values [p.u.] for some power flows in the IEEE 118-bus system

Power flow	Input nodal power range of variation: [90–110%]					
	S_{min}		$S_{100\%}$		S_{max}	
$S_{1\text{–}2}$	−0.1764	−0.1498j	−0.1236	−0.1304j	−0.0706	−0.1105j
$S_{1\text{–}3}$	−0.4516	−0.1982j	−0.3864	−0.1706j	−0.3212	−0.1420j
$S_{4\text{–}5}$	−1.1622	−0.2872j	−1.0321	−0.2679j	−0.9024	−0.2471j
$S_{3\text{–}5}$	−0.7701	−0.1697j	−0.6810	−0.1449j	−0.5920	−0.1191j
$S_{5\text{–}6}$	0.7559	0.0212j	0.8846	0.0411j	1.0138	0.0611j
	…		…		…	
$S_{61\text{–}62}$	0.0890	−0.1728j	0.2549	−0.1386j	0.4224	−0.1035j
$S_{63\text{–}59}$	1.0671	0.6363j	1.5175	0.6748j	1.9740	0.7075j
$S_{63\text{–}64}$	−1.9740	−0.7075j	−1.5175	−0.6748j	−1.0671	−0.6363j
$S_{64\text{–}61}$	−0.1336	0.0666j	0.3052	0.1399j	0.7507	0.1986j
$S_{38\text{–}65}$	−3.1305	−0.8305j	−1.8141	−0.5701j	−0.5064	−0.1667j
	…		…		…	
$S_{114\text{–}115}$	−0.0430	−0.0124j	0.0135	0.0061j	0.0702	0.0168j
$S_{68\text{–}116}$	1.6571	−0.9956j	1.8413	−0.6636j	2.0257	−0.6345j
$S_{12\text{–}117}$	0.1812	0.0426j	0.2015	0.0520j	0.2219	0.0615j
$S_{75\text{–}118}$	0.0811	0.1398j	0.4019	0.2359j	1.0174	0.3322j
$S_{76\text{–}118}$	−0.6546	−0.1966j	−0.0683	−0.0970j	0.2376	0.0252j

Fig. 5 System voltage inadequacies with hard voltage constraint for the IEEE 118-bus system when input nodal powers are trapezoidal FNs with (90, 95, 105, and 110%) of the base case values. *Solid line* sum operator; *dotted line* max operator

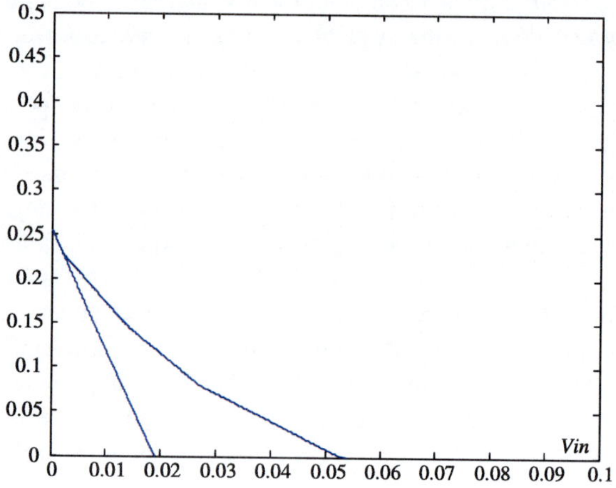

Fig. 6 System voltage inadequacies with soft voltage constraint for the IEEE 118-bus system when input nodal powers are trapezoidal fuzzy numbers with (90, 95, 105, and 110%) of the base case values. *Solid line* sum operator; *dotted line* max operator

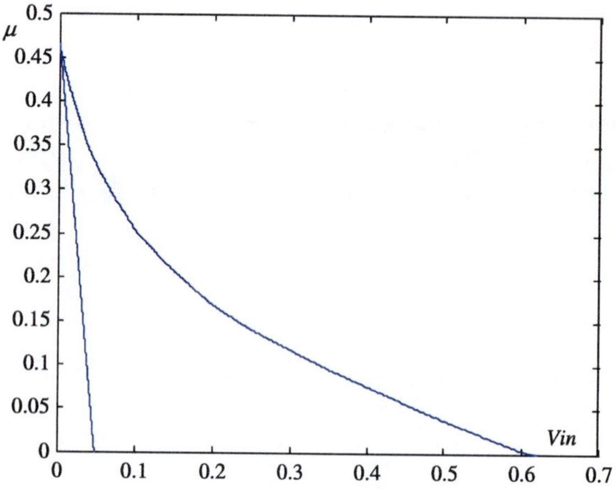

lowest voltage is approximately 0.02 p.u. (bus 21) below the hard constraint of 0.95 p.u. The total under-voltage violation with some non-zero possibility is slightly bigger than 0.05. In addition, there will be an under-voltage violation with 0.25 possibility.

The same indices are shown again in Fig. 6, but now for the soft under-voltage constraint case (shown with a dotted line in Fig. 3). Figure 7 shows results for the system current inadequacies with soft and hard over-current constraints. Because there is no information in the original data about the line capacities, we have assumed that each line is loaded at 50% in the base case. This is not a realistic assumption but is simple allows us to show properties of system inadequacies. The hard constraint was set at each element capacity. The soft constraint was represented as a line that goes from zero at each element capacity to one at 20% above the capacity.

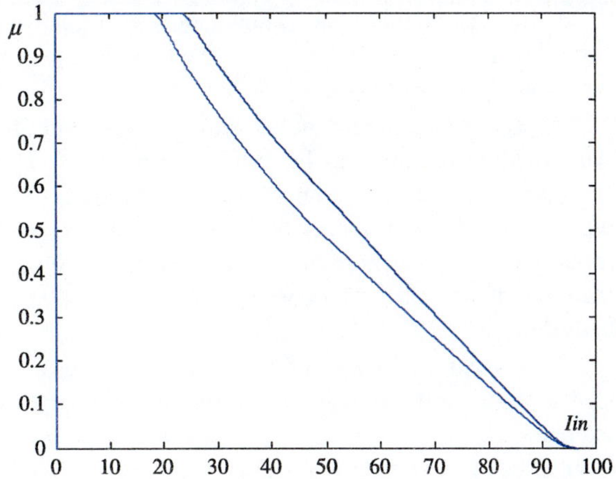

Fig. 7 System current inadequacies for the IEEE 118-bus system for when input nodal powers are trapezoidal fuzzy numbers with (90, 95, 105, and 110%) of the base case values, aggregated with the sum operator. *Solid line* hard constraint; *dotted line* soft constraint

As can be seen from Fig. 7, there is a 1.0 possibility that some elements in the system will be overloaded. This means that given the input power variation there is always a combination of loads and generation that results in overload of at least one of the elements in the system. This is a result of the assumption that we have made regarding the element capacities, namely, some of the elements in the base case scenario are lightly loaded and, consequently, we have assumed small capacities. With such capacities only a slight change in power distribution results in overloading of those elements. Because we used fuzzy summation to aggregate the results system-wide, the system over-current inadequacy is a big number.

5 Affine Arithmetic Power Flow

5.1 Elements of AA

Affine arithmetic (AA), introduced by Comba and Stolfi [13], is a method for range analysis widely appreciated for its ability to manipulate sources of error both external, such as imprecise or missing input data, uncertainty in the mathematical modeling, and internal, such as round-off and truncation errors. This model is similar to standard interval mathematics (IM) but, in addition, it keeps track of correlations between the input and computed quantities. This extra information allows providing much tighter bounds in the computing process avoiding the probability to generate the error explosion problem observed in long interval computations [14].

In AA a partially unknown quantity x is represented by an affine form $\hat{x}$ which is a first degree polynomial of the form:

$$\hat{x} = x_0 + x_1\varepsilon_1 + x_2\varepsilon_2 + \cdots + x_n\varepsilon_n \tag{19}$$

In the Eq. 19 the x_i are known real coefficients and, in particular, x_0 is the central value of the affine form $\hat{x}$, while the other coefficients x_i represent the corresponding partial deviations.

The symbol, ε_i, called noise symbol, is a symbolic variable whose values are unknown but assumed to lie in the interval $[-1, 1]$. Each ε_i stands for an independent source of uncertainty that contributes to the total uncertainty of the quantity x. The source may be external, if it is originated by uncertainty in some input quantities, or internal, if it originated by round-off and truncation errors in the computation of $\hat{x}$. The corresponding coefficient x_i gives the magnitude of that contribution.

The key feature of the AA model is that the same noise symbol may contribute to the uncertainty of two or more quantities arising in the evolution of an expression.

In order to evaluate a formula with AA, it is necessary to replace each elementary operation on real quantities by a corresponding operation on their affine forms, returning an affine form.

In particular given a general operation: $z \rightarrow f(x, y)$ the corresponding AA operation $\hat{z} \rightarrow f(\hat{x}, \hat{y})$ is a procedure that computes an affine form for $z = f(x, y)$ that is consistent with affine forms $\hat{x}, \hat{y}$.

If the operation f is an affine function of its argument x and y then the affine representation for z can be obtained by expanding and rearranging into an affine combination the noise symbols ε_i. This is the case for example for the following basic operations:

$$\hat{x} \pm \hat{y} = (x_0 \pm y_0) + (x_1 \pm y_1)\varepsilon_1 + (x_2 \pm y_2)\varepsilon_2 + \cdots + (x_n \pm y_n)\varepsilon_n \tag{20}$$

$$\alpha\hat{x} = (\alpha x_0) + (\alpha x_1)\varepsilon_1 + (\alpha x_2)\varepsilon_2 + \cdots + (\alpha x_n)\varepsilon_n \quad \forall \alpha \in \mathrm{R} \tag{21}$$

$$\hat{x} \pm \lambda = (x_0 \pm \lambda) + x_1\varepsilon_1 + x_2\varepsilon_2 + \cdots + x_n\varepsilon_n \quad \forall \lambda \in \mathrm{R} \tag{22}$$

On the other hand if f is a non-affine operation z cannot be expressed exactly as an affine combination of the noise symbols ε_i:

$$\begin{aligned} \hat{z} &= f(\hat{x}, \hat{y}) = f(x_0 + x_1\varepsilon_1 + x_2\varepsilon_2 + \cdots + x_n\varepsilon_n, y_0 + y_1\varepsilon_1 + y_2\varepsilon_2 + \cdots + y_n\varepsilon_n) \\ &= f^*(\varepsilon_1, \varepsilon_2, \ldots, \varepsilon_n) \end{aligned} \tag{23}$$

In this case the problem leads to the identification of an affine function:

$$f^a(\varepsilon_1, \varepsilon_2, \ldots, \varepsilon_n) = z_0 + z_1\varepsilon_1 + \cdots + z_n\varepsilon_n \tag{24}$$

that approximate the function $f^*(\varepsilon_1, \varepsilon_2, \ldots, \varepsilon_n)$ reasonably well over its domain jointly with an extra term $z_k\varepsilon_k$ that represents the error introduced by this approximation:

$$\hat{z} = f^a(\varepsilon_1, \varepsilon_2, \ldots, \varepsilon_n) + z_k\varepsilon_k = z_0 + z_1\varepsilon_1 + \cdots + z_n\varepsilon_n + z_k\varepsilon_k x \tag{25}$$

The last term represents the residual or approximation error:

$$e^*(\varepsilon_1, \varepsilon_2, \ldots, \varepsilon_n) = f^*(\varepsilon_1, \varepsilon_2, \ldots, \varepsilon_n) - f^a(\varepsilon_1, \varepsilon_2, \ldots, \varepsilon_n) \tag{26}$$

The noise symbol ε_k must be distinct from all other noise symbols that already appeared in the same computation, and the coefficient $\hat{z}_k$ must be an upper bound on the absolute magnitude of e^* :

$$|z_k| > \max\{|e^*(\varepsilon_1, \varepsilon_2, \ldots, \varepsilon_n)| : (\varepsilon_1, \varepsilon_2, \ldots, \varepsilon_n) \in U\} \tag{27}$$

The affine approximation function f^a could assume different forms, depending on the desired degree of accuracy and the available computational resources.

A good trade-off between these goals could be reached by employing as approximation function an affine combination of the input forms $\hat{x}$ and $\hat{y}$:

$$f^a(x) = \alpha\hat{x} + \beta\hat{y} + \varsigma \tag{28}$$

Where the unknown function coefficients can be identified thanks to the Chebyshev approximation theory for univariate functions, which furnishes an algorithm for finding the optimum coefficients α, β, and ζ of the affine approximation function.

5.2 An AA-Based Approach for Power Flow Analysis

AA can be effectively adopted to represent sources of uncertainty affecting the systems state variables in power flow analysis [15]. With AA each state variable (i.e., the voltage magnitude of the load buses and the voltage angles of all nodes but the slack) is expressed by a central value and a set of partial deviations. These deviations are associated to as many noise symbols as those which describe the effect of the various phenomena affecting the system state variables. Without loss of generality, we assumed as driving phenomena the effects of active and reactive power uncertainty of the load nodes and the effect of active power uncertainty of the generation nodes.[2]

Thus the affine forms representing the power system state variables are:

$$\begin{cases} V_i = V_{i,0} + \sum_{j\in n_P} V^P_{i,j}\varepsilon_{Pj} + \sum_{k\in n_Q} V^Q_{i,k}\varepsilon_{Qk} & \text{for } i \in n_Q \\ \delta_i = \delta_{i,0} + \sum_{j\in n_P} \delta^P_{i,j}\varepsilon_{Pj} + \sum_{k\in n_Q} \delta^Q_{i,k}\varepsilon_{Qk} & \text{for } i \in n_P \end{cases} \tag{29}$$

where ε_{Pj} is the noise symbol representing the uncertainty due to the active power injection at the jth bus; ε_{Qk} is the noise symbol representing the uncertainty due to

[2] Further noise symbols describing other uncertainty sources (i.e. network modeling errors) and/ or more complex correlations between the affine forms could be assumed without loss of generalization. For example, after detailed statistical analysis of the historical load profiles, it could be possible to share the same noise symbols for statistically dependent loads.

the reactive power injection at the kth bus; $V_{i,0}$ is the central value of the ith bus voltage magnitude; $\delta_{i,0}$ is the central value of the ith bus voltage angle; $V_{i,j}^{P}$ is the partial deviations of the ith bus voltage magnitude due to the active power injected at the jth bus; $V_{i,j}^{Q}$ is the partial deviations of the ith bus voltage magnitude due to the reactive power injected at the jth bus; $\delta_{i,j}^{P}$ is the partial deviations of the ith bus voltage angle due to the active power injected at the jth bus; $\delta_{i,j}^{Q}$ is the partial deviations of the ith bus voltage angle due to the reactive power injected at the jth bus.

The central values of the affine forms Eq. 29 are calculated by solving a conventional power flow problem in correspondence of the "nominal operating point" defined by:

$$\begin{cases} P_i^{SP} = \mathrm{mid}\left(\left[P_{i,\min}^{SP}, P_{i,\max}^{SP}\right]\right) = \frac{P_{i,\max}^{SP} - P_{i,\min}^{SP}}{2} & \text{for } i \in n_P \\ Q_i^{SP} = \mathrm{mid}\left(\left[Q_{i,\min}^{SP}, Q_{i,\max}^{SP}\right]\right) = \frac{Q_{i,\max}^{SP} - Q_{i,\min}^{SP}}{2} & \text{for } i \in n_Q \end{cases} \tag{30}$$

A first estimation of the partial deviations of the affine forms Eq. 29 are calculated by (1) linearizing the power flow equations at the "nominal operating point" with respect to the uncertain parameters and (2) computing the corresponding partial derivatives (sensitivity coefficients). The partial deviations can then be estimated as:

$$\begin{cases} V_{i,j}^{P} = \left.\frac{\partial V_i}{\partial P_j}\right|_{x_0} \Delta P_j V_{i,k}^{Q} = \left.\frac{\partial V_i}{\partial Q_k}\right|_{x_0} \Delta Q_k & \text{for } j \in nP\ k, i \in n_Q \\ \delta_{i,j}^{P} = \left.\frac{\partial \delta_i}{\partial P_j}\right|_{x_0} \Delta P_j \delta_{i,k}^{Q} = \left.\frac{\partial \delta_i}{\partial Q_k}\right|_{x_0} \Delta Q_k & \text{for } i,j \in nP\ k \in n_Q \end{cases} \tag{31}$$

It is worth to note that if the equation system contains only affine expressions (linear system of equations) the obtained affine forms would be the exact solution. As the power flow equations contain non-linear expressions, the obtained affine forms are usually an underestimation of the exact result [16]. Thus, to guarantee the inclusion of the solution domain each partial deviation is multiplied by an amplification coefficient [15].

Starting from this initial affine solution, we propose a "domain contraction" based method for narrowing its bounds.

The proposed solution algorithm first expresses the power flow Eq. 1 by using AA-based computing:

$$\begin{cases} \widehat{Q}_i = Q_{i,0} + \sum_{j \in nP} Q_{i,j}^{P} \varepsilon_{Pj} + \sum_{k \in nQ} Q_{i,k}^{Q} \varepsilon_{Qk} + \sum_{h \in nN} Q_{i,h} \varepsilon_h & \text{for } i \in n_Q \\ \widehat{P}_i = P_{i,0} + \sum_{j \in nP} P_{i,j}^{P} \varepsilon_{Pj} + \sum_{k \in nQ} P_{i,k}^{Q} \varepsilon_{Qk} + \sum_{h \in nN} P_{i,h} \varepsilon_h & \text{for } i \in n_P \end{cases} \tag{32}$$

where $\widehat{Q}_i, \widehat{P}_i$ are the affine forms representing the active and reactive power injected in the ith bus; ε_h are new noise symbols introduced in the course of the computation due to the presence of non-affine operations (n_N denotes the list of the new noise symbols); $Q_{i,0}$, $Q_{i,j}^{P}$, $Q_{i,j}^{Q}$, $Q_{i,h}$, $P_{i,0}$, $P_{i,j}^{P}$, $P_{i,j}^{Q}$, $P_{i,h}$ are the central values and the partial deviations of the affine forms representing the active and reactive powers injected in the ith bus (they are real numbers determined in the course of computations).

The obtained affine forms can be arranged according to the following matrix formalism:

$$\begin{bmatrix} \widehat{Q}_1 \\ \dots \\ \widehat{Q}_{N_Q} \\ \widehat{P}_1 \\ \dots \\ \widehat{P}_{N_P} \end{bmatrix} = \begin{bmatrix} Q_{1,0} \\ \dots \\ Q_{N_Q,0} \\ P_{1,0} \\ \dots \\ P_{N_P,0} \end{bmatrix} + \begin{bmatrix} Q^P_{1,1} & \dots & Q^P_{1,N_P} & Q^Q_{1,1} & \dots & Q^Q_{1,N_Q} \\ \dots & \dots & \dots & \dots & \dots & \dots \\ Q^P_{N_Q,1} & \dots & Q^P_{N_Q,N_P} & Q^Q_{N_Q,1} & \dots & Q^Q_{N_Q,N_Q} \\ P^P_{1,1} & \dots & P^P_{1,NP} & P^Q_{1,1} & \dots & P^Q_{1,N_Q} \\ \dots & \dots & \dots & \dots & \dots & \dots \\ P^P_{N_P,1} & \dots & P^P_{N_P,N_P} & P^Q_{N_P,1} & \dots & P^Q_{N_P,N_Q} \end{bmatrix} \begin{bmatrix} \varepsilon_{P_1} \\ \dots \\ \varepsilon_{P_{N_P}} \\ \varepsilon_{Q_1} \\ \dots \\ \varepsilon_{Q_{NQ}} \end{bmatrix}$$
$$+ \begin{bmatrix} Q_{1,1} & \dots & Q_{1,n_N} \\ \dots & \dots & \dots \\ Q_{N_Q,1} & \dots & Q_{N_Q,n_N} \\ P_{1,1} & \dots & P_{1,nN} \\ \dots & \dots & \dots \\ P_{N_P,1} & \dots & P_{N_P,n_N} \end{bmatrix} \begin{bmatrix} \varepsilon_1 \\ \dots \\ \dots \\ \dots \\ \dots \\ \varepsilon_{n_N} \end{bmatrix} \tag{33}$$

and in a more general form:

$$f(X) = AX + B \tag{34}$$

where:

$$\boldsymbol{A} = \begin{bmatrix} Q^P_{1,1} & \dots & Q^P_{1,N_P} & Q^Q_{1,1} & \dots & Q^Q_{1,N_Q} \\ \dots & \dots & \dots & \dots & \dots & \dots \\ Q^P_{N_Q,1} & \dots & Q^P_{N_Q,N_P} & Q^Q_{N_Q,1} & \dots & Q^Q_{N_Q,N_Q} \\ P^P_{1,1} & \dots & P^P_{1,NP} & P^Q_{1,1} & \dots & P^Q_{1,N_Q} \\ \dots & \dots & \dots & \dots & \dots & \dots \\ P^P_{N_P,1} & \dots & P^P_{N_P,N_P} & P^Q_{N_P,1} & \dots & P^Q_{N_P,N_Q} \end{bmatrix} \tag{35}$$

$$\boldsymbol{X} = \begin{bmatrix} \varepsilon_{P_1} \\ \dots \\ \varepsilon_{P_{N_P}} \\ \varepsilon_{Q_1} \\ \dots \\ \varepsilon_{Q_{N_Q}} \end{bmatrix} \tag{36}$$

$$\boldsymbol{B} = \begin{bmatrix} Q_{1,0} \\ \dots \\ Q_{N_Q,0} \\ P_{1,0} \\ \dots \\ P_{N_P,0} \end{bmatrix} + \begin{bmatrix} Q_{1,1} & \dots & Q_{1,n_N} \\ \dots & \dots & \dots \\ Q_{N_Q,1} & \dots & Q_{N_Q,n_N} \\ P_{1,1} & \dots & P_{1,n_N} \\ \dots & \dots & \dots \\ P_{N_P,1} & \dots & P_{N_P,n_N} \end{bmatrix} \begin{bmatrix} \varepsilon_1 \\ \dots \\ \dots \\ \dots \\ \dots \\ \varepsilon_{n_N} \end{bmatrix} \tag{37}$$

where N_P is the number of buses in which the active power is specified (i.e., load, generation and slack buses) and N_Q is the number of buses in which the reactive power is fixed (i.e., load buses); A is a matrix of real coefficients; X is the vector to contract (the initial value of each component of X is [−1, 1]); B is an interval vector (since the new noise symbols ε_h vary in the interval [−1, 1] and it is not possible to contract them since they represent internal noise introduced by AA-based computing).

The power flow solution could then be obtained by contracting the vector X such that:

$$f = AX + B = f^{\mathrm{SP}} \tag{38}$$

where f^{SP} is the interval vector defining the specified range of the active and reactive powers:

$$f^{\mathrm{SP}} = \begin{bmatrix} \left[Q^{SP}_{1\,\min}, Q^{SP}_{1\,\max}\right] \\ \cdots \\ \left[Q^{SP}_{N_Q\,\min}, Q^{SP}_{N_Q\,\max}\right] \\ \left[P^{SP}_{1\,\min}, P^{SP}_{1\,\max}\right] \\ \cdots \\ \left[P^{SP}_{N_P\,\min}, P^{SP}_{NP\,\max}\right] \end{bmatrix} \tag{39}$$

The problem (38) can be formalized as:

$$AX = C \tag{40}$$

where the interval vector C is:

$$C = f^{\mathrm{SP}} - B \tag{41}$$

The form 41 derives directly by the application of AA for uncertainty representation in power flow analysis. It represents an alternative to the traditional and widely used linearization formalism adopted in Interval Newton method:

$$J\Delta x = -f(x_0) \tag{42}$$

where the Jacobian matrix J is an interval matrix while $f(x_0)$ is a real vector. The adoption of the formalism 38 asks for the inversion of the interval matrix J, which is a very complex issue to address. As pointed out by many authors [17, 18], the inversion of the interval matrix J, which requires the solution of a linear systems of equation with interval coefficient, represents the main impediment in using classic IM in power flow analysis. On the contrary, since the application of the formalism 41 does not require any kind of interval matrix inversion, it could be highly suitable to address the problem under study. The application of this formalism for the solution of non-linear systems of equations in the presence of data uncertainty was studied in detail by Kolev [18].

According to these studies, the problem 41 can be effectively addressed by solving the following $2(n_P + n_Q)$ constrained *linear* optimization problems:

$$\begin{cases} \min / \max(\varepsilon_{Qk}, \varepsilon_{Pj}) & \text{for } k \in n_Q,\ j \in n_P \\ \text{s.t.} \ -1 \le \varepsilon_{Qk} \le 1 & -1 \le \varepsilon_{Pj} \le 1 \\ \inf(C_i) \le \sum_{j \in n_P} A_{ij}\varepsilon_{Pj} + \sum_{k \in n_Q} A_{ik}\varepsilon_{Qk} \le \sup(C_i) & \text{for } i = 1, 2, \ldots, n_P + n_Q \end{cases} \tag{43}$$

The solution of these problems has been extensively explored in literature and it does not represent a computationally intensive issue [15].

The final power flow solution is then obtained as:

$$\begin{cases} V_i = V_{i,0} + \sum_{j \in nP} V^P_{i,j}\left[\varepsilon_{Pj,\min}, \varepsilon_{Pj,\max}\right] + \sum_{k \in nQ} V^Q_{i,k}\left[\varepsilon_{Qk,\min}, \varepsilon_{Qk,\max}\right] & \text{for } i \in n_Q \\ \delta_i = \delta_{i,0} + \sum_{j \in nP} \delta^P_{i,j}\left[\varepsilon_{Pj,\min}, \varepsilon_{Pj,\max}\right] + \sum_{k \in nQ} \delta^Q_{i,k}\left[\varepsilon_{Qk,\min}, \varepsilon_{Qk,\max}\right] & \text{for } i \in n_P \end{cases} \tag{44}$$

The AA-based solution strategy can be easily adapted to account for the effect of reactive power limits on generator voltage settings and properly model the generators' voltage regulators. For this purpose, the voltage modules at each PV bus should be assumed as new input variables of the power flow problem and further noise symbols ε_{Vi}, taking into account the uncertainty of these new input variables, should be introduced in the computation.

5.3 *Consideration for an AA-Based OPF*

The previously described AA-based solution strategy allows the analyst to calculate the affine forms representing the power flow state variables once the voltage magnitudes and the active power at the generator buses and the active and reactive powers at the load buses are known. In particular, it leads to identify an affine approximation of the following vectorial mapping:

$$\tilde{Y} = F(V_{\text{gen}}, f^{\text{SP}}) \tag{45}$$

where $\tilde{Y}$ is the vector of the affine forms representing the unknown power flow state variables (V_i for $i \in n_Q$ and δ_j for $j \in n_P$); V_{gen} is the vector of the fixed voltage magnitudes at the generator buses.

Once $\tilde{Y}$ has been identified it is straightforward to calculate the affine forms representing the reactive power generated at the generation buses:

$$Q^i_{\text{gen}} = G_i(\tilde{Y}) = G_i\left(F\left(V_{\text{gen}}, f^{\text{SP}}\right)\right) \quad i \in \text{PV} \tag{46}$$

Thanks to this feature, the AA-based solution methodology could be integrated in optimal power flow studies in order to evaluate the solution robustness and to check the constraints consistency. This is very useful in addressing the problem of voltage regulation in the presence of data uncertainty.

The solution of this complex problem asks for the identification of the generator voltage modules which minimizes a cost function (accounting for both technical and economical issues) and satisfies the constraints on the generators operation. To check the constraints consistency the proposed AA-based solution methodology could be conveniently adopted. The employment of AA leads to solve the power flow problem and to calculate the corresponding intervals of the reactive power generated at each PV buses for each candidate solution.

In particular, the optimal voltage regulation settings can be obtained by solving the following constrained non-linear optimization problem:

$$\min_{\mathbf{V}_{\mathbf{gen}}} f \tag{47}$$

$$Q^i_{\text{gen}} = G_i\left(F\left(V_{\text{gen}}, f^{\text{SP}}\right)\right) \tag{48}$$

$$\left(\inf\left(Q^i_{\text{gen}}\right) - Q_i^{\min}\right) v^i_a \leq 0 \tag{49}$$

$$\left(Q_i^{\max} - \sup\left(Q^i_{\text{gen}}\right)\right) v^i_b \leq 0 \tag{50}$$

$$V^i_{\text{gen}} = V^i_0 + v^i_a - v^i_b \tag{51}$$

$$v^i_a, v^i_b \geq 0 \quad i \in \text{PV} \tag{52}$$

where v^i_a, v^i_b are two auxiliary variables representing the changes in the *i*th generator bus voltage due to reactive power limits.

- f is the cost function that should be minimized.
- inf(.), sup(.) are interval operators returning, respectively, the lower and upper bound of an interval.

In order to account for the effect of reactive power limits on generator voltage settings and properly model the generators' voltage regulators, constraints 49–52 are introduced [1]. These constraints ensure that all the generators will be operating at their terminal voltage settings, as long as the reactive power is within its limits; in this case, the two variables v^i_a and v^i_a will be equal to zero to satisfy Eqs. 49 and 52. If the reactive power output of any of the generators hits its maximum limit constraints 49 and 50 will force v^i_b to have a positive value, therefore reducing the voltage at this generator bus according to Eq. 51. Similarly, if the lower limit of reactive power output for any generator is reached, v^i_a will have a positive value, hence increasing the voltage at this generator bus. Note that v^i_a and v^i_a may still have a zero value even if reactive power limits are reached; these variables only simulate the loss of voltage control due to limits.

5.4 Simulation Study

This section discusses the application of the AA-based methodology to the power flow analysis of the IEEE 118-bus test system in the presence of data uncertainties. The power flow solution bounds obtained by the AA-based technique are compared to those calculated using a Monte Carlo simulation with a uniform distribution. For the latter, 5,000 different values of the input variables within the assumed input bounds were randomly selected, and a conventional power flow solution was obtained for each one; this procedure yielded the desired interval solutions defined by the largest and the smallest values of the bus voltage magnitudes and angles as well as line flows.

Without loss of generality, a $\pm 10\%$ tolerance on load and generator powers was assumed.

Based on the assumed load and generator power bounds to represent input data uncertainty, the AA-based methodology was applied to estimate the bounds of the power flow solution. The computed solution was compared with that obtained by using the Monte Carlo approach. The corresponding profiles are shown in Figs. 8, 9, with Fig. 8 depicting the bus voltages magnitude bounds; Fig. 9 shows the bus voltages angle bounds.

Observe that the AA-based methodology gives fairly good approximations of the power flow solution bounds when compared to the benchmark intervals obtained with the Monte Carlo approach; this is mainly due to the intrinsic characteristic of AA that keeps track of correlations between the power systems state variables.

Notice also that the solution bounds are slightly conservative, which is due to the fact that AA yields "worst case" bounds, which take into account any uncertainties in the input data as well as all internal truncation and round-off errors. This is to be expected, since the random, uniformly distributed variation of parameters (with mean equal zero) assumed in the Monte Carlo approach tends to underestimate the worst case variations. This can be considered an advantage of

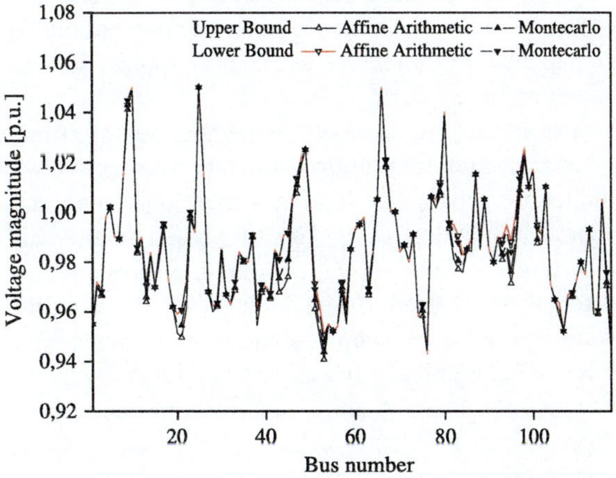

Fig. 8 Upper and lower bounds computed for the bus voltage magnitude

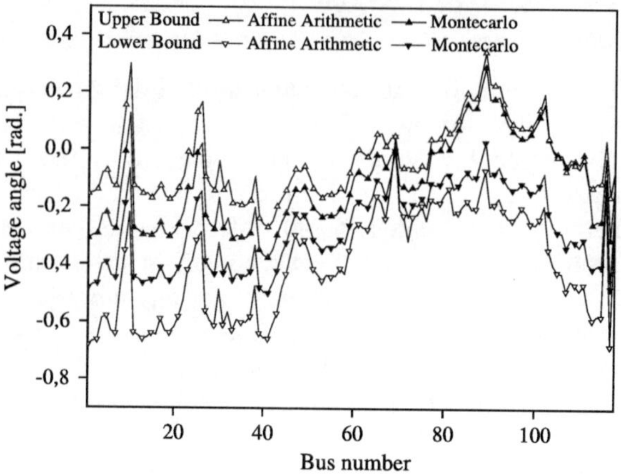

Fig. 9 Upper and lower bounds computed for the bus voltage angle

the AA approach, since no assumptions regarding the probability distribution of load and generator power variations are required.

6 Conclusions

This chapter analyzed and discussed two advanced methodologies for reliable power flow analysis in the presence of data uncertainty allowing to better handle uncertainty compared to the traditional and widely used sampling approaches.

The first one (namely the Boundary Power Flow) is used to find extreme solutions for system voltages, currents and power flows given ranges of values of nodal powers as input variables. These solutions are referred to as boundary values and are obtained by following an optimization procedure. It uses information about the gradients obtained from the inverse system Jacobian matrix, but it is not like conventional gradient based procedures because the objective functions are not explicitly defined. The procedure is simple but the posed problem is difficult. It is computationally intensive and is best applied selectively on a subset of variables which are of interest.

The second one is based on the adoption of affine arithmetic. The employment of this methodology allows analysts to effectively overcome some of the main limitations characterizing the traditional interval arithmetic based solution approaches. In details, the latter are based on interval Newton methods that require inverting an interval matrix and thus presenting a major impediment for its practical application. Based on the AA formalism, the power flow solution bounds were shown to be simply obtained by solving a power flow plus two straight forward LP problems. It was shown with the help of tests run on a realistic power system that using AA allows addressing effectively the “wrapping effect” and the “dependency problem” of interval arithmetic, that leads to a better characterization of the effects of input data uncertainty in power flow solutions, and a more realistic approximation of the solution domain compared to the typical "hyper box" form obtained with interval approaches.

References

1. El-Samahy I, Bhattacharya K, Cañizares CA, Anjos MF, Pan J (2008) A procurement market model for reactive power services considering system security. IEEE Trans Power Syst 23(1):137–149
2. Verbic G, Cañizares CA (2006) Probabilistic optimal power flow in electricity markets based on a two-point estimate method. IEEE Trans Power Syst 21(4):1883–1893
3. Wan Y-H, Parsons BK (1993) Factors relevant to utility integration of intermittent renewable technologies. NREL, August 1993
4. Al-Othman AK, Irving MR (2006) Analysis of confidence bounds in power system state estimation with uncertainty in both measurements and parameters. Electr Power Syst Res 76(12):1011–1018
5. Vaccaro A, Villacci D (2009) Radial power flow tolerance analysis by interval constraint propagation. IEEE Trans Power Syst 24(1):28–39
6. Allan RN, Leite da Silva AM (1981) Probabilistic load flow using multilinearisations. Proc IEE 128(Pt. C 5):280–287
7. Dimitrovski A, Tomsovic K (2004) Boundary load flow solutions. IEEE Trans Power Syst 19(1):348–355
8. Dimitrovski A, Tomsovic K (2004) Slack bus treatment in load flow solutions with uncertain nodal powers. In: Proc. of the 8th Intl. conference on probabilistic methods in power systems (PMAPS)-2004, Ames, Iowa
9. Dimitrovski A, Tomsovic K (2003) Risk assessment using boundary load flow solutions. In: Proc of the 2003 intl. conference on intelligent system applications to power systems (ISAP), Lemnos, Greece, pp 1–6
10. Ajjarapu V, Christy C (1992) The continuation power flow: a tool for steady state voltage stability analysis. IEEE Trans Power Syst 7(1):416–423
11. Task Force on Probabilistic Aspects of Reliability Criteria (2004) Probabilistic security assessment for power system operations, reliability, risk, and probability applications subcommittee. IEEE PES General Meeting, Denver, 6–10, pp 212–220
12. Merrill HM, Wood AJ (1990) Risk and uncertainty in power system planning presented at the 10th PSCC, Graz
13. de Figueiredo LH, Stolfi J (1997) Self-validated numerical methods and applications, Brazilian Mathematics Colloquium monograph, IMPA. Rio de Janeiro, Brazil
14. de Figueiredo LH, Stolfi J (2004) Affine arithmetic: concepts and applications. Numer Algorithms 37(1–4):147–158
15. Vaccaro A, Canizares C, Villacci D (2010) An affine arithmetic based methodology for reliable power flow analysis in the presence of data uncertainty. IEEE Trans Power Syst 25(2):624–632
16. Grabowski D, Olbrich M, Barke E (2008) Analog circuit simulation using range arithmetics. In: Proc. 2008 conf. Asia and South Pacific design automation, Seoul, Korea, pp. 762–767, January 2008
17. Kolev L, Nenov I (2000) A combined interval method for global solution of nonlinear systems. In: Proc. XXIII int. conf. fundamentals of electronics and circuit theory SPETO 2000, Gliwice, Poland, pp 365–368
18. Kolev L (1997) A general interval method for global nonlinear DC analysis. In: Proc. 1997 European Conf. circuit theory and design ECCD'97, vol 3. Technical University of Budapest, Budapest, Hungary, pp 1460–1462

Index